AGRICULTURAL PRODUCTIVITY:
Measurement and Sources of Growth

Studies in Productivity and Efficiency

Series Editors:

Books in the series:

Fox, Kevin J.:
Efficiency in the Public Sector
Ball, V. Eldon and Norton, George W.:
Agricultural Productivity:
Measurement and Sources of Growth

AGRICULTURAL PRODUCTIVITY:
Measurement and Sources of Growth

edited by

V. Elton Ball
Economic Research Service,
U.S. Department of Agriculture, Washington, DC

and

George W. Norton
Department of Agricultural and Applied Economics,
Virginia Polytechnic Institute and State University,
Blacksburg, VA

Kluwer Academic Publishers
Boston/Dordrecht/London

Distributors for North, Central and South America:
Kluwer Academic Publishers
101 Philip Drive
Assinippi Park
Norwell, Massachusetts 02061 USA
Telephone (781) 871-6600
Fax (781) 871-6528
E-Mail <kluwer@wkap.com>

Distributors for all other countries:
Kluwer Academic Publishers Group
Post Office Box 322
3300 AH Dordrecht, THE NETHERLANDS
Telephone 31 786 576 000
Fax 31 786 576 474
E-mail <services@wkap.nl>

Electronic Services <http://www.wkap.nl>

Library of Congress Cataloging-in-Publication Data

A C.I.P. Catalogue record for this book is available from the Library of Congress.

Printed on acid-free paper.

Printed in the United States of America

Contents

Contributing Authors

Albert K. A. Acquaye – Department of Agricultural and Resource Economics, University of California, Davis, California.

Mary Ahearn – Economic Research Service, U.S. Department of Agriculture, Washington, D.C.

Julian M. Alston – Department of Agricultural and Resource Economics, University of California, Davis, California.

V. Eldon Ball – Economic Research Service, U.S. Department of Agriculture, Washington, D.C.

Vince Breneman – Economic Research Service, U.S. Department of Agriculture, Washington, D.C.

Jean-Pierre Butault – Institut National de la Recherche Agronomique, France.

Kelly A. Chaston – Department of Economics, Davidson College, Davidson, North Carolina.

Robert E. Evenson – Department of Economics, Yale University, New Haven, Connecticut.

Rolf Färe – Department of Economics, Oregon State University, Corvallis, Oregon.

Ronald G. Felthoven – Department of Agricultural and Resource Economics, University of California, Davis, California.

Frank M. Gollop – Department of Economics, Boston College, Boston, Massachusetts.

Bruce L. Gardner – Department of Agricultural and Resource Economics, University of Maryland, College Park, Maryland.

Munisamy Gopinath – Department of Agricultural and Resource Economics, Oregon State University, Corvallis, Oregon.

Donald W. Goss – Texas Agricultural Experiment Station, Temple, Texas.

Shawna Grosskopf – Department of Economics, Oregon State University, Corvallis, Oregon.

Arthur Grube – Office of Pesticide Programs, U.S. Environmental Protection Agency.

F. Hernandez-Sancho – Department of Applied Economics, University of Valencia, Valencia, Spain.

Wallace E. Huffman – Economics Department, Iowa State University, Ames, Iowa.

Dale W. Jorgenson – Department of Economics, Harvard University, Cambridge, Massachusetts.

Robert L. Kellogg – National Resources Conservation Service, Washington, D.C.

Gregory Livanis – Department of Food and Resource Economics, University of Florida, Gainesville, Florida.

Catherine J. Morrison Paul – Department of Agricultural and Resource Economics, University of California, Davis, California.

Charles B. Moss – Department of Food and Resource Economics, University of Florida, Gainesville, Florida.

Richard F. Nehring – Economic Research Service, U.S. Department of Agriculture, Washington, D.C.

Doris Newton – Economic Research Service, U.S. Department of Agriculture, Washington, D.C.

George W. Norton – Department of Agriculture and Applied Economics, Virginia Tech, Blacksburg, Virginia.

Christopher J. O'Donnell – School of Economic Studies, University of New England, Armidale, Australia.

Philip G. Pardey – International Food Policy Research Institute, Washington, D.C.

Steven Plotkin – Natural Resources Conservation Service, Amherst, Massachusetts.

D.S. Prasada Rao – School of Economic Studies, University of New England, Armidale, Australia.

C. Richard Shumway – Department of Agricultural Economics, Washington State University, Pullman, Washington.

Agapi Somwaru – Economic Research Service, U.S. Department of Agriculture, Washington, D.C.

Jet Yee – Economic Research Service, U.S. Department of Agriculture, Washington, D.C.

Acknowledgments

The authors thank V. Kerry Smith, Richard Perrin, and Tom Cox for careful reviews of and comments on earlier drafts of chapters in this volume. We also appreciate the comments by Jim Ohmeke, David Schimmelpfennig, Brent Moulton, Barbara Craig, Colin Thirtle, Spiro Stefanou, and Deborah Aiken on earlier versions of the chapters when they were presented at the USDA/ERS/NC 208 conference on Agricultural Productivity Analysis: Data, Methods, and Measures, Washington, D.C., March 9-10, 2000. Chapter 13 in this volume is based on closing comments by Dale Jorgenson, Bruce Gardner, Bob Evenson, and Dick Shumway at that conference, and we appreciate their willingness to synthesize the comments in writing.

The editorial and production assistance of Mary Holliman, Bruce Wallace, and Dot Wnorowski was invaluable in producing the book manuscript. The editorial assistance of Bob Russell and Carolyn O'Neil of Kluwer is also gratefully acknowledged. Finally, we thank the Farm Foundation and USDA/ERS for their financial support of the conference that led to most of the chapters in this volume, and Virginia Tech for supporting production of the book.

Foreword

Dale W. Jorgenson

This is an important and timely volume for at least three reasons. First, productivity is a very topical issue. The resurgence of the U.S. economy since 1995 has outrun all but the most optimistic expectations. The growth slowdown beginning in 2001 has touched off a strenuous debate about whether the improvements in America's economic performance are permanent or transitory. The focus of this debate has been on the sustainability of rapid productivity growth.

Second, agriculture has been a very important contributor to productivity growth for the U.S. economy for decades. An updated version of my calculations with Kevin Stiroh (Jorgenson and Stiroh, 2000) attributes 21% of growth in U.S. productivity to agriculture for the period 1958-99, far in excess of agriculture's 1.3% contribution to the gross domestic product. And productivity accounts for well over half the growth in agricultural output, while only 12.6% of U.S. GDP growth can be attributed to productivity.

Finally, the Economic Research Service of the U.S. Department of Agriculture (USDA) has been the leader in introducing "production theory" methods of productivity measurement into the federal statistical system. These methods have established a secure foundation for empirical research on the determinants of the very impressive productivity gains in agriculture. Among the many factors to be considered are research and development, agricultural extension, education, infrastructure investment, and conservation and environmental programs.

The essence of the production theory approach is to resolve each of the issues that arise in productivity measurement by appealing to the theory of production. Frank Gollop, Barbara Fraumeni, and I (1987) have presented this approach in detail and have integrated productivity measures for individual sectors, such as agriculture, into measures for the economy as a whole (Jorgenson, Gollop and Fraumeni, 1987). Gollop (2000) shows how this approach has replaced the "welfare" approach that initially dominated productivity measurement.

Michael Trueblood and Vernon Ruttan (1995) have pointed out that the leadership role of the USDA in our federal productivity statistics reflects a long tradition in the Economic Research Service. In 1960 the ERS was the first federal agency to publish official estimates of total factor productivity. It was not until 1983, more than two decades later, that the agency responsible for productivity statistics, the Bureau of Labor Statistics, published official estimates for the U.S. economy.

Under the leadership of Eldon Ball, the ERS was the first federal agency to adopt the "production theory" approach in its official farm sector production and efficiency statistics in 1991. These estimates of agricultural productivity updated the estimates presented in a pioneering paper by Ball (1985). Ball, Gollop, Kelly-Hawke, and Swinand (1999) have successfully disaggregated these estimates to the state level. An update of these state-level estimates by Ball, Butault, and Nehring is presented in this volume.

The first hallmark of the production approach is the use of gross output, including the value of capital, labor, and intermediate goods, as the measure of agricultural output. As a result, inputs of intermediate goods, obviously crucial to agricultural production, are treated symmetrically with capital and labor inputs. The alternative approach is to employ value added or even value added net of depreciation as a measure of agricultural output, an approach that is far inferior to gross output in modeling the behavior of agricultural producers. Unsurprisingly, empirical research in agricultural economics has come to rely on gross output, rather than on value added.

The second hallmark of the production theory approach is careful attention to the quality of inputs — labor, capital, and intermediate goods. This has long been a very important focus for research on agricultural productivity, beginning with the path-breaking work of Zvi Griliches (1960). The current challenge is to link the composition of inputs more directly to the underlying developments in agricultural technology. Genetic research has been of critical importance in development of new technology for decades. In order to deal with the rapid advance of biotechnology, it will be necessary to disaggregate varieties of plant and animal breeding stock down to genetically distinguishable types.

Construction of aggregates for output, as well as for capital, labor, and intermediate inputs, requires careful attention to index number issues. As a consequence of the many contributions of Erwin Diewert (1976), index number issues have become part of every agricultural economist's toolkit. Techniques for aggregation of capital goods based on hedonic price indices have also been thoroughly integrated into measures of agricultural productivity through the construction of measures of capital input. These techniques will become increasingly important as advances in biotechnology are incorporated into agricultural productivity statistics.

The third hallmark of the production theory approach is to derive measures of productivity for the economy as a whole by aggregating over such individual sectors as agriculture. This aggregation makes it possible to identify the relative importance of productivity growth in each sector and leads to the conclusion that agriculture is a leading contributor to growth in aggregate productivity. To resolve the issue of the sustainability of productivity growth for the U.S. economy, it is essential to analyze trends in technology at the sectoral level. I have outlined this approach for information technology in my Presidential Address (2001) to the American Economic Association.

This volume describes the state of the art in the modeling of agricultural production and the impact of agricultural policy. While much has been accomplished in understanding the determinants of agricultural productivity, projecting future productivity growth in agriculture remains a daunting task. The challenge that lies ahead is to push the description of agricultural technology down to the level of genetic types of both plants and animals, thus opening the way to understanding the potential of biotechnology for enhancing the growth of the U.S. economy and economies around the world.

References

Ball, V. E. 1985. "Output, Input, and Productivity Measurement in U.S. Agriculture." *American Journal of Agricultural Economics*, 67(3)(August):475-86.

Ball, V. E., F. M. Gollop, A. Kelly-Hawke, and G. Swinand. 1999. "Patterns of State Productivity Growth in the U.S. Farm Sector: Linking State and Aggregate Models." *American Journal of Agricultural Economics*, 81(1)(February):164-79.

Ball, V. E., J. P. Butault, and R. Nehring. (this volume). "U.S. Agriculture, 1960-96: A Multilateral Comparison of Total Factor Productivity."

Diewert, W. E. 1976. "Exact and Superlative Index Numbers." *Journal of Econometrics*, 39(1)(May):115-46.

Gollop, F. M. 2000. "The Cost of Capital and the Measurement of Productivity." *Econometrics and the Cost of Capital*, ed. Lawrence J. Lau. Cambridge: The MIT Press, pp. 85-110.

Griliches, H. Z. 1960. "Measuring Inputs in Agriculture: A Critical Survey." *Journal of Farm Economics*, 42(5)(December):1411-27.

Jorgenson, D. W. 2001. "Information Technology and the U.S. Economy." *American Economic Review*, 91(1)(March):1-32.

Jorgenson, D. W., and K. J. Stiroh. 2000. "Raising the Speed Limit: U.S. Economic Growth in the Information Age." *Brookings Papers on Economic Activity*, 1:125-211.

Jorgenson, D. W., F. M. Gollop, and B. M. Fraumeni. 1987. *Productivity and U.S. Economic Growth*. Cambridge: Harvard University Press.

Trueblood, M. A., and V. W. Ruttan. 1995. "A Comparison of Multifactor Productivity Calculations of the U.S. Agricultural Sector." *Journal of Productivity Analysis*, 6:321-31.

1

Introduction and Overview

V. Eldon Ball and George W. Norton

INTRODUCTION

The rise in agricultural productivity has long been chronicled as the single most important source of economic growth in the U.S. farm sector (Kendrick and Grossman, 1980; Ball, 1985; Jorgenson, Gollop, and Fraumeni, 1987; Ball et al., 1997). In a 1992 study, Jorgenson and Gollop (1992) estimated that productivity growth over the 1947-85 period accounted for 82% of output growth in agriculture compared to only 13% in the non-farm sector. Moreover the estimated rate of productivity growth in agriculture (1.58%) was substantially greater than in the non-farm economy (0.44%). This singularly important role of productivity change in economic growth in agriculture has stimulated interest in trying to explain productivity change and in ascertaining that it is measured properly.

While measured productivity growth for the aggregate farm sector provides a useful summary statistic of the sector's overall contribution to the economy, it can mask significant state or regional differences and trends (Ball et al., 1999). Not only is state productivity more variable than national productivity, but evidence indicates that productivity growth in the U.S. farm sector is wholly a function of productivity trends in the individual states (Ball et al., 1999). Interstate shifts in production activity and resource reallocations have little impact. These results make clear that disaggregated analysis of farm sector productivity growth is essential for understanding the sources of farm sector growth.

This disaggregation requires a consistent set of state and aggregate farm sector accounts conforming to the economic theory of production. Empirical methods for measuring productivity and assessing its impacts in a theoretically sound manner have advanced in recent years. The U.S. agricultural sector provides an excellent setting for applying these advances, due to the wealth of data available, not only at the national level but now at the state level as well through efforts at and supported by the Economic Research Service of the U.S. Department of Agriculture (USDA/ERS).

Data also are becoming available on environmental risks associated with chemical use in modern agriculture. Productivity measures that take into account the non-marketed environmental effects associated with chemical use can provide a more complete picture of changes in agricultural productivity growth rates.

The purpose of this book is to present recent research on agricultural productivity measurement, drawing on the latest theories and methods. The chapters also describe and utilize new data sets recently made available at USDA/ERS. The book is organized around three main topics. Part 1 contains four papers that present multi-lateral comparisons of total factor productivity, three applying parametric and one non-parametric methods. Part 2 uses USDA/ERS data to estimate differences in economic efficiency by state, and the sources of agricultural productivity differences, especially the effects of infrastructure and research and extension (R&E). These three papers give special attention to benefits of R&D that spill across state lines.

Part 3 includes four papers that explore the linkage between productivity growth and the environment. One provides a set of environmental risk indicators that can be included in models to account for environmental and human health effects associated with agricultural growth. The second introduces an environmental performance index to evaluate the environmental performance of the U.S. agricultural sector. The third explores the effects of groundwater regulation on productivity growth in agriculture, and the fourth uses a cost function model to estimate the costs of reducing environmental risks from pesticide use.

The book concludes with an assessment by several experts in the field of where we most profitably might go from here in terms of agricultural productivity measurement and associated research. This introduction provides an overview of each chapter. It also highlights the lessons learned from the papers, including the strengths and limitations of the methods employed. Guidance is provided for future agricultural productivity assessments.

Overview of the Book

The first chapter in Part 1 by Ball, Butault, and Nehring, presents a set of state and aggregate farm-sector accounts developed at USDA that are consistent with the economic theory of production. Estimates are provided of the growth rates and relative levels of agricultural productivity for the 48 contiguous states for the 1960-96 period. An index number approach is used. The index number procedure, proposed independently by Eltetö and Köves (1964) and Szulc (1964) yields transitive multilateral comparisons that deviate the least from binary comparisons. They find that the range of levels of productivity has narrowed over time across states, although the pattern of convergence has been far from uniform. They test the catch-up hypothesis that implies that states that lag particularly far behind the technology leaders experience the most rapid productivity gains. They find strong statistical support for this hypothesis. They also find a positive interaction between capital accumulation and productivity growth. The data and variables provided in the paper also form the basis for several of the subsequent papers in this volume, and are available at the USDA/ERS web site at http://preview.ers.usda.gov/data/stateproductivity/.

In chapter 3, Acquaye, Alston, and Pardey use annual data from 1949 to 1991 for 48 states to account for changes in the composition of input and output aggregates over time and space. They employ index number procedures as well, and obtain new evidence on changes in inputs, outputs, and productivity in U.S. agriculture. Their measures change significantly when they use state-specific rather than national prices and when they allow for changes in the composition of the aggregates, especially labor and capital. State-specific annual-average productivity growth rates ranged from .79 to 3.23, compared with a national aggregate rate of 1.9 percent per annum. Over the period 1960 to 1990, they obtain similar national but quite different state-specific estimates from those presented in chapter 2. Their analysis points to the importance of how the capital aggregate is defined in determining these differences.

Chapter 4, by Rao, O'Donnell, and Ball, focuses on the use of EKS index numbers for multilateral comparisons, while exploring the feasibility of employing the concept of minimum spanning trees. The EKS index number procedure, discussed in chapter 2, is routinely applied in multilateral comparisons. One of the main problems with an EKS index is that it considers all pair-wise binary comparisons to be equally reliable. This reliability problem arises because not all states produce all commodities or have similar input mixes. If the overlap between commodities produced in the two states is low, the comparison is less meaningful. The chapter describes a method of generalizing EKS indices by assigning weights to different pair-wise comparisons. The empirical part of the chapter focuses on the construction of minimum spanning trees (introduced by Hill (1999) and used to compute purchasing power parities) for outputs and inputs, and then uses them to derive sets of alternative output, input, and productivity comparisons.

Chapter 5, by Moss, Livanis, Breneman, and Nehring, considers the importance of the spatial dimension of farmland values for productivity measurement. It sorts out the relative importance of productivity versus urban influences in explaining changes in farmland values using county level data. Accurate productivity measurement requires input qualities to be homogeneous. Therefore the authors conduct an hedonic analysis to estimate farmland value, holding urbanization pressure constant.

Part 2 presents methods for measuring economic efficiency and sources of productivity growth. The parametric approach to measuring technical and allocative efficiency typically involves estimating systems of equations that include a cost frontier and/or a set of first derivative equations. O'Donnell in chapter 6 points out that functions of the estimated parameters of these systems may be inconsistent with important curvature conditions (concavity, monotonicity) implied by microeconomic theory. He illustrates how Bayesian methods can be used to impose inequality constraints on a system of equations derived from a translog shadow cost frontier to overcome this problem. The estimated parameters are used to obtain levels of technical, allocative, and cost efficiency in 48 states. Florida is found to be the most technically- and cost-efficient state, while South Dakota is the most allocatively efficient. Point estimates of levels of technical and cost efficiency average 64 percent and 56 percent across states, while average allocative efficiency is 86 percent.

The two chapters that focus on sources of change in agricultural productivity use the state-level data described in chapter 2 and new series on public R&D, extension, and infrastructure. States are heterogeneous in agro-climatic conditions and resource bases, and their competitiveness in agricultural products can be shifted by changes in such public investments. In chapter 7, Huffman, Ball, Gopinath, and Somwaru apply a cost function approach to assess impacts of these investments on the cost of production and input demand for Midwestern agriculture. The own-price elasticities estimated for capital services, land, hired labor, and materials are –0.86, -0.82, -0.78, and –0.63. Own-state public agricultural research, spill-in research, and highway infrastructure all reduced total variable costs. In chapter 8, Yee, Huffman, Ahearn, and Newton, using the same data, examine the contributions of these same variables in a model with total factor productivity (TFP) as the dependent variable. They apply the model to 48 states and emphasize the analysis of spillovers of agricultural research and of weather. Public R&D and highways are found to have significant positive effects on TFP, but the results for extension are mixed.

The first chapter of Part 3, by Kellogg et al., presents indicators of how changes in pesticide use in agriculture have changed the potential risks to human health and the environment from pesticide contamination of water leaving farm fields. These indicators are designed to provide relative estimates of potential risk that are based on pesticide use and the factors that are known to be important determinants of pesticide loss from farm fields, such as the potential of soils to leach or allow pesticide runoff, the chemical properties of the pesticides, annual rainfall and its relationship to leaching and runoff, and changes in cropping patterns. The other three papers in this section of the book make use of these indicators in their analyses. The framework for constructing the indicators consists of 4700 resource polygons representing the intersection of 48 states, 280 watersheds at the 6-digit Hydrologic Unit level, and 1,400 combinations of climate and soil groups. Twelve major crops are included in the analysis. Estimates of pounds of pesticides applied, mass loss, and annual concentrations leaving the farm field (edge of field and bottom of root zone) were obtained for pesticides used on each of the 12 crops in each of the resource polygons for each year from 1960 through 1997. Indicators of potential risk are constructed from estimates of annual concentrations that exceed "safe" thresholds for chronic exposure to four target groups: humans, fish, crustaceans, and algae. The spatial and temporal trends of these indicators should track the change in actual risk to human health and the environment from agricultural use of pesticides.

The availability of these indicators facilitates the adjustment of productivity measures to account for environmental risk. Conceptually, proper measures of productivity growth should consider air and water as scarce inputs, just as they do labor, capital, and materials. In chapter 10, Ball, Fare, Grosskopf, Hernandez-Sancho, and Nehring introduce an environmental performance index and apply it to the U.S. agricultural sector. They use distance functions and exploit the USDA/ERS output and input data set described in chapter 2 and the environmental data set from chapter 9. The index is the ratio of distance functions that is modified to provide a summary of the degree to which agriculture succeeds in producing maximum "good" outputs with minimal "bad" outputs. Their overall environmental performance index is generally less than one before 1975 and generally greater than one after 1975, suggesting a clear-cut pattern of improvement in the latter half of the sample period.

In chapter 11, Chaston and Gollop demonstrate the relative ease with which changing environmental quality can be incorporated into formal productivity modeling. State-specific total resource productivity (TRP) growth rates are calculated and compared with the corresponding total factor productivity (TFP) growth rates. Because of a rapid rise in groundwater pollution from U.S. farms through 1977, TRP growth is below conventional TFP growth through that year. Regulatory-induced environmental improvements after 1977 lead to TRP exceeding TFP after that year. The importance of environmental regulation in contributing to TRP is empirically assessed in a cost-function model. While annual TFP growth rates across states decreased by an average 0.50 percent over the period 1960-96, annual TRP growth increased by an average of 0.45 points, caused largely by a 0.95 point increase in TRP due to regulatory-induced reductions in chemical use that improved water quality.

In chapter 12, Ball, Felthovan, Nehring, and Morrison-Paul use the same data to estimate cost-based shadow values associated with environmental risk reduction. These shadow values not only can be useful for estimating a more complete measure of productivity growth, but can be used with their input-specific components to impute the social benefit required to justify environmental legislation that limits the environmental risk factors stemming from chemical applications in agriculture. The authors explore the changes in input and output composition necessary to accommodate meaningful reductions in environmental risk as a result of reduced leaching and runoff of pesticides. The costs of reducing risks involve higher levels of most inputs except land, and a movement toward animal and away from crop production.

LESSONS AND LIMITATIONS

In 1980, the American Agricultural Economics Association and the U.S. Department of Agriculture created a task force to recommend to USDA a set of changes in data and methods that, if they were implemented, would provide more accurate assessment of agricultural productivity growth. As the authors of the concluding chapter in this volume note, nearly all those changes have been made, which is a credit both to USDA and to certain individuals. In a book on agricultural productivity measurement, from a workshop 16 years ago, Capalbo and Antle called for the generation of a time-series/cross-sectional national database and for studies that incorporate non-market environmental costs in calculating TFPs. Chapters in sections 1 and 3 below illustrate that these calls were heard.

As the available data have improved in quantity and quality, the range of questions that have been addressed has grown, but a number of important questions remain unanswered. The discussion below highlights some of what has been learned in recent years, as well as key gaps in our knowledge with respect to agricultural productivity measurement and sources.

The median agricultural productivity growth rate over the past several decades has averaged almost two percent per year, with a range across states from less than one percent to around three percent. These results are obtained both by Ball et al., and by Acquaye et al. Such growth rates are remarkable when compared with those in the non-agricultural sectors. Moreover, agricultural productivity growth rates among states appear to be converging over time. The chapters by Huffman et al. and by Yee et al., as well as the results of previous studies, indicate that agricultural research, education, and infrastructure play important roles in generating these productivity gains. Unfortunately, the results do little to explain growth rate differences between the agricultural and non-agricultural sectors.

Adjusting inputs for changes in their quality has served to explain a significant proportion of growth in output, but productivity growth remains the dominant source of economic growth in agriculture, unlike the case for most other sectors in the U.S. economy. What is different between agriculture and most other sectors to explain this difference? Three factors that seem to stand out are the relative importance of publicly funded research in producing new technologies, the relative importance of economies of scale, and perhaps the insufficiently measured embodiment of new technologies in capital and intermediate inputs in agriculture. However, additional research is needed to explain the sharp differences in sources of growth between the agricultural and most non-farm sectors.

Another lesson from research results presented in this book is the sharp reduction in estimated productivity growth rates that occurs when changes in environmental quality are accounted for, such as degree of groundwater pollution. If groundwater pollution differences are not considered, productivity growth from a total resource accounting perspective is biased upward for the two decades prior to 1977 and biased downward for the two decades since then.

Parametric and non-parametric methods for multilateral comparisons of agricultural productivity growth generated generally similar results, but for some states gave significant differences, especially for those states with very different output or input sets. The minimum spanning tree method in particular seemed to give results that made less intuitive sense.

It is clear that much progress has been made in methods for measuring and assessing the sources of agricultural productivity growth over the past few years. Much remains to be learned, however, and fortunately, due to efforts of several of the contributors to this volume, state level production accounts and other data are now available that can be used to increase our understanding of productivity-related issues.

Notes

[1] The underlying economic principles for developing state and national accounts are equivalent, but their application leads to an important difference: state production accounts must explicitly recognize interstate and inter-farm transactions in farm goods while aggregate U.S. farm sector accounts must ignore these transfers. In the context of aggregate accounts, intermediate goods transfers are internal and self-canceling transactions.

References

Ball, V. E. 1985. "Output, Input, and Productivity Measurement in U.S. Agriculture, 1948-79." *American Journal of Agricultural Economics*, 67:475-86.

Ball, V. E., J. C. Bureau, R. Nehring, and A. Somwaru. 1997. "Agricultural Productivity Revisited." *American Journal of Agricultural Economics*, 79:1045-63.

Ball, V. E., F. M. Gollop, A. Kelly-Hawke, and G. P. Swinand. 1999. "Patterns of State Productivity Growth in the U.S. Farm Sector: Linking State and Aggregate Models." *American Journal of Agricultural Economics*, 81:164-79.

Capalbo, S. M., and J. M. Antle. 1988. "Agricultural Productivity Measurement and Explanation." Washington: Resources for the Future.

Eltetö, O., and P. Köves. 1964. "On a Problem of Index Number Computation Relating to International Comparisons." *Szatistikai Szemle*, 42:507-18.

Hill, R. J. 1999. "International Comparisons using Spanning Trees." In *International and Interarea Comparisons of Income, Output and Prices*, ed. A. Heston and R. E. Lipsey. Chicago: University of Chicago Press, 109-20.

Jorgenson, D., F. Gollop, and B. Fraumeni. 1987. *Productivity and U.S. Economic Growth.* Cambridge: Harvard University Press.

Jorgenson, D., and F. Gollop. 1992. "Productivity Growth in U.S. Agriculture: A Postwar Perspective." *American Journal of Agricultural Economics*, 74:745-50.

Kendrick, J. W., and E. S. Grossman. 1980. *Productivity in the United States, Trends and Cycles*. Baltimore: The Johns Hopkins University Press.

Szulc, B. 1964. "Indices for Multiregional Comparison." *Przeglad Statystyczny*, 3:239-54.

Part I.
Production Accounts and Productivity of U.S. Agriculture

2

U.S. Agriculture, 1960-96: A Multilateral Comparison of Total Factor Productivity

V. Eldon Ball, Jean-Pierre Butault, and Richard F. Nehring

Abstract

Increased productivity is the main contributor to growth in U.S. agriculture. This chapter provides estimates of the growth and relative levels of productivity for the 48 contiguous states from 1960 to 1996. For the full 27-year period, every state exhibits a positive and generally substantial average annual rate of productivity growth. There is considerable variance, however, and the wide disparity in growth rates resulted in substantial changes in the rank order of states. For each year, the coefficient of variation (CV) of productivity levels is calculated. These CVs are used to show that the range of levels of productivity has narrowed over time, although the pattern of convergence was far from uniform. The fact that some states grew faster than others and yet the cross section dispersion decreased implies that the states that grew most rapidly were those with lower initial levels of productivity. This result is consistent with Gerschenkron's notion of the advantage of relative backwardness. The states that were particularly far behind the productivity leaders had the most to gain from the diffusion of technical knowledge and proceeded to grow most rapidly. We also observe a positive relationship between capital accumulation and productivity growth, implying embodiment of technology in capital.

Introduction

The rise in agricultural productivity has long been chronicled as the single most important source of economic growth in the U.S. farm sector. Though their methods differ in important ways, the major sectoral productivity studies by Kendrick and Grossman (1980) and Jorgenson, Gollop, and Fraumeni (1987) share this common conclusion.[1] In a recent study, Jorgenson and Gollop (1992) find that productivity growth over the 1947-85 period accounted for 82 percent of the economic growth in agriculture, compared with only 13 percent in the private non-farm economy.[2] Moreover, the rate of productivity growth over this period in agriculture (1.58 percent) was nearly four times the corresponding rate in the private non-farm economy (0.44 percent).

The U.S. Department of Agriculture (USDA) has been monitoring the industry's productivity performance for decades. In fact, the USDA in 1960 was the first agency to introduce multifactor productivity measurement into the Federal statistical program. Today, the U.S. Department of Agriculture's Economic Research Service (ERS) routinely produces total factor productivity measures for the aggregate farm sector from production accounts that distinguish multiple outputs and inputs, adjust for quality change in each input category,[3] and recognize that some farm production (e.g., breeding livestock) is both an investment good as well as an agricultural output.[4]

A properly constructed measure of productivity growth for the aggregate farm sector is certainly important. It provides a useful summary statistic indicating how economic welfare is being advanced through productivity gains in agriculture, but it may mask important state-specific or regional trends. A recent study by Ball et al. (1999) focuses on agricultural productivity at both sector and state levels. A model accounting for interstate transactions in farm goods links sector-wide and state-specific measures of productivity growth. They conclude that there is much more volatility across states than can be inferred from productivity measures for the aggregate farm sector. The results also indicate that productivity growth in the U.S. farm sector is wholly a function of the productivity trends in individual states. Interstate shifts in production activity and resource reallocations have had little effect.

In the present study, we adopt an index number approach to estimate the growth and relative levels of productivity for the states for the period 1960 to 1996. Recent developments in economic theory have improved our knowledge about which index number procedures are most attractive for making productivity comparisons. The distinguishing feature of these procedures is that they possess many properties considered desirable in classical index number analysis, and they exactly represent production structures that have attractive properties.

Problems arise, however, when more than two states are compared simultaneously. The use of binary indices to compare each possible state pair gives results that may not satisfy Fisher's (1922) circularity test. Eltetö and Köves (1964) and Szulc (1964) have proposed a method that achieves transitivity while minimizing the deviations from the binary comparisons. The procedure they developed is used in this study to provide estimates of the relative levels of productivity. Changes in these levels are compared across states over time. We find that the range of levels of productivity (as measured by the coefficient of variation) narrowed over the study period, although the pattern of convergence was far from uniform. This narrowing is remarkable given the wide variation in state growth rates.

To account for the observed convergence in productivity levels, we consider two hypotheses, which are not mutually exclusive. The first is the catch-up hypothesis, which simply states that those states that lagged furthest behind the leading states in terms of levels of productivity, benefit the most from the diffusion of technical knowledge and, hence, should exhibit the most rapid rates of productivity growth. The second hypothesis is that technological innovation is embodied in capital and intermediate inputs. If the input measures do not correct for changes in input quality, then a positive relation should be observed between the rate of productivity growth and the rates of growth of capital and intermediate inputs.

Support is found for the two hypotheses. First, we find a strong inverse relationship between the rates of growth of productivity and the initial levels of productivity. Second, our results support the existence of a positive interaction between capital accumulation and productivity growth. The relation between growth in materials input and productivity growth is positive, as predicted, but statistically insignificant.

METHODOLOGY

A productivity index is generally defined as an output index divided by an input index. The measures of productivity presented in this paper are formed from Fisher indices of outputs and inputs. Let $p^i \gg 0_N$ denote the positive price vector for state $i, i = 1, \Lambda, M$. The corresponding quantity vector is $x^i \geq 0_N$. The Fisher quantity index is defined as:

$$Q_F^{jk}(p^j, p^k, x^j, x^k) \equiv \left[\frac{p^k \cdot x^j}{p^k \cdot x^k} \frac{p^j \cdot x^j}{p^j \cdot x^k} \right]^{1/2}, \qquad (1)$$

where the superscripts j and k can be interpreted as time periods or as states. Diewert (1976) showed that equation (1) can be derived from a homogeneous quadratic production or input requirements function (or aggregator function). The homogeneous quadratic aggregator function can provide a second-order approximation to an arbitrary twice differentiable, linear homogeneous function.

The direct application of equation (1) to the $M(M-1)/2$ possible pairs of states yields a matrix of binary comparisons that may not satisfy the transitivity condition. To eliminate this problem, we apply the multilateral Eltetö and Köves (1964) and Szulc (1964) method, which defines the index for state j relative to state k as the unweighted geometric mean of M binary Fisher indices:

$$Q_{EKS}^{jk} \equiv \left(\prod_{i=1}^{M} Q_F(p^j, p^i, x^j, x^i) \cdot Q_F(p^i, p^k, x^i, x^k) \right)^{1/M}, \forall j, k. \quad (2)$$

The multilateral Eltetö-Köves-Szulc index defined by equation (2) satisfies transitivity while minimizing the deviations from the binary Fisher indices.

The binary Fisher indices, which are the building blocks of the multilateral Eltetö-Köves-Szulc indices, are based on prices and quantities of commodities common to both states in the comparison. Even so, these binary indices sometimes rely on a very small number of commodities. In this study, we construct direct binary Fisher indices if the commodities common to both states represent a minimum percentage of the value of production in both states. Below this percentage, the Eltetö-Köves-Szulc indices are constructed using indirect binary comparisons through other states.

The indirect binary indices are calculated using a chain-link method. Adjacent states are selected on the basis of the shortest possible path (i.e., fewest number of states) without falling below this predefined threshold. This method is similar to the chain-link method used in inter-temporal comparisons. The only difference is that there is no natural ordering of the data points (such as chronological ordering).

Production Accounts

ERS has constructed state and aggregate accounts for the farm sector. Output of the sector is defined as gross production leaving the farm, as opposed to real value added. Inputs are not limited to capital and labor, but include intermediate inputs as well. Both state and aggregate accounts view all of agriculture within their respective boundaries as if it were a single farm. Output includes all off-farm deliveries but excludes intermediate goods produced and consumed on the farm. The difference is that output in the aggregate accounts is defined as deliveries to final demand and intermediate demands in the non-farm sector. State output accounts include these deliveries plus interstate shipments to intermediate farm demands. The text in this section provides an overview of the sources and methods used to construct the annual production accounts for the 1960-96 period for each of the 48 contiguous states.

Output

The development of a measure of output begins with disaggregated data for physical quantities and market prices of crops and livestock compiled for each state by ERS's Resource Economics Division.[5] The output quantity for each crop and livestock category consists of quantities of commodities sold off the farm, additions to inventory, and quantities consumed as part of final demand in farm households during the calendar year. As discussed above, off-farm sales in the aggregate accounts are defined only in terms of output leaving the sector. Off-farm sales in the state accounts include sales to the farm sector in other states as well.

The price corresponding to each disaggregated output reflects the value of that output to the sector; i.e., subsidies are added and indirect taxes are subtracted from market values. Using equation (2), we construct multilateral indices of crop and livestock output for the forty-eight contiguous states for the period 1960 to 1996. Indices of crop output and livestock output for this period and these states are found on the U.S. Department of Agriculture, Economic Research Service website: http://preview.ers.usda.gov/data/stateproductivity/. All values are presented relative to Alabama in 1996. Average annual rates of growth are presented for the full 1960 to 1996 period, and for six sub-periods.[6]

Intermediate Inputs

Intermediate inputs consist of goods used in production during the calendar year, whether withdrawn from beginning inventories or purchased from outside the farm sector, or (in the case of the state production accounts) from farms in other states. The inclusion and treatment of open-market purchases of feed, seed, and livestock inputs require little discussion. These inputs should enter both state and aggregate farm sector intermediate goods accounts. However, the treatment of withdrawals from producers' inventories requires elaboration.

Inventories enter the measurement of output, intermediate input, and capital input. Beginning inventories of crops and livestock represent capital inputs and are treated as such in the discussion of capital later. Additions to these inventories represent deliveries to final demand and, therefore, are treated as part of output. Goods withdrawn from inventory are symmetrically defined as intermediate goods and, therefore, must enter the farm input accounts.

Data on current dollar consumption of petroleum fuels, natural gas, and electricity in agriculture are compiled for each state for the 1960 to 1996 period. Prices of individual fuels are taken from the Energy Information Administration's Monthly Energy Review. The index of energy consumption is formed implicitly as the ratio of total expenditures (less state and federal excise tax refunds) to the corresponding price index.

Pesticides and fertilizers have undergone significant changes in input quality over the study period. Since input price and quantity series used in a study of productivity must be denominated in constant-efficiency units, we construct price indices for fertilizers and pesticides from hedonic regression results. A price index of fertilizers is formed by regressing the prices of single-nutrient and multi-grade fertilizer materials on the proportion of nutrients contained in the materials.[7] Price differences across pesticides are assumed due to differences in physical characteristics such as toxicity, persistence in the environment, and leaching potential.[8] The corresponding quantity indices are formed implicitly as the ratio of the value of each aggregate to its price index.

There remain several purchased inputs that collectively account for some 15 percent of the input service flow. We compute price and implicit quantity indices of purchased services such as contract labor services, custom machine services (less income from machine hire), machine and building maintenance and repairs, and irrigation from public sellers of water. Indices of intermediate input are constructed by aggregating across each category of intermediate inputs described above. These indices and their average annual rates of growth are found at http://preview.ers.usda.gov/data/stateproductivity/.

Capital Input

This study requires measures of capital input and capital service prices for each state. Construction of these series begins with estimating the capital stock and rental price for each asset type for each state. The perpetual inventory method is used to develop capital stocks from data on investment. Implicit rental prices for each asset are based on the correspondence between the purchase price of the asset and the discounted value of future service flows derived from that asset.

Capital Stocks

Under the perpetual inventory method, capital stock at the end of each period, K_t, is measured as the sum of all past investments, each weighted by its relative efficiency, d_t:

$$K_t = \sum_{\tau=0}^{\infty} d_\tau I_{t-\tau} . \qquad (3)$$

We assume that the relative efficiency of capital goods declines with age, giving rise to the need for replacement of productive capacity. The proportion of investment to be replaced at age τ is equal to the decline in efficiency from age $\tau - 1$ to age τ:

$$m_\tau = -(d_\tau - d_{\tau-1}), \tau = 1, \Lambda, t. \tag{4}$$

These proportions represent mortality rates for capital goods of different ages. Replacement requirements in period t can be expressed as a weighted sum of past investments:

$$R_t = \sum_{\tau=1}^{\infty} m_\tau I_{t-\tau}, \tag{5}$$

where the weights are the mortality rates.

Taking the first difference of expression equation (3) and substituting from equations (4) and (5), we can write:

$$K_t - K_{t-1} = I_t - R_t. \tag{6}$$

The change in capital stock in any period is equal to the acquisition of investment goods less replacement requirements.

To estimate replacement requirements, we must introduce an explicit description of the decline in efficiency. The relative efficiency of an asset τ years of age is given by:

$$\begin{aligned} d_\tau &= (L - \tau)/(L - \beta\tau), 0 \le \tau \le L \\ d_\tau &= 0, \tau > L, \end{aligned} \tag{7}$$

where L is the service life of the asset and β is a curvature or decay parameter.[9]

Little empirical evidence is available to suggest a precise value of β. However, two studies provide evidence that suggests efficiency decay occurs more rapidly in the later years of service. Utilizing data on expenditures for maintenance and repairs of farm tractors covering the period 1958-74, Penson, Hughes, and Nelson (1977) found that efficiency loss was, in fact, very small in the early years of service and increased rapidly as the end of the asset's service life approached. More recently, Romain, Penson, and Lambert (1987) compared the explanatory power of alternative capacity depreciation patterns for farm tractors in a model of investment behavior that also included the price of capital services. They observed that the concave depreciation pattern better reflected actual investment decisions.

Taken together, these studies suggest that possible values of β should be restricted to the zero-one interval. Ultimately, the β values chosen for this study were 0.50 for durable equipment and 0.75 for structures. It is assumed that the efficiency of a structure declines slowly over most of the service life until a point is reached where the cost of repairs exceeds the increased service flows derived from the repairs, at which point the structure is allowed to depreciate rapidly. The decay parameter for durable equipment assumes that the decline in efficiency was more uniformly distributed over the asset's service life.

Investment as used in this study is composed of different types of capital goods. Each type of capital good is a homogeneous group of assets for which the service life L is a random variable reflecting quality differences, maintenance schedules, etc. For each asset type, there exists some mean service life $\bar{L}$ around which there exists some distribution of actual service lives. In order to determine the amount of capital available for production, the different service lives and their frequency of occurrence must be determined. It is assumed that this distribution can accurately be depicted by the normal distribution truncated at a point two standard deviations before and after the mean.[10]

Once the frequency of occurrence of a particular service life is determined, the efficiency function for that service life is calculated using the assumed value of β. This process is repeated for all possible service lives. An aggregate efficiency function is then constructed as a weighted sum of the individual efficiency functions using as weights the frequency of occurrence. This function reflects not only changes in efficiency, but also the discard distribution around the mean service life of the asset.

Beginning inventories of crops and livestock are treated as capital inputs.[11] We estimate the stock of inventories using the perpetual inventory method, assuming zero replacement.

PRICE OF CAPITAL SERVICES

Firms add to the capital stock so long as the present value of the net revenue generated by an additional unit of capital exceeds the purchase price of the asset. Following Coen (1975), this requirement can be stated algebraically as:

$$\sum_{t=1}^{\infty}\left(p\frac{\partial y}{\partial K} - w\frac{\partial R_t}{\partial K}\right)(1+r)^{-t} > w, \tag{8}$$

where p is the price of output, w is the price of an additional unit of capital, and r is the real discount rate.

To maximize net present value, firms add to the capital stock until this equation holds as an equality. For this equality to hold:

$$p\frac{\partial y}{\partial K} = rw + r\sum_{t=1}^{\infty} w\frac{\partial R_t}{\partial K}(1+r)^{-t} = c. \tag{9}$$

The expression for c is the implicit rental price of capital corresponding to the mortality distribution m. The rental price consists of two components. The first term, rw, represents the opportunity cost of invested funds. The second term, $r\sum_{t=1}^{\infty} w\frac{\partial R_t}{\partial K}(1+r)^{-t}$, is the present value of all future replacements required to maintain the productive capacity of the capital stock.

Let us define F as the present value of the stream of capacity depreciation on one unit of capital according to the mortality distribution m:

$$F = \sum_{t=1}^{\infty} m_t (1+r)^{-t}. \tag{10}$$

Since replacement at time t is equal to capacity depreciation at time t:

$$\sum_{t=1}^{\infty} \frac{\partial R_t}{\partial K}(1+r)^{-t} = \sum_{t=1}^{\infty} F^t = F/(1-F) \tag{11}$$

and

$$c = \frac{rw}{(1-F)}. \tag{12}$$

The real rate of return r in the above expression is calculated as the nominal yield on investment grade corporate bonds less the rate of inflation as measured by the implicit deflator for gross domestic product.[12] An *ex ante* rate is obtained by expressing observed real rates as an ARIMA process.[13] We then calculate F holding r constant for that particular vintage of capital goods. In this way, implicit rental prices c are calculated for each asset type.

Indices of capital input in each state are constructed by aggregating over the different capital assets using as weights the asset-specific rental prices. Service prices for capital input are formed implicitly as the ratio of the total current dollar value of capital service flows to the quantity index. The resulting measure of capital input for each state is adjusted for changes in input quality (See the USDA/ERS website cited above).

Land Input

To obtain a constant-quality land stock, we compile data on land area and average value (excluding buildings) per acre in each Agricultural Statistics District in each state. We further disaggregate land input into irrigated and dry cropland, grazing land, and other land in 11 western states. The land area in each district and use category is reported in the quinquennial Census of Agriculture (U.S. Department of Commerce). USDA's National Agricultural Statistics Service annually updates state estimates of total land in farms. For the years intermediate to the censuses, percentages in each district and use category are interpolated. Land values per acre are used to aggregate across the different land categories in each state. The service flows from public lands were estimated by means of grazing fees paid using data from the U.S. Department of Interior's Bureau of Land Management and USDA's Forest Service.

Finally, the differences in the relative efficiency of land across states prevent the direct comparison of observed prices. We construct relative prices of land in each state based on hedonic regressions. Indices of land service flows are formed implicitly as the ratio of the value of service flows to the price index.

Labor Input

The USDA labor accounts, for the aggregate farm sector, incorporate the demographic cross-classification of the agricultural labor force developed by Jorgenson, Gollop, and Fraumeni (1987).[14] Matrices of hours worked and compensation per hour have been developed for laborers cross-classified by sex, age, education, and employment class—employee versus self-employed and unpaid family workers.

The task for this study is to develop a set of similarly formatted but otherwise demographically distinct matrices of labor input and labor compensation by state. These matrices are developed using the RAS procedure popularized by Jorgenson, Gollop, and Fraumeni (1987, pp. 72-76) by combining the aggregate farm sector matrices initially produced in that study but updated through 1996 with state-specific demographic information available from the decennial Census of Population (U.S. Department of Commerce).[15] The result is state-by-year matrices of hours worked and hourly compensation with cells cross-classified by sex, age, education, and employment class and with each matrix controlled to the USDA hours worked and compensation totals.

Labor compensation (opportunity cost) data for self-employed and unpaid family workers are not available. As a result, for each state and year, self-employed and unpaid family workers in each state are imputed the mean wage earned by hired workers with the same demographic characteristics.

Indices of labor input are constructed for each state over the 1960-96 period using the demographically cross-classified hours and compensation data. Labor hours having higher marginal productivity (wages) are given higher weights in forming the index of labor input than are hours having lower marginal productivities. Doing so explicitly adjusts indices of labor input for quality change in hours as originally defined by Jorgenson and Griliches (1967).[16]

Total Factor Productivity

In a table on the USDA/ERS website http://www.ers.usda.gov/ we present indices of total output for the 48 states for the period 1960 to 1996. These indices are constructed from disaggregated industry data described earlier. In another table on that website, we present indices of total factor input. Indices of total factor productivity for each state and year are formed as the ratio of the output index to the input index. These indices are presented both on the website and in table 2.1 below, along with their percentage rates of growth. Finally, in table 2.2, we rank the states by their relative level of productivity in 1996. We also include in the table their rank in 1960 and the average annual rate of growth from 1960 to 1996.

One remarkable similarity exists across all states. For the full 1960-96 period, every state exhibited a positive and generally substantial average annual rate of productivity growth. There is considerable variance, however. The median rate of productivity growth was 1.94 percent per year. Nearly half the states (nineteen of forty-eight) had productivity growth rates averaging more than 2 percent per year. Only two states (Oklahoma and Wyoming) had an average rate of growth less than 1 percent per year. The reported annual rates of growth ranged from 0.94 percent for Wyoming to 2.84 percent for Louisiana.

The wide disparity in growth rates over the 1960-96 period resulted in substantial changes in the rank order of states. For example, between 1960 and 1996, Connecticut rose from twentieth to first, Georgia rose from fourteenth to third, and North Carolina rose from twenty-second to fourth. In contrast, Iowa fell from first to fifth, Wisconsin fell from third to fourteenth, and Colorado fell from fourth to twenty-first in terms of levels of productivity. West Virginia was last throughout the period. Moreover, its productivity relative to Connecticut fell from one-half in 1960 to one-third in 1996.

Table 2.1. Total Factor Productivity Relative to 1996 Level for Alabama

Year	AL	AR	AZ	CA	CO	CT	DE	FL
1960	0.511	0.484	0.556	0.612	0.654	0.549	0.595	0.701
1961	0.534	0.520	0.597	0.638	0.676	0.600	0.645	0.784
1962	0.514	0.530	0.573	0.656	0.649	0.595	0.640	0.840
1963	0.578	0.545	0.610	0.673	0.649	0.632	0.663	0.727
1964	0.582	0.584	0.578	0.709	0.666	0.633	0.668	0.712
1965	0.590	0.618	0.595	0.691	0.664	0.657	0.732	0.799
1966	0.552	0.600	0.599	0.736	0.703	0.698	0.671	0.824
1967	0.557	0.584	0.590	0.715	0.701	0.741	0.824	0.879
1968	0.570	0.622	0.616	0.769	0.730	0.720	0.796	0.812
1969	0.593	0.614	0.637	0.782	0.757	0.738	0.892	0.909
1970	0.594	0.639	0.629	0.770	0.785	0.735	0.861	0.860
1971	0.671	0.656	0.617	0.791	0.795	0.778	0.903	0.890
1972	0.663	0.664	0.635	0.808	0.787	0.723	0.979	0.933
1973	0.646	0.693	0.665	0.839	0.813	0.699	0.996	1.015
1974	0.670	0.643	0.692	0.849	0.769	0.696	0.919	1.056
1975	0.752	0.827	0.667	0.861	0.812	0.742	0.973	1.154
1976	0.726	0.732	0.668	0.865	0.816	0.719	1.001	1.184
1977	0.727	0.794	0.656	0.905	0.846	0.753	0.932	1.154
1978	0.698	0.773	0.646	0.880	0.804	0.750	0.868	1.082
1979	0.725	0.821	0.682	0.905	0.811	0.720	0.920	1.012
1980	0.642	0.683	0.670	0.957	0.800	0.718	0.807	1.122
1981	0.793	0.871	0.737	0.975	0.798	0.743	0.919	1.118
1982	0.832	0.883	0.717	1.024	0.799	0.886	0.952	1.109
1983	0.766	0.735	0.677	0.951	0.814	0.768	0.960	1.142
1984	0.790	0.869	0.699	1.025	0.862	0.807	0.915	1.128
1985	0.854	0.891	0.721	1.085	0.899	0.881	1.044	1.168
1986	0.829	0.888	0.766	1.093	0.900	0.965	1.053	1.249
1987	0.827	0.938	0.781	1.210	0.917	1.101	1.031	1.228
1988	0.852	0.974	0.725	1.075	0.918	1.119	1.079	1.380
1989	0.858	0.944	0.831	1.164	0.922	1.015	1.086	1.381
1990	0.873	0.937	0.762	1.222	0.970	1.129	1.198	1.357
1991	0.973	0.940	0.799	1.084	1.039	1.107	1.248	1.386
1992	0.935	1.081	0.825	1.227	1.062	1.215	1.288	1.431
1993	0.893	0.989	0.842	1.175	1.103	1.171	1.278	1.484
1994	1.017	1.098	0.824	1.204	1.069	1.190	1.297	1.532
1995	0.925	1.045	0.821	1.061	1.059	1.327	1.146	1.438
1996	1.000	1.184	0.925	1.146	1.083	1.509	1.197	1.504
Average annual growth rates:								
1960-96	0.019	0.025	0.014	0.017	0.014	0.028	0.019	0.021
1960-66	0.013	0.036	0.013	0.031	0.012	0.040	0.020	0.027
1966-69	0.023	0.008	0.020	0.020	0.025	0.019	0.095	0.033
1969-73	0.022	0.030	0.011	0.018	0.018	-0.013	0.028	0.028
1973-79	0.019	0.028	0.004	0.013	0.000	0.005	-0.013	0.000
1979-89	0.017	0.014	0.020	0.025	0.013	0.034	0.017	0.031
1989-96	0.022	0.032	0.015	-0.002	0.023	0.057	0.014	0.012

Year	GA	IA	ID	IL	IN	KS	KY	LA
1960	0.560	0.712	0.525	0.599	0.510	0.636	0.496	0.386
1961	0.609	0.747	0.558	0.636	0.533	0.630	0.552	0.409
1962	0.605	0.749	0.587	0.641	0.557	0.616	0.552	0.429
1963	0.660	0.797	0.603	0.679	0.595	0.599	0.576	0.478
1964	0.667	0.803	0.598	0.657	0.557	0.625	0.631	0.477
1965	0.677	0.795	0.595	0.723	0.606	0.660	0.581	0.489
1966	0.673	0.815	0.588	0.666	0.560	0.629	0.561	0.503
1967	0.757	0.844	0.679	0.751	0.623	0.670	0.655	0.546
1968	0.713	0.855	0.694	0.719	0.650	0.713	0.625	0.583
1969	0.729	0.832	0.695	0.737	0.678	0.759	0.677	0.536
1970	0.775	0.830	0.706	0.675	0.639	0.750	0.671	0.594
1971	0.863	0.896	0.744	0.804	0.727	0.832	0.699	0.593
1972	0.840	0.898	0.737	0.784	0.680	0.832	0.693	0.606
1973	0.822	0.891	0.729	0.757	0.677	0.840	0.682	0.579
1974	0.918	0.790	0.728	0.635	0.581	0.754	0.734	0.583
1975	0.960	0.806	0.744	0.826	0.681	0.794	0.701	0.667
1976	0.896	0.790	0.758	0.743	0.726	0.769	0.781	0.665
1977	0.841	0.847	0.730	0.772	0.717	0.836	0.877	0.688
1978	0.881	0.894	0.733	0.762	0.717	0.705	0.748	0.663
1979	0.912	0.894	0.751	0.816	0.725	0.764	0.804	0.707
1980	0.770	0.874	0.775	0.711	0.698	0.687	0.766	0.606
1981	0.914	0.981	0.808	0.837	0.729	0.742	0.906	0.699
1982	1.036	0.903	0.812	0.858	0.800	0.790	0.946	0.774
1983	0.949	0.754	0.843	0.626	0.616	0.736	0.851	0.710
1984	1.015	0.920	0.865	0.823	0.821	0.826	1.005	0.784
1985	1.077	1.036	0.859	0.983	0.897	0.890	1.106	0.735
1986	0.984	1.052	0.914	0.944	0.850	0.893	0.963	0.744
1987	1.030	1.014	0.990	0.943	0.903	0.921	0.937	0.750
1988	1.110	0.880	0.954	0.737	0.724	0.885	0.854	0.855
1989	1.141	1.078	0.964	1.037	0.953	0.848	1.008	0.703
1990	1.120	1.068	1.078	1.037	0.968	0.950	0.943	0.825
1991	1.238	1.071	1.012	0.952	0.874	0.958	1.010	0.790
1992	1.309	1.251	1.145	1.175	1.109	1.022	1.085	0.906
1993	1.200	0.978	1.254	1.065	1.041	0.975	1.055	0.825
1994	1.409	1.390	1.177	1.214	1.155	1.089	1.115	0.986
1995	1.302	1.186	1.145	0.976	0.997	0.915	1.066	0.932
1996	1.398	1.299	1.218	1.093	1.040	1.032	1.181	1.074
Average annual growth rates:								
1960-96	0.025	0.017	0.023	0.017	0.020	0.013	0.024	0.028
1960-66	0.031	0.022	0.019	0.018	0.015	-0.002	0.020	0.044
1966-69	0.026	0.007	0.055	0.034	0.064	0.063	0.063	0.021
1969-73	0.030	0.017	0.012	0.007	0.000	0.026	0.002	0.019
1973-79	0.017	0.000	0.005	0.012	0.012	-0.016	0.027	0.033
1979-89	0.022	0.019	0.025	0.024	0.027	0.011	0.023	-0.001
1989-96	0.029	0.027	0.033	0.008	0.013	0.028	0.023	0.061

continued

Table 2.1., *continued*

Year	MA	MD	ME	MI	MN	MO	MS	MT
1960	0.477	0.468	0.593	0.384	0.592	0.498	0.398	0.423
1961	0.514	0.487	0.663	0.418	0.619	0.523	0.442	0.380
1962	0.531	0.480	0.687	0.443	0.581	0.516	0.440	0.471
1963	0.558	0.490	0.696	0.447	0.646	0.556	0.509	0.485
1964	0.571	0.509	0.698	0.485	0.601	0.536	0.525	0.498
1965	0.577	0.541	0.685	0.471	0.626	0.573	0.519	0.536
1966	0.608	0.512	0.680	0.465	0.646	0.540	0.525	0.512
1967	0.629	0.594	0.743	0.476	0.656	0.576	0.531	0.519
1968	0.635	0.589	0.792	0.513	0.691	0.631	0.581	0.560
1969	0.654	0.641	0.789	0.522	0.671	0.561	0.588	0.530
1970	0.700	0.636	0.808	0.530	0.700	0.576	0.590	0.528
1971	0.719	0.642	0.876	0.548	0.748	0.660	0.635	0.552
1972	0.685	0.655	0.894	0.596	0.727	0.635	0.632	0.550
1973	0.672	0.648	0.838	0.581	0.782	0.641	0.657	0.520
1974	0.713	0.650	0.735	0.570	0.655	0.544	0.629	0.521
1975	0.742	0.688	0.854	0.674	0.688	0.605	0.739	0.617
1976	0.702	0.695	0.845	0.600	0.619	0.575	0.698	0.604
1977	0.705	0.680	0.803	0.717	0.837	0.708	0.750	0.528
1978	0.698	0.699	0.784	0.679	0.779	0.673	0.698	0.555
1979	0.647	0.684	0.813	0.688	0.794	0.708	0.753	0.485
1980	0.641	0.613	0.725	0.695	0.766	0.610	0.628	0.513
1981	0.694	0.666	0.735	0.739	0.840	0.764	0.776	0.621
1982	0.674	0.718	0.907	0.763	0.839	0.685	0.863	0.646
1983	0.694	0.684	0.832	0.702	0.751	0.589	0.737	0.634
1984	0.736	0.768	0.830	0.754	0.855	0.671	0.810	0.531
1985	0.785	0.838	0.893	0.873	0.895	0.806	0.889	0.404
1986	0.852	0.813	0.941	0.851	0.928	0.775	0.756	0.679
1987	0.882	0.832	0.904	0.835	0.938	0.749	0.852	0.703
1988	0.926	0.831	0.996	0.785	0.802	0.692	0.853	0.472
1989	0.957	0.825	0.900	0.904	1.005	0.799	0.789	0.670
1990	0.921	0.850	0.963	0.942	1.011	0.761	0.822	0.686
1991	1.003	0.848	0.952	0.889	1.001	0.762	0.847	0.753
1992	0.933	0.885	1.038	0.966	1.061	0.889	1.026	0.694
1993	0.989	0.870	1.011	0.973	0.832	0.786	0.868	0.816
1994	0.887	0.908	0.944	1.014	1.113	0.896	0.972	0.697
1995	0.888	0.850	0.988	1.002	1.049	0.779	0.965	0.797
1996	1.033	0.954	1.208	0.981	1.132	0.933	1.034	0.707
Average annual growth rates:								
1960-96	0.021	0.020	0.020	0.026	0.018	0.017	0.027	0.014
1960-66	0.040	0.015	0.023	0.032	0.014	0.014	0.046	0.032
1966-69	0.025	0.075	0.049	0.039	0.013	0.013	0.038	0.011
1969-73	0.007	0.003	0.015	0.027	0.039	0.033	0.028	-0.005
1973-79	-0.006	0.009	-0.005	0.028	0.002	0.017	0.023	-0.012
1979-89	0.039	0.019	0.010	0.027	0.024	0.012	0.005	0.032
1989-96	0.011	0.021	0.042	0.012	0.017	0.022	0.039	0.008

Year	NC	ND	NE	NH	NJ	NM	NV	NY
1960	0.522	0.437	0.557	0.442	0.581	0.450	0.559	0.603
1961	0.545	0.343	0.536	0.498	0.599	0.462	0.551	0.646
1962	0.578	0.539	0.561	0.538	0.605	0.489	0.584	0.639
1963	0.591	0.493	0.563	0.538	0.608	0.490	0.632	0.684
1964	0.643	0.516	0.575	0.542	0.599	0.474	0.621	0.681
1965	0.572	0.570	0.588	0.544	0.634	0.493	0.621	0.729
1966	0.602	0.519	0.651	0.578	0.631	0.531	0.647	0.729
1967	0.661	0.525	0.654	0.600	0.665	0.516	0.594	0.746
1968	0.615	0.591	0.651	0.604	0.661	0.542	0.623	0.737
1969	0.669	0.600	0.704	0.598	0.668	0.544	0.687	0.735
1970	0.695	0.527	0.674	0.588	0.674	0.573	0.668	0.761
1971	0.720	0.699	0.727	0.654	0.685	0.559	0.700	0.776
1972	0.755	0.641	0.726	0.672	0.637	0.592	0.657	0.720
1973	0.809	0.644	0.709	0.674	0.672	0.609	0.700	0.730
1974	0.817	0.564	0.624	0.633	0.694	0.556	0.637	0.713
1975	0.888	0.656	0.714	0.694	0.655	0.579	0.668	0.739
1976	0.846	0.651	0.693	0.656	0.657	0.571	0.654	0.706
1977	0.790	0.611	0.790	0.644	0.660	0.610	0.679	0.729
1978	0.815	0.704	0.743	0.629	0.668	0.573	0.601	0.734
1979	0.764	0.642	0.774	0.654	0.635	0.556	0.638	0.767
1980	0.783	0.557	0.708	0.593	0.604	0.584	0.651	0.759
1981	0.869	0.853	0.819	0.627	0.664	0.607	0.663	0.759
1982	0.885	0.839	0.793	0.652	0.707	0.641	0.757	0.833
1983	0.840	0.780	0.717	0.700	0.679	0.635	0.753	0.783
1984	0.901	0.831	0.838	0.676	0.714	0.662	0.778	0.773
1985	0.957	0.923	0.929	0.694	0.805	0.754	0.758	0.889
1986	0.961	0.965	0.928	0.696	0.796	0.721	0.726	0.907
1987	1.003	0.905	0.956	0.763	0.862	0.692	0.757	0.936
1988	1.093	0.488	0.961	0.815	0.838	0.742	0.734	0.930
1989	1.130	0.736	0.981	0.750	0.840	0.771	0.787	0.939
1990	1.208	0.925	1.012	0.740	0.870	0.752	0.794	0.955
1991	1.283	0.906	1.057	0.818	0.969	0.782	0.782	0.970
1992	1.327	1.114	1.117	0.905	0.949	0.799	0.700	0.991
1993	1.343	0.853	1.055	0.925	0.963	0.807	0.804	0.993
1994	1.451	0.972	1.141	0.944	0.970	0.856	0.789	0.994
1995	1.345	0.891	1.028	0.845	0.962	0.869	0.800	0.983
1996	1.386	1.000	1.122	0.924	1.080	0.969	0.855	1.042
Average annual growth rates:								
1960-96	0.027	0.023	0.019	0.020	0.017	0.021	0.012	0.015
1960-66	0.024	0.029	0.026	0.045	0.014	0.028	0.024	0.032
1966-69	0.035	0.048	0.026	0.011	0.019	0.008	0.020	0.003
1969-73	0.048	0.018	0.002	0.030	0.002	0.028	0.005	-0.002
1973-79	-0.010	-0.001	0.015	-0.005	-0.009	-0.015	-0.016	0.008
1979-89	0.039	0.014	0.024	0.014	0.028	0.033	0.021	0.020
1989-96	0.029	0.044	0.019	0.030	0.036	0.033	0.012	0.015

continued

Table 2.1., *continued*

Year	OH	OK	OR	PA	RI	SC	SD	TN
1960	0.460	0.490	0.479	0.500	0.424	0.456	0.613	0.387
1961	0.472	0.485	0.498	0.535	0.464	0.476	0.587	0.424
1962	0.493	0.438	0.512	0.528	0.508	0.514	0.637	0.421
1963	0.521	0.453	0.523	0.569	0.565	0.537	0.639	0.437
1964	0.520	0.500	0.543	0.584	0.605	0.544	0.620	0.463
1965	0.537	0.575	0.564	0.607	0.585	0.574	0.662	0.467
1966	0.569	0.513	0.601	0.578	0.597	0.563	0.661	0.426
1967	0.560	0.507	0.612	0.654	0.573	0.622	0.707	0.448
1968	0.610	0.550	0.620	0.628	0.559	0.541	0.728	0.461
1969	0.592	0.553	0.662	0.663	0.598	0.605	0.685	0.485
1970	0.619	0.564	0.669	0.683	0.647	0.586	0.678	0.484
1971	0.675	0.549	0.700	0.694	0.661	0.663	0.749	0.512
1972	0.646	0.587	0.738	0.678	0.594	0.617	0.777	0.499
1973	0.578	0.650	0.757	0.672	0.633	0.641	0.738	0.507
1974	0.628	0.643	0.802	0.656	0.570	0.646	0.694	0.514
1975	0.706	0.692	0.831	0.665	0.689	0.793	0.693	0.585
1976	0.705	0.647	0.820	0.678	0.611	0.702	0.567	0.574
1977	0.708	0.725	0.763	0.684	0.583	0.698	0.773	0.603
1978	0.682	0.612	0.731	0.719	0.612	0.723	0.744	0.596
1979	0.749	0.700	0.748	0.740	0.597	0.744	0.739	0.613
1980	0.701	0.659	0.822	0.701	0.603	0.644	0.703	0.546
1981	0.676	0.692	0.802	0.785	0.600	0.801	0.778	0.667
1982	0.745	0.807	0.798	0.786	0.790	0.875	0.836	0.679
1983	0.624	0.730	0.819	0.742	0.755	0.723	0.752	0.522
1984	0.827	0.733	0.875	0.847	0.800	0.868	0.917	0.694
1985	0.911	0.761	0.875	0.925	0.968	0.896	0.911	0.747
1986	0.864	0.780	0.948	0.940	0.964	0.802	0.972	0.711
1987	0.868	0.757	0.991	0.946	1.046	0.917	0.981	0.758
1988	0.761	0.757	1.020	0.923	1.141	0.942	0.811	0.709
1989	0.852	0.764	0.967	0.934	0.973	0.943	0.985	0.703
1990	0.933	0.746	1.007	0.942	0.929	0.875	1.087	0.693
1991	0.807	0.717	1.011	0.901	0.896	0.944	1.103	0.699
1992	0.994	0.746	0.979	1.043	0.956	1.083	1.207	0.775
1993	0.900	0.735	1.112	1.014	0.864	0.958	1.033	0.730
1994	1.022	0.775	1.053	1.059	0.810	1.185	1.225	0.836
1995	0.895	0.677	0.968	1.044	0.836	1.020	1.003	0.747
1996	0.884	0.699	0.990	1.112	0.851	1.100	1.213	0.775
Average annual growth rates:								
1960-96	0.018	0.010	0.020	0.022	0.019	0.024	0.019	0.019
1960-66	0.035	0.008	0.038	0.024	0.057	0.035	0.013	0.016
1966-69	0.014	0.025	0.032	0.046	0.001	0.024	0.012	0.043
1969-73	-0.006	0.040	0.034	0.003	0.014	0.014	0.018	0.011
1973-79	0.043	0.012	-0.002	0.016	-0.010	0.025	0.000	0.032
1979-89	0.013	0.009	0.026	0.023	0.049	0.024	0.029	0.014
1989-96	0.005	-0.013	0.003	0.025	-0.019	0.022	0.030	0.014

Year	TX	UT	VA	VT	WA	WI	WV	WY
1960	0.478	0.480	0.423	0.560	0.554	0.684	0.248	0.449
1961	0.482	0.492	0.450	0.626	0.581	0.723	0.258	0.454
1962	0.472	0.523	0.460	0.629	0.612	0.754	0.251	0.466
1963	0.507	0.531	0.420	0.660	0.641	0.746	0.261	0.505
1964	0.505	0.543	0.487	0.668	0.674	0.786	0.273	0.496
1965	0.535	0.576	0.482	0.665	0.681	0.790	0.274	0.490
1966	0.539	0.559	0.457	0.685	0.735	0.781	0.263	0.502
1967	0.518	0.617	0.505	0.715	0.782	0.794	0.297	0.540
1968	0.554	0.616	0.504	0.748	0.778	0.821	0.295	0.526
1969	0.534	0.606	0.521	0.762	0.776	0.779	0.299	0.517
1970	0.563	0.636	0.533	0.806	0.765	0.807	0.305	0.527
1971	0.544	0.648	0.548	0.837	0.829	0.865	0.323	0.561
1972	0.582	0.638	0.562	0.829	0.860	0.837	0.322	0.531
1973	0.644	0.655	0.566	0.851	0.825	0.805	0.315	0.518
1974	0.607	0.636	0.588	0.804	0.866	0.793	0.317	0.543
1975	0.671	0.630	0.601	0.834	1.027	0.810	0.335	0.554
1976	0.646	0.633	0.606	0.843	0.989	0.754	0.315	0.565
1977	0.680	0.601	0.613	0.813	0.954	0.890	0.304	0.544
1978	0.602	0.577	0.616	0.832	0.956	0.839	0.348	0.542
1979	0.636	0.591	0.630	0.800	0.911	0.837	0.363	0.510
1980	0.579	0.592	0.574	0.754	0.960	0.831	0.370	0.516
1981	0.674	0.695	0.657	0.761	1.000	0.860	0.356	0.570
1982	0.702	0.679	0.663	0.858	0.999	0.900	0.339	0.546
1983	0.634	0.679	0.641	0.815	1.058	0.818	0.382	0.539
1984	0.629	0.663	0.742	0.781	1.043	0.897	0.422	0.548
1985	0.704	0.698	0.754	0.858	1.055	0.930	0.464	0.545
1986	0.686	0.768	0.736	0.863	1.089	0.978	0.472	0.626
1987	0.678	0.792	0.784	0.901	1.177	0.993	0.455	0.576
1988	0.690	0.816	0.845	0.947	1.253	0.910	0.430	0.561
1989	0.738	0.821	0.899	0.918	1.170	1.067	0.451	0.559
1990	0.721	0.834	0.862	0.924	1.259	1.027	0.487	0.633
1991	0.783	0.843	0.848	0.983	1.145	1.080	0.462	0.646
1992	0.814	0.878	0.914	1.116	1.260	1.057	0.496	0.634
1993	0.818	0.923	0.863	1.118	1.308	1.055	0.490	0.669
1994	0.851	0.876	0.943	1.108	1.265	1.125	0.519	0.579
1995	0.778	0.892	0.878	0.992	1.266	1.077	0.537	0.614
1996	0.778	0.913	0.916	1.102	1.287	1.137	0.485	0.630
Average annual growth rates:								
1960-96	0.014	0.018	0.021	0.019	0.023	0.014	0.019	0.009
1960-66	0.020	0.026	0.013	0.034	0.047	0.022	0.010	0.019
1966-69	-0.003	0.027	0.044	0.036	0.018	-0.001	0.042	0.010
1969-73	0.047	0.019	0.021	0.028	0.015	0.008	0.014	0.001
1973-79	-0.002	-0.017	0.018	-0.010	0.017	0.006	0.023	-0.003
1979-89	0.015	0.033	0.036	0.014	0.025	0.024	0.022	0.009
1989-96	0.007	0.015	0.003	0.026	0.014	0.009	0.010	0.017

Table 2.2. States Ranked by 1996 Level of Productivity

State	1996		1960		Average Annual Growth of Productivity, 1960-96	
	Rank	Level	Rank	Level	Rank	Growth
CT	1	1.509	20	0.549	2	0.0284
FL	2	1.504	2	0.701	17	0.0212
GA	3	1.398	14	0.560	6	0.0254
NC	4	1.386	22	0.522	3	0.0271
IA	5	1.299	1	0.712	37	0.0167
WA	6	1.287	19	0.554	10	0.0234
ID	7	1.218	21	0.525	11	0.0234
SD	8	1.213	6	0.613	27	0.0190
ME	9	1.208	11	0.593	22	0.0198
DE	10	1.197	10	0.595	24	0.0194
AR	11	1.184	29	0.484	7	0.0249
KY	12	1.181	27	0.496	9	0.0241
CA	13	1.146	7	0.612	35	0.0174
WI	14	1.137	3	0.684	42	0.0141
MN	15	1.132	12	0.592	32	0.0180
NE	16	1.122	17	0.557	23	0.0195
PA	17	1.112	25	0.500	13	0.0222
VT	18	1.102	15	0.560	28	0.0188
SC	19	1.100	36	0.456	8	0.0244
IL	20	1.093	9	0.599	38	0.0167
CO	21	1.083	4	0.654	43	0.0140
NJ	22	1.080	13	0.581	36	0.0172
LA	23	1.074	46	0.386	1	0.0284
NY	24	1.042	8	0.603	39	0.0152
IN	25	1.040	24	0.510	21	0.0198
MS	26	1.034	44	0.398	4	0.0265
MA	27	1.033	33	0.477	15	0.0215
KS	28	1.032	5	0.636	45	0.0134
AL	29	1.000	23	0.511	29	0.0186
ND	30	1.000	40	0.437	12	0.0230
OR	31	0.990	31	0.479	19	0.0202
MI	32	0.981	47	0.384	5	0.0261
NM	33	0.969	37	0.450	16	0.0213
MD	34	0.954	34	0.468	20	0.0198
MO	35	0.933	26	0.498	34	0.0174
AZ	36	0.925	18	0.556	41	0.0142
NH	37	0.924	39	0.442	18	0.0205
VA	38	0.916	43	0.423	14	0.0215
UT	39	0.913	30	0.480	33	0.0179
OH	40	0.884	35	0.460	31	0.0181
NV	41	0.855	16	0.559	46	0.0118
RI	42	0.851	41	0.424	25	0.0193

continued

Table 2. 2., *continued*

State	1996		1960		Average Annual Growth of Productivity, 1960-96	
	Rank	Level	Rank	Level	Rank	Growth
TX	43	0.778	32	0.478	44	0.0135
TN	44	0.775	45	0.387	26	0.0193
MT	45	0.707	42	0.423	40	0.0143
OK	46	0.699	28	0.490	47	0.0098
WY	47	0.630	38	0.449	48	0.0094
WV	48	0.485	48	0.248	30	0.0186

Figure 2.1 provides details for the intervening years. It plots for each year the coefficient of variation (the ratio of the standard deviation to the mean) of productivity levels for all forty-eight states. We use these coefficients to show that there was some narrowing of the range of levels of productivity between 1960 and 1987. This narrowing is a remarkable result, given the wide variation

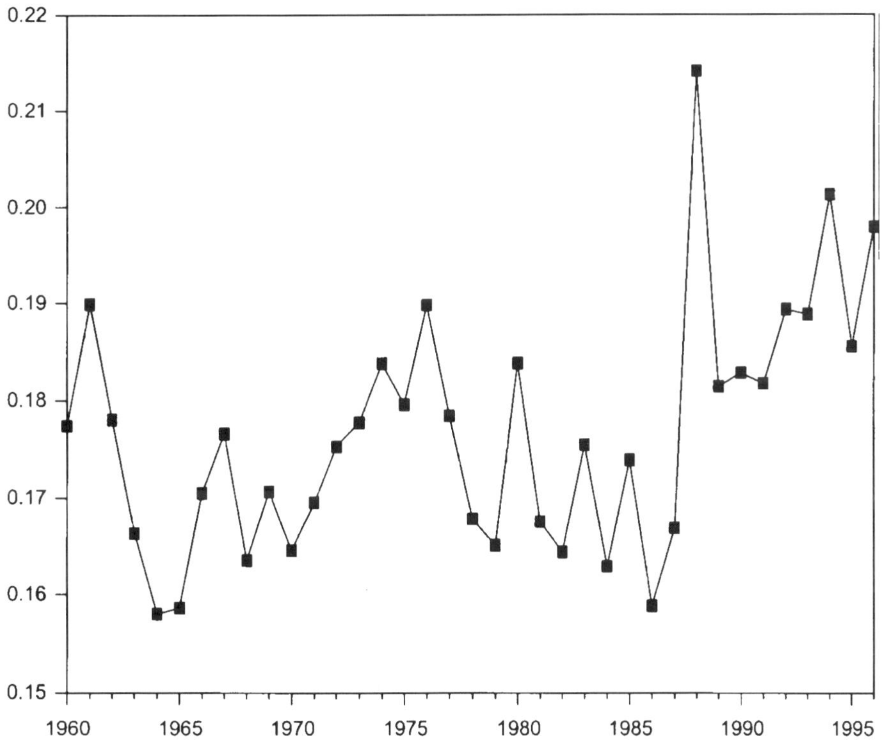

Figure 2.1. Coefficients of variation of state productivity.

in productivity growth rates. The fact that some states grew more rapidly than others and yet the cross-section dispersion decreased implies that the states that grew most rapidly were those with lower initial levels of productivity, a finding consistent with Gerschenkron's (1952) notion of the advantages of relative backwardness. Those states that were particularly far behind had the most to gain from the diffusion of technical information and proceeded to grow most rapidly.

There was a sharp increase in dispersion between 1988 and 1996. This increased dispersion was, in part, a consequence of the "industrialization" of agriculture, characterized by the expanding presence of very large, vertically integrated firms. The industrialization phenomenon has been especially apparent in the Southeastern United States, with accompanying relative gains in productivity.

A second factor was extreme variability in growing conditions. The drought of 1988 was the worst since 1956, with one-third of the nation experiencing a severe to extreme drought, as defined by the Palmer Drought Index. Also during 1988, the Southwestern United States experienced above-normal precipitation. The effect of such extremes on productivity can be seen in table 2.1. Twenty-three states showed increases in productivity in 1988, while twenty-four states exhibited decreases in productivity.

In 1993, when severe spring and summer flooding affected the upper Midwest, a major drought gripped the Southeastern United States. In fact, while floodwaters coursed down the Mississippi River in late July and August, adjacent land in the Bootheel of Missouri was parched by drought. According to table 9, twenty-nine of forty-eight states had negative productivity growth.

In the West, above-normal precipitation has been the rule during recent wet seasons (roughly October-April). This wet pattern follows the long-running drought of 1986-87 to 1991-92 and a recurrence of drought in 1993-94. The summer of 1994, which featured record corn and soybean production in the Midwest, was one of the hottest, driest summers on record in the West.

The Southern High Plains, including western Texas, were parched by the drought of 1994, and have yet to fully recover. But nearly-ideal growing conditions prevailed in the Midwest during the 1992, 1994, and 1996 seasons. In 1996, forty-three of forty-eight states posted positive productivity growth. Such a large number of states recorded gains in productivity in only three of the thirty-seven years we studied.

Analysis of Differences in Productivity

In this section, we turn to a regression framework to test two hypotheses concerning technology convergence. The first is the catch-up hypothesis, which states simply that those states that lag furthest behind the technology leaders should exhibit the most rapid rates of growth in productivity. Taking each state as an observation, this hypothesis implies that the rate of growth of productivity is inversely correlated with the initial level of productivity.

The second hypothesis is that technological innovation is embodied in capital and intermediate inputs. If the input measures do not fully reflect the changes in input quality, then this hypothesis suggests that the rate of growth of productivity will be positively correlated with growth of capital and intermediate inputs. Again, we can treat each state as an observation to test this hypothesis.

To investigate both hypotheses, we employ the basic specification:

$$\hat{TFP}_t^i = \beta_0 + \beta_1 \ln TFP_t^i + \beta_2 \left(\frac{\hat{K}}{L}\right)_t^i + \beta_3 \left(\frac{\hat{M}}{L}\right)_t^i + \varepsilon_{it}, i = 1, \Lambda, M, t = 1, \Lambda, T, \quad (13)$$

where TFP_t^i is the relative level of productivity of state i at the beginning of each period, $\left(\frac{K}{L}\right)_t^i$ and $\left(\frac{M}{L}\right)_t^i$ are relative factor intensities, and ε_{it} is a stochastic disturbance. The circumflexes (^) denote time derivatives or relative rates of change. Three-year and 5-year averages are used for the rates of growth to reduce random noise.

In equation (13), we also include the Palmer Drought Index as an explanatory variable. While the coefficient on the Palmer Index may have an interpretation, its role in the current estimation is to remove noise that could affect the estimation of the other coefficients.

The data consist of time series observations on each of several cross sections or states. We include state dummy variables to control for state-specific effects, such as the stock of infrastructure and expenditures on research and extension. Time dummy variables are included to allow productivity growth rates to vary by period (e.g., in response to unevenness in the flow of new technologies).

The regression results, shown in table 2.3, confirm the catch-up hypothesis, showing a highly significant inverse relation between the rate of productivity convergence by state and its initial level of productivity. The results for the embodiment hypothesis are mixed. Coefficient estimates for growth of the capital-labor ratio are positive and significant at the 1-percent level. These results suggest that embodiment of technology in capital was an important source of productivity growth.

The coefficient estimates for the materials-labor ratio are positive, as predicted, but statistically insignificant. We conclude that the input indices fully reflect the improvements in their quality.

Table 2.3. Regressions of Total Factor Productivity Growth on Relative Productivity Level and Growth in Factor Intensities, 1960-1996

Independent Variables	Three-Year Averages $\hat{TFP}$	Five-Year Averages $\hat{TFP}$
Constant	-0.042**	-0.007
	(-5.58)	(1.29)
ln *TFP*	-0.189**	-0.185**
	(25.12)	(25.71)
$\left(\hat{K/L}\right)$	0.105**	0.106**
	(4.60)	(4.55)
$\left(\hat{M/L}\right)$	0.023	0.019
	(1.09)	(0.86)
R^2	0.505	0.546
$\sqrt{MSE}$	0.023	0.014
F Value	16.79**	19.14**
Sample Size	1547	1453

Note: t-ratios are shown in parentheses below the coefficient estimate. The Palmer Drought Index and state and time dummy variables are included in the specification, but the results are not shown. The observations are based on 3- or 5-year averages, as indicated.

* Significant at the 5 percent level; ** significant at the 1 percent level.

Summary and Conclusion

In this study, we estimate the growth and relative levels of productivity for the 48 contiguous states for the period 1960 to 1996. For the full 1960-96 period, every state exhibits a positive and generally substantial average annual rate of productivity growth. There is considerable variance, however. The median rate of productivity growth was 1.94 percent per year, while average growth rates ranged from 0.94 percent for Wyoming to 2.84 percent for Louisiana.

The wide disparity in growth rates resulted in substantial changes in the rank order of states. For each year, we compute the coefficient of variation of productivity levels for all 48 states. We use these coefficients to show that the range of levels of productivity has narrowed over time, although the pattern of convergence was far from uniform. The fact that some states grew faster than others and yet the cross section dispersion decreased implies that the states that grew most rapidly were those with lower initial levels of productivity. This result is consistent with Gerschenkron's (1952) notion of the advantages of relative backwardness. The states that were particularly far behind the productivity leaders had the most to gain from the diffusion of technical information and proceeded to grow most rapidly. Finally, we observed a positive relation between productivity growth and growth of the capital-labor ratio, implying embodiment of technology in capital.

Notes

1. Jorgenson, Gollop, and Fraumeni (1987), table 6.7, pp. 198-200.

2. Jorgenson and Gollop (1992), pp. 746 and 750.

3. See Jorgenson and Griliches (1967) for a discussion of input quality.

4. For a description of USDA methods, see Ball (1985) and Ball et al. (1997).

5. The estimates of output are product based. The product approach proceeds by valuing final output of all agricultural products, but only those products. Thus, agricultural products from establishments that produce agricultural goods as a minor or secondary activity are valued, while goods and services that might be produced on farms but are secondary to agriculture are excluded.

6. The six sub-periods are not chosen arbitrarily but represent cyclical peaks. Since the data reported for each sub-period are average annual growth rates, the unequal lengths of the sub-periods do not affect comparisons among the sub-periods.

7. The observations consist of average prices for each of the chemicals. When average data are used rather than actual observations of prices, the disturbance terms are likely to be heteroskedastic. Efficient parameter estimates are obtained by applying weighted least squares. Expenditure shares are used as weights.

8. The pesticides model is estimated in double logarithmic form. It is well known that the antilog of $\hat{\delta}_t$ is not an unbiased estimate of δ_t, which means that price indices based on the dummy variable method are biased. A standard bias correction is to add one-half of the coefficient's standard error to the estimated coefficient. See Teekens and Koerts (1972).

9. The decay function defined by equation 8 incorporates many of the commonly used forms of capacity depreciation as special cases. The upper limit of β is unity, a value that corresponds to the one-hoss shay form of depreciation. As the value of β approaches zero, decay occurs at an increasing rate over time. If β is zero, the function corresponds to the formula for straight-line depreciation. Finally, if β is negative, decay occurs most rapidly in the early years of service, corresponding to accelerated forms of depreciation such as geometric decay.

10. Mean service lives for each asset type are taken from *Fixed Reproducible Tangible Wealth in the United States, 1925-1994* (U.S. Department of Commerce).

11. Net additions to inventory during the calendar year are considered a component of output.

12. The nominal rate was taken to be the average yield on Moody's BAA bonds over all maturities.

13. Observed real rates are expressed as an AR(1) process. We use this specification after examining the correlation coefficients for auto correlation, and partial and inverse auto correlation, and after performing the unit root and white noise tests. We centered each time series by subtracting its sample mean. The analysis was performed on the centered data.

14. See Ball et al. (1997).

15. For 1960 and 1970, see "Characteristics of the Population," *Census of the Population, Volume 1*; for 1980, see "Detailed Population Characteristics," *Census of the Population,* Chapter D. The Bureau of the Census did not tabulate state-specific demographic data for agriculture in 1990. Consequently, the 1990 marginal matrices described in the text were constructed from 1980 marginals adjusted for 1980-90 trends from series updated from those initially reported in Jorgenson, Gollop, and Fraumeni (1987) and controlled to USDA totals by employment class.

16. For a discussion of the theoretical basis for adjusting labor input for compositional shifts in the labor force, see Jorgenson and Gollop (1992).

References

Ball, V. E. 1985. "Output, Input, and Productivity Measurement in U.S. Agriculture, 1948-79." *American Journal of Agricultural Economics,* 67:475-86.

Ball, V. E., J. C. Bureau, R. Nehring, and A. Somwaru. 1997. "Agricultural Productivity Revisited." *American Journal of Agricultural Economics*, 79(1997):1045-63.

Ball, V. E., F. M. Gollop, A. Kelly-Hawke, and G. P. Swinand. 1999. "Patterns of State Productivity Growth in the U.S. Farm Sector: Linking State and Aggregate Models." *American Journal of Agricultural Economics,* 81:164-79.

Coen, R. 1975. "Investment Behavior, the Measurement of Depreciation, and Tax Policy." *American Economic Review*, 65:59-74.

Denison, E. 1979. *Accounting for Slower Economic Growth.* Washington, D.C: The Brookings Institution.

Diewert, W. E. 1976. "Exact and Superlative Index Numbers." *Journal of Econometrics*, 4: 115-46.

Eltetö, O. and P. Köves. 1964. "On a Problem of Index Number Computation Relating to International Comparisons." *Szatistikai Szemle,* 42:507-18.

Federal Reserve Board. *Moody's BAA Seasoned Corporate Bonds*. Washington, D.C., monthly.

Fisher, I. 1922. *The Making of Index Numbers*. Boston: Houghton Mifflin.

Gerschenkron, A. 1952. "Economic Backwardness in Historical Perspective." *The Progress of Underdeveloped Areas*, ed. Bert F. Hoselitz. Chicago: University of Chicago Press.

Jorgenson, D., and F. Gollop. 1992. "Productivity Growth in U.S. Agriculture: A Postwar Perspective." *American Journal of Agricultural Economics*, 74:745-50.

Jorgenson, D., and Z. Griliches. 1967. "The Explanation of Productivity Change." *Review of Economic Studies*, 34(July):249-83.

Jorgenson, D., F. Gollop, and B. Fraumeni. 1987. *Productivity and U.S. Economic Growth.* Cambridge, Mass.: Harvard University Press.

Kendrick, J. W., and E. S. Grossman. 1980. *Productivity in the United States, Trends and Cycles*. Baltimore: The Johns Hopkins University Press.

Palmer, W. D. 1965. "Meteorological Drought." U.S. Department of Commerce, Weather Bureau, Research Paper No. 45.

Penson, J. B., D. W. Hughes, and G. L. Nelson. 1977. "Measurement of Capacity Depreciation Based on Engineering Data." *American Journal of Agricultural Economics*, 59:321-29.

Romain, R., J. Penson, and R. Lambert. 1987. "Capacity Depreciation, Implicit Rental Prices and Investment Demand for Farm Tractors in Canada." *Canadian Journal of Agricultural Economics*, 35:373-78.

Szulc, B. 1964. "Indices for Multiregional Comparison." *Przeglad Statystyczny*, 3:239-54.

Teekens, R., and J. Koerts. 1972. "Some Statistical Implications of the Transformation of Multiplicative Models." *Econometrica*, 40:793-819.

U.S. Department of Agriculture. *Agricultural Land Values Survey*, annual.

U.S. Department of Agriculture. *Agricultural Prices*. Washington, D.C., annual.

U.S. Department of Agriculture. 1979. *Producer-Owned Grain Stocks*.

U.S. Department of Commerce, Bureau of the Census. "Geographic Area Series, State and County Data." *Census of Agriculture*, volume 1. 1959, 1964, 1969, 1974, 1978, 1982, 1987, and 1992.

U.S. Department of Commerce, Bureau of the Census. 1960 and 1970. "Characteristics of Population." *Census of Population*, volume 1.

U.S. Department of Commerce, Bureau of the Census. 1980. "Detailed Population Characteristics." *Census of Population*, Chapter D.

U.S. Department of Energy. Energy Information Administration. *Monthly Energy Review*, monthly.

U.S. Department of Commerce. 1999. *Fixed, Reproducible, Tangible Wealth of the United States, 1925-1994.*

3

A Disaggregated Perspective on Post-War Productivity Growth in U.S. Agriculture: Isn't That Spatial?

Albert K.A. Acquaye, Julian M. Alston, and Philip G. Pardey[1]

Abstract

Annual data from 1949 through 1991 for 48 states are used to account for changes in the composition of input and output aggregates over space and time, and thereby to obtain new evidence on changes in inputs, outputs, and productivity in U.S. agriculture. The measures change significantly when we use state-specific rather than national prices and when we allow for changes in the composition of the aggregates, especially of labor and capital inputs. The national picture is not representative of spatial sub-aggregates. For instance, state-specific annual average productivity growth rates for 1949-1991 ranged from 0.79 percent (Nevada) to 3.23 percent (Georgia) compared with a national aggregate rate of 1.90 percent per annum. We compare our estimates and those reported by Ball et al. (1999). The national estimates were similar but some substantial differences were found in state-level productivity growth, accounted for by differences in the underlying patterns of measured output and especially input quantities.

Introduction

> As I understand it, we are interested in "productivity" because we are interested in understanding . . . . the forces that affect "output" because we hope, ultimately, to be able to affect them for the better. We approach this task first by trying to take into account the "obvious" factors: changes in labor and capital (and other materials if our output measures are gross). We measure these inputs as best we can, aggregate them using some sensible weighting procedure, and get a "total input" index. We compare this index with our output index and call any discrepancy "productivity." Crudely speaking then, the "productivity" indices measure the changes

in output that have not been accounted for by the analyst's input measures. It is a measure of our ignorance, of the unknown, and of the magnitude of the task that is still ahead of us.

— Griliches (1961, p. 446)

Every agricultural economist knows that U.S. agricultural output and productivity have grown rapidly over the past 100 years, and especially over the past 50 years. Most of us know less about the timing and magnitude of the important changes in agricultural output and productivity, and the direction and magnitude of changes in aggregate input use, and few know anything much about the changing composition of agricultural outputs and inputs – apart from knowing that agriculture uses much less labor and more purchased inputs than it once did. Even less is known about the spatial pattern of input, output, and productivity change within the United States.

Our quantitative understanding is limited in particular because the quantities of interest are hard to measure. Because quantifying output or inputs involves aggregating over heterogeneous goods and services, using prices that are not always observed on the appropriate units, index number and aggregation problems arise. The purpose of this paper is twofold. First, we show the implications of different indexing and aggregation procedures, taking into account the composition of input and output aggregates, with particular reference to quality issues and the spatial dimension. Specifically, we show the implications of carefully accounting for price variation across qualities or states for measures of aggregate national and regional output and input growth. Second, the spatial patterns are intrinsically interesting, and there are important differences among states or regions, as well as between individual states and the national aggregate. We wish to quantify these differences, to enrich our understanding of the state-specific economic history of inputs, outputs, and productivity in American agriculture.

In studies of U.S. agricultural productivity, the preponderance of past work has studied national trends in multi-factor productivity (MFP). The first of these measures was provided by Barton and Cooper (1948), with Loomis and Barton (1961), and Kendrick (1961) extending them to include from 1866 to 1957. The USDA published a Laspeyres productivity index (including coverage of the 10 USDA sub-regions) up until 1990, including estimates back to 1870. A number of studies (e.g., Diewert 1976) have shown that a chain-linked index, specifically an approximation of a Divisia index, is to be preferred. Hence, the types of index number formulation have changed. For instance, Ball (1985) provided alternative measures using Törnqvist-Theil indices (recently extended and updated by Ahearn et al. 1998), and recent studies, such as Ball et al. (1997), have used Fisher Ideal indices.[2]

Another dimension for change has been the spatial aggregate. Craig and Pardey (1990b) were among the first to compile state-level input, output, and productivity indices. Like Craig and Pardey, Huffman and Evenson (1992, 1993) reported Törnqvist-Theil productivity indices, including estimates for the individual states.[3] Craig and Pardey also used the state-level data to compute improved national input, output, and productivity measures, but they did not exploit the panel nature of the data to measure productivity multilaterally (i.e., the states were treated separately). In a separate study, however, Craig and Pardey (1990a) reported an explicitly multilateral index of U.S. agricultural productivity based on state-level data. More recently, Ball and Nehring (1998) provided state-level multilateral indices. Most recently, Ball et al. (1999) analyzed U.S. agricultural productivity patterns, emphasizing the implications of interstate transactions in farm goods.

A third dimension of change in methods and measures has been in accounting for changes in the composition of aggregates, including changes in quality of inputs and outputs. That input and output quality changes affect measures of agricultural productivity is not a new idea (e.g., Griliches 1963), but until relatively recently it has been neglected in most of the empirical work. Jorgenson and Griliches (1967) and Jorgenson, Gollop, and Fraumeni (1987) highlighted the changing composition of labor in their studies of sectoral productivity. The same issue was featured in the empirical work by Craig and Pardey (1990b) in their analysis of U.S. agricultural productivity, and emphasized by Jorgenson and Gollop (1992). Craig and Pardey also paid particular attention to changes in the quality of land and capital. At a 1993 NC-208 conference in Orlando, the quality issue was raised in papers by Ball (1994) and Pardey, Craig, and Deininger (1994) who treated different elements in different ways. The discussion at that conference highlighted the quality issue and some controversy about how to address it, among other things. Subsequently, the issue of quality adjustments and the related spatial aspects were the focus of papers by Craig and Pardey (1996a, 1996b), and quality adjustment has been a feature of other studies, even when that was not the main issue, e.g., Ahearn et al. (1998) and Ball et al. (1999).

The estimates of agricultural input, output, and productivity growth in this paper are different from any previously published estimates because they combine various elements of the innovations discussed above. The methods to account for the composition of quantity aggregates and quality change, and the use of state-specific prices, follow Craig and Pardey (1996a, b). Like Craig and Pardey (1996b) we provide estimates of growth of agricultural inputs, outputs, and productivity over the period 1949-1991, for the United States as a whole and for 48 individual states. However, our main results are based on the Fisher Ideal index rather than the Törnqvist-Theil index, some data have been revised, and some different specific choices were made about aggregation and the treatment of Conservation Reserve Program (CRP) acreage. In addition, we also report some estimates by USDA regions. Finally, we provide a preliminary comparison of our productivity estimates with those of Ball et al. (1999) to show the consequences of alternative, perhaps equally reasonable, treatments of the data.[4]

Data

In table 3.1 we give a summary description of the input and output data coverage in this data set. To capture changes in the composition and quality of inputs we use disaggregated data that distinguish among 58 types of inputs. Annual data for 30 categories of farm operator, and for hired and family labor, were used in forming the labor aggregate, a total of 32 categories. Data on days worked off-farm by farm operators were used to adjust the measure of operator labor, and account for the substantial but uneven shift toward part-time farming. State-specific wages were obtained for the hired and family labor while implicit operator wages were developed using national income earned by rural farm males, categorized by age and educational attainment.

The land input is sub-divided into three basic types: namely, service flow from pasture and rangeland, non-irrigated cropland, and irrigated cropland. This measure of land quantity differs from the more traditional measure in that it excludes non-grazed forest and woodlands (areas which, although in farms, are not in agriculture), and includes Federally-owned land rented or leased for rangeland grazing purposes. Also included in the cropland measures are land acres idled for whatever reason. The price weights used for aggregation of the land input are annual, state-specific, cash rents for each of the three land types.

We identified as many separate capital classes as possible. The twelve different capital inputs in this data set include seven classes of physical capital and five classes of biological capital. In addition, we adjusted inventories of the physical capital classes to reflect quality change over time, depending on the nature of the data available and the service flow profile of each capital type. For all farm machinery, automobiles, and trucks, inventories on farm were converted to equivalent head counts of a new machine, by incorporating information on the average age of machines. For tractors and combines, additional information on the productive characteristics of machines on farms was used to adjust for quality as well, which led to inventories measured in equivalent numbers of a new, numeraire (55 horsepower) tractor or combine (medium capacity). For only one capital class, buildings, the quantities were implicit quantities derived by dividing the total rental value of buildings on farms by a price index developed for the purpose.Rents for capital items were taken

Table 3.1. Input, Output, and Spatial Details

Inputs/Outputs	Subcategory	Examples (where applicable)
Inputs		
Land	Cropland	
	Irrigated cropland	
	Pasture and grassland	
Labor	Family labor	
	Hired labor	
	Operator labor	Thirty classes characterized by: Education—0-7 years; 8 years; 1-3 years of high school, 4 years of high school, 1-3 years of college, 4 years or more of college. Age—25-34, 35-44, 45-54, 55-64, or 65 or more years of age.
Capital	Physical capital	Automobiles, combines, mowers and conditioners, pickers and balers, tractors, and trucks.
	Biological capital	Breeding cows, chickens, ewes, milking cows, and sows.
	Purchased capital	Electricity, purchased feed, fuel, hired machines, pesticides, nitrogen, phosphorus, potash, repairs, seeds, and miscellaneous purchases.
Outputs		
Crops	Field crops	Barley, corn, cotton, flax, oats, peanuts, rice, rye, sugar beets, sugarcane, sorghum, soybeans, tobacco, tomatoes for processing, wheat.
	Fruits and Nuts	Almonds, apples, apricots, avocados, cherries, cranberries, grapefruit, grapes, lemon, oranges, pears, peaches, pecans, strawberries, walnuts.
	Vegetables	Beans, broccoli, carrots, cauliflower, celery, cucumber, fresh tomatoes, lettuce, onions, peas, potatoes, sweet corn (fresh, and for processing).
	Greenhouse and Nursery Products	
	Livestock	Broilers, cattle, eggs, hogs, honey, milk, sheep, turkeys, wool.
	Machines rented out	
	Returns from CRP	

Region	Name and abbreviation	States within region
	Northeast 1 (NE1)	Connecticut, Maine, Massachusetts, New Hampshire, Rhode Island, and Vermont.
	Northeast 2 (NE2)	Delaware, Maryland, New Jersey, New York, and Pennsylvania.
	Cornbelt (CB)	Illinois, Indiana, Iowa, Missouri, and Ohio.

continued

Table 3.1, *continued*

Region	Name and abbreviation	States within region
	Lake states (LS)	Michigan, Minnesota, and Wisconsin.
	Northern plains (NP)	Kansas, Nebraska, North Dakota, and South Dakota.
	Appalachian (AP)	Kentucky, North Carolina, Tennessee, Virginia, and West Virginia.
	Southeast (SE)	Alabama, Florida, Georgia, and South Carolina.
	Delta states (DS)	Arkansas, Louisiana, and Mississippi.
	Southern plains (SP)	Oklahoma and Texas.
	Mountain (M)	Arizona, Colorado, Idaho, Montana, New Mexico, Nevada, Utah, and Wyoming.
	Pacific (P)	California, Oregon, and Washington.

to be specific fractions of the purchase price, fractions that varied among capital types. Purchase prices were assumed to reflect the expected present value of real capital services over the lifetime of the specific type of capital. The fraction used to construct rents from purchase prices depended upon the lifespan, depreciation, and age of each capital class (e.g., tractors as a class versus combines as a class). Rents for physical capital inputs were based on the national value of a representative machine, whereas rents for biological capital inputs were based on state-specific value per head. A constant real interest rate of four percent was used for all rent calculations.

Eleven types of purchased inputs are included in this data set. Apart from fertilizers (measured as quantities of elemental nitrogen, phosphorus, and potash), the purchased input quantities were implicit quantities derived from state-specific expenditure totals. The miscellaneous category was pre-aggregated, and included a list of disparate inputs, such as fencing, irrigation fees, hand tools, veterinary services, and insurance costs, among others. State-specific prices were available only for electricity; all other prices were national prices or price indices based on national prices paid by farmers.

The output data cover (in the disaggregated form) 55 output categories, including nine livestock commodities, 15 field crops, 25 fruits and vegetables, three tree nuts, implicit quantities of machines rented out by farmers, implicit quantities of greenhouse and nursery products, and Conservation Reserve Program (CRP) acreage. As a departure from Craig and Pardey (1996a, b) and other studies, CRP acreage is treated as an output for which producers receive revenue equal to the CRP payments from the government. That is, the measures of output (and hence productivity) incorporate the environmental benefits produced by farmers and sold to the government as CRP acreage. Because data were not available, horses and mules are not included in this data set either as biological capital inputs or as outputs, even though we know they have been important.

The prices used as weights to form an aggregate output are state-specific prices received by farmers for all commodities, except machines for hire and greenhouse and nursery products. For these composite commodities a price index based on national prices for their sub-components was constructed, to be used to deflate reported current dollar values. A somewhat questionable, but standard, choice was to use the (policy-distorted) prices paid and received by farmers as the relevant prices for aggregating inputs and outputs. As a counterpart to that decision, we included set-aside acreage within the inputs used by farmers. It is not clear to us how others have dealt with this issue.

Aggregation Procedures for Inputs and Outputs

Bias from the procedure used to aggregate inputs and outputs can be kept to a minimum by choosing an appropriate index, carefully selecting value weights for all inputs and outputs, and disaggregating inputs and outputs as finely as possible. The effects of these choices are illustrated next.[5]

Effects of Choosing an Inappropriate Index

We compare two fixed-weight indices (the Laspeyres and Paasche indices) with two Divisia approximations (the Fisher Ideal and Törnqvist-Theil indices). Since the computation uses ratios (or logarithms) of quantities, the Törnqvist-Theil index cannot be applied when observations take a value of zero (which becomes more of a problem when the data are more finely disaggregated). It seems that a common treatment is to exclude those goods for which *any* of the observations is zero. We adopt this approach and, to facilitate comparisons among the indices, we use the same (substantially) reduced data for every index at this stage.[6] But in the main analysis we use the entire data set to compute Fisher Ideal indices. The national multifactor productivity (MFP) indices for U.S. agriculture were created by dividing each output quantity index by the corresponding input quantity index.[7] The Laspeyres index understated productivity growth because it overstated growth in inputs more seriously than outputs (and conversely the Paasche index overstated productivity growth). In 1991, the indices of MFP were 279.8 (Paasche), 160.8 (Laspeyres), 223.7 (Törnqvist-Theil), and 222.1 (Fisher Ideal) relative to a base of 100 in 1949. The Törnqvist-Theil and Fisher Ideal indices were almost indistinguishable throughout the entire period. Between these two, the Fisher Ideal index is preferred (Diewert 1992).

Effects of Using the Wrong Value Weights

Economic theory prescribes the use of appropriate prices as weights for the aggregation of quantities. Most studies of productivity have used pre-aggregated national data and, more often than not, national average input (and output) prices have been used in constructing the indices, perhaps because disaggregated quantity, or state-specific price data were not readily available. Indices derived using such pre-aggregated data may suffer from aggregation bias, which is reduced by using state-specific prices and quantities.[8]

To illustrate the effects of aggregation bias, regional quantity indices were computed in two ways. *Pre-aggregated* input (and output) quantity indices were computed using average national unit values, obtained by dividing the national value of each agricultural input (output) by the corresponding total national input use (output production) for each year as the price weights. *Disaggregated* productivity indices were formed using state-specific prices, where possible. In general, the use of the wrong (national-average rather than state-specific) prices led to biased regional productivity indices. Figure 3.1 presents the two productivity indices for each of the 11 regions. At the regional level, the bias was negligible, but substantial biases were found in three regions, overstating productivity growth in the Northeast 1 region (where agriculture became unimportant during the period) and understating productivity growth in the Northeast 2 and Pacific regions (reflecting distortions in estimates of agriculture).

Disaggregated Outputs and Inputs

As shown formally by Star (1974), one is safe in using pre-aggregated data only if all inputs (outputs) in the class are growing at the same rate or are perfect substitutes for one another. If, for example, the rate of growth of the higher-priced inputs (outputs) exceeds the rate of growth of the lower-priced inputs (outputs), the estimated rate of growth of the group will be biased downwards when pre-aggregated data are used. Hence, growth rates of agricultural productivity will tend to be overstated if the quantities of higher-priced (higher-quality) inputs are growing relatively quickly.

The composition of land used in U.S. agriculture underwent some substantial changes from 1949 to 1991.[9] Figure 3.2a shows the national agricultural land quantity index measured as either a sum of acres of land in agriculture or as the rental prices of the different categories of land in the aggregation procedure. The difference between the two indices is a measure of the implicit changes in the quality of the national land quantity index. Figure 3.2b shows that the effects of adjusting for changes in the composition of the farm labor force are relatively small at the national level, but some more striking contrasts are re-

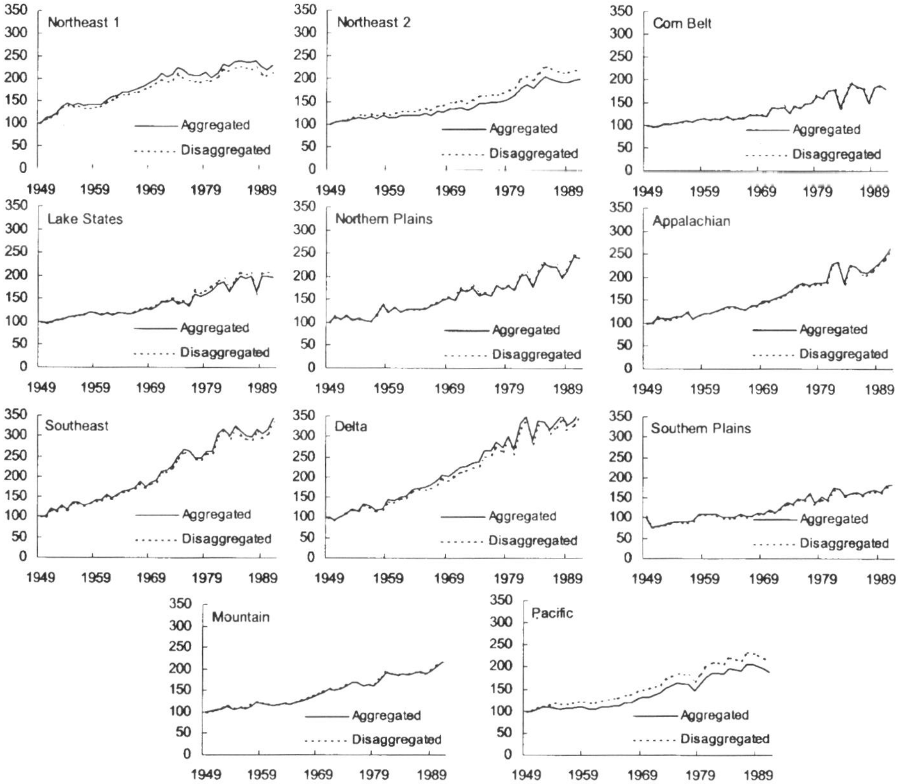

Figure 3.1. Comparison of regional productivity indices using national -average and state-specific prices.

vealed in certain states (e.g., in Florida, the average quality of labor increased dramatically during 1949-91).

To illustrate the effects of quality adjustment on partial factor productivity measures, we use the graphical technique developed by Hayami and Ruttan (1971). Figure 3.3 plots regional data without quality adjustments in panel a, and with quality adjustments in panel b. In these figures, the horizontal axis measures labor productivity, and the vertical axis measures land productivity. All of the productivity paths move in a northeasterly direction starting in 1949 and ending in 1991. The diagonals indicate constant factor ratios. As a region's productivity locus crosses a diagonal from left to right, an increase is indicated in the number of acres per hour in that region. The productivity gains were especially dramatic for labor. Output per hour was much lower than the national average in the South in the immediate postwar period and yet there was

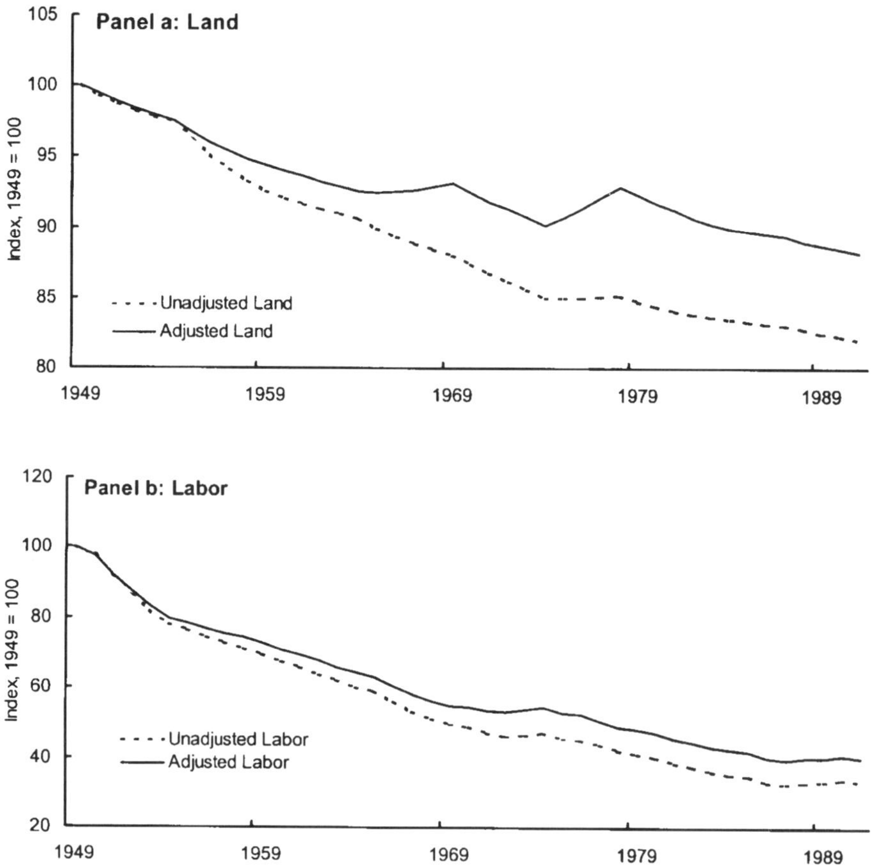

Figure 3.2. National land and labor indices, quality-unadjusted, and adjusted.

virtually no difference in the more-recent years. A longer productivity locus indicates a greater percentage change in productivity in that region (e.g., the Delta region). The quality adjustment reduced measured productivity gains for both factors, as evidenced by the shorter path of the quality-adjusted locus in panel b. In addition, quality adjustment of the labor and land inputs reduced the dissimilarities among the regional productivity paths, in terms of their starting and ending points; they are generally closer together.

Input, Output, and Productivity Measures

In 1949, the agricultural sector accounted for 7.6 percent of national gross domestic product (GDP), but since then has shrunk in many relative senses (to

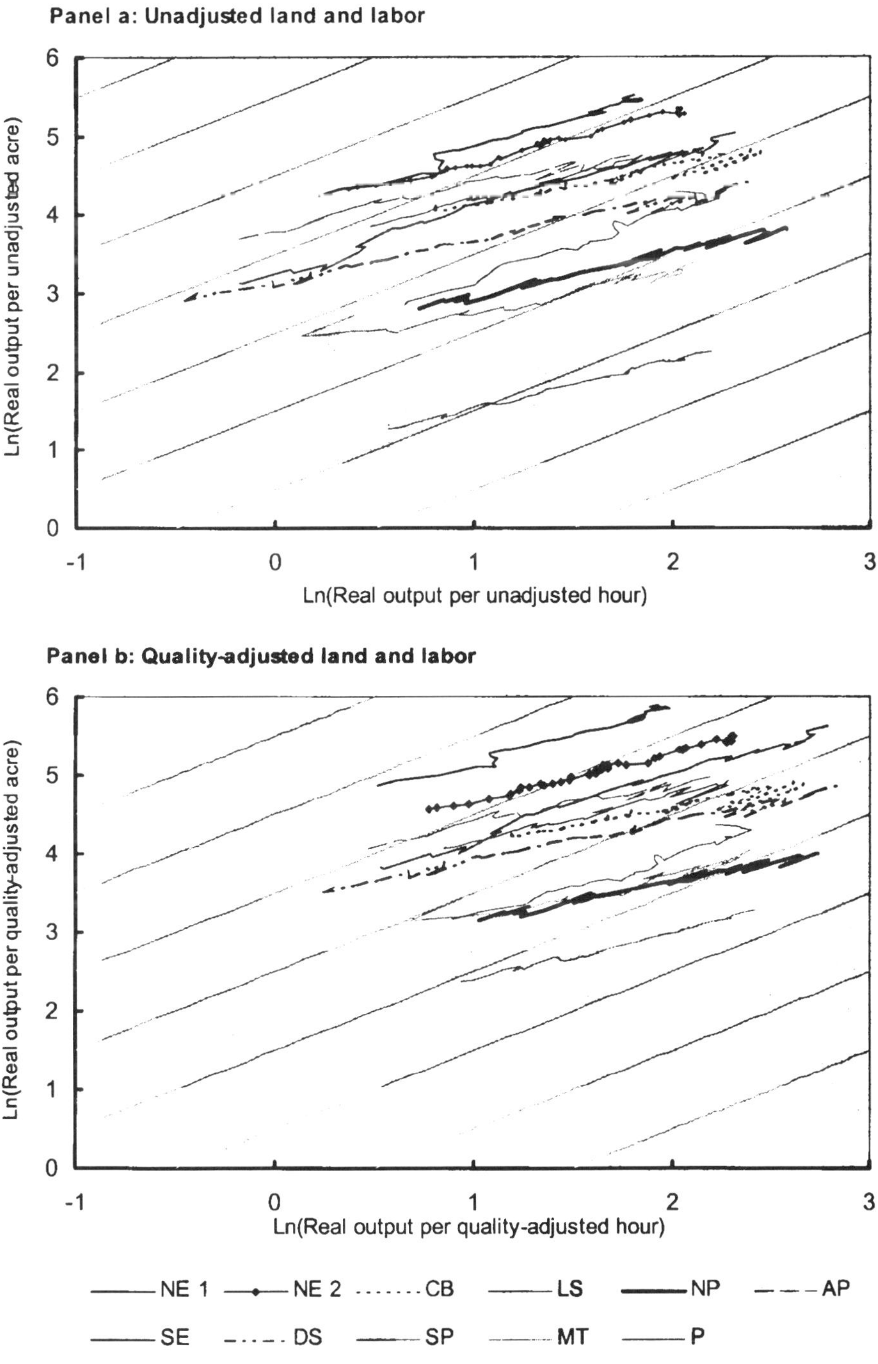

Figure 3.3. Effects of quality adjustment on regional land and labor productivity measures.

1.4 percent of GDP in 1990), while continuing to grow in absolute terms. Not all states have shared in that growth or developed in the same way. Appendix A includes detailed tables on inputs, outputs, and productivity patterns.

Inputs

The area of land used in U.S. agriculture decreased from 1.43 billion acres in 1949 to 1.17 billion acres in 1991. The area of grassland and cropland decreased both in total and for most states from 1949 to 1991; notably, Iowa and Illinois increased their area of cropland. The area of irrigated cropland grew in total and in all but four states: Utah, Nevada, Arizona, and Connecticut, and California's share fell from about 25 percent to 16 percent from 1949 to 1991.

Labor use in agriculture changed substantially. In 1949, 20.5 billion hours of labor were employed in the 48 contiguous states compared with 6.8 billion hours in 1991. The composition of the labor input also changed dramatically. Table 3.2 shows the shares of total labor hours accounted for by the various labor sub-classes in 1949 and 1991. A "+" for a labor sub-class indicates an increase in its share of total labor hours over the 43 years; a "-" indicates a decrease. We see a predominance of "+" for hired labor and for better-educated and older operators. Whereas the total hours of hired, family, and operator labor decreased, the share of hired labor in total labor hours increased markedly from 19 percent to 31 percent between 1949 and 1991. In 1949, the largest proportion of hours by farm operators were by operators with less than eight years of education; in 1991, the largest proportion had four years of high-school education.

Quantities of combines and tractors increased steadily between 1949 and the early 1980s, but after that the numbers gradually fell.[10] In contrast, the use of agricultural chemicals increased steadily up until 1961, and then rose more rapidly to a peak in the early 1980s. Most of the tractors and combines were in the Midwestern and Plains states, and in general, the Northeastern and Mountain states used the least-powerful tractors, and medium-capacity combines.

There was considerable variation among states in their patterns of input use and growth rates. Table 3.3 presents the highest and lowest average annual growth rates in input and output quantities for the period 1949-1991, among the states, as well as the averages across states and the national figures. Total input use in agriculture declined by about 0.19 percent annually on average, during the period 1949-1991.

Figure 3.4 shows distributions of state-specific growth rates of aggregate input among the 48 states over the 1949 to 1991 period. Panel a presents linearized distributions of state-specific input growth rates for each USDA production region. Panel b is a frequency distribution of the 48 state-specific input

Table 3.2. Share of Labor Sub-Class Hours in Total Labor Hours: 1949 and 1991

Labor subclass	1949	1991	Change
	(percentage)		
Hired Labor	19.80	31.15	+
Family Labor	16.95	12.38	-
Operators with 4+ years of college education and			
65+ years old	0.10	1.14	+
55 - 64 years old	0.23	1.57	+
45 - 54 years old	0.29	1.93	+
35 - 44 years old	0.31	2.68	+
25 - 34 years old	0.20	0.71	+
Operators with 1 - 3 years of college education and			
65+ years old	0.17	1.63	+
55 - 64 years old	0.40	2.16	+
45 - 54 years old	0.47	3.17	+
35 - 44 years old	0.52	4.13	+
25 - 34 years old	0.40	1.76	+
Operators with 4 years of high-school education and			
65+ years old	0.26	3.65	+
55 - 64 years old	0.65	5.75	+
45 - 54 years old	1.02	5.63	+
35 - 44 years old	2.06	4.01	+
25 - 34 years old	2.33	2.76	+
Operators with 1 - 3 years of high-school education and			
65+ years old	0.48	1.67	+
55 - 64 years old	1.31	2.02	+
45 - 54 years old	2.00	1.67	-
35 - 44 years old	2.52	0.87	-
25 - 34 years old	1.82	0.56	-
Operators with 8 years of education and			
65+ years old	1.90	1.38	-
55 - 64 years old	3.99	1.26	-
45 - 54 years old	4.88	0.61	-
35 - 44 years old	4.26	0.26	-
25 - 34 years old	2.40	0.12	-
Operators with 0 - 7 years of education and			
65+ years old	4.28	1.11	-
55 - 64 years old	6.90	1.13	-
45 - 54 years old	7.23	0.66	-
35 - 44 years old	6.35	0.32	-
25 - 34 years old	3.47	0.14	-

Table 3.3. Average Annual Growth Rates

	National	State growth rates: Minimum	Maximum	Range	Standard Deviation
			(percentage)		
MFP	1.90	0.79	3.23	2.44	0.56
Output	1.72	-0.63	3.27	3.90	1.02
Input	-0.19	-2.34	1.30	3.64	0.92
Outputs					
Field Crops	1.79	-6.76	3.82	10.58	2.12
Fruits and Nuts	1.80	-4.52	4.91	9.43	2.23
Livestock	1.43	-2.89	4.47	7.36	1.57
Vegetables	1.69	-4.56	5.58	10.14	2.40
Greenhouse and Nursery	3.84	0.45	7.01	6.56	1.76
Inputs					
Land	-0.30	-3.44	0.47	3.90	0.96
Labor	-2.17	-4.26	-0.25	4.02	0.98
Capital	0.37	-1.59	2.57	4.17	0.82
Purchased Inputs	2.27	-1.64	4.05	5.69	1.38

	States with annual average growth rates: Above U.S. Average	More than 1 percent below	More than 1 percent above
		(number)	
MFP	18	1	5
Output	21	11	6
Input	22	9	2
Outputs			
Field Crops	18	17	3
Fruits and Nuts	9	28	5
Livestock	21	15	8
Vegetables	14	19	8
Greenhouse and Nursery	18	19	13
Inputs			
Land	21	13	2
Labor	27	6	1
Capital	21	12	7
Purchased Inputs	20	12	0

growth rates; it can be seen that the preponderance of states (i.e., 31) reduced their total input use. Each regional distribution in panel a is represented as a line with a maximum and a minimum, representing the highest and lowest state-specific growth rates within the region, and a central tendency, the growth rate of the regional index. In addition, comparing among regions, we can see the range among all 48 states; the vertical line indicates the growth rate of the national index. While the national index of input use fell, 17 of the 48 contiguous states increased their input use (on average), and in two states (Florida and California) input use grew over one percent per annum faster than the national average. Some regions are relatively homogeneous (Corn Belt, Lake States, Northern Plains, and Southern Plains), and this is reflected in their relatively low variance in growth rates of total input use. The large spread within the Northeast 2 and Southeast regions reflects their diversity. Input use grew fastest generally in the Pacific region, and contracted fastest in the Northeast 1 region.

The patterns of change differed among input categories. While land and labor became less important in the production of agricultural output over the 1949-91 period, our adjustments for changes in the composition of land and labor over time reduced the measured rates of decline in the use of these inputs. At the national level, quality-adjusted land and labor use declined over the 1949 to 1991 period. Labor use fell by 2.17 percent per annum; land by 0.30 percent per annum. Capital and purchased inputs grew; purchased inputs grew the fastest, at 2.27 percent per annum. The use of quality-adjusted capital inputs grew in the first decade by about two percent per annum; little in the following decades. The use of purchased inputs, on the other hand, increased throughout the 1949-91 period, increasing from 19 percent of the total cost of production in 1949 (calculated as a sum of the value of all inputs measured in this study, using state-specific prices and quantities) to 31 percent in 1991. The cost shares of land, labor, and capital all fell from 1949 to 1991.

Figure 3.5 shows frequency distributions of state-specific input growth rates by input category. Some states increased their amount of land in agricultural production, but most declined slowly around the national growth rate of -0.30 percent per annum. Although the number of acres in agriculture in Nebraska fell by almost three million acres, the quality improvement of the land (as a result of expanded irrigation) meant that the measured quantity of land grew by 0.47 percent per annum, the national maximum. During the 1949-1991 period, labor use in U.S. agriculture declined in all 48 contiguous states. The reduction in labor use was less dramatic in the West. Over the entire period, labor use contracted most quickly in Mississippi and most slowly in California, where labor use has increased during the period since 1975. Although the 48-state aggregate measure of capital use grew by 0.37 percent per year, capital use fell in one-third of the 48 states (including all of the states in the

Panel a

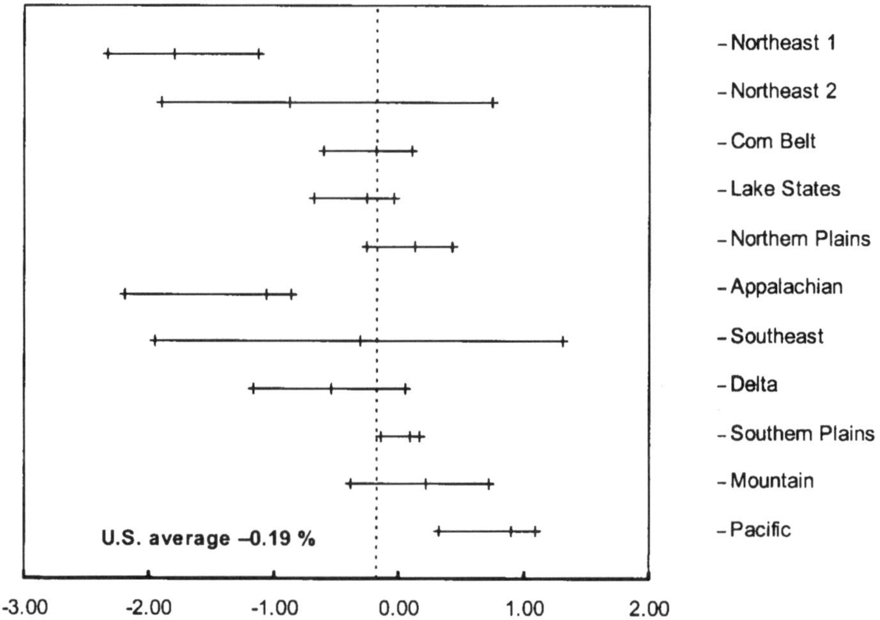

Panel b

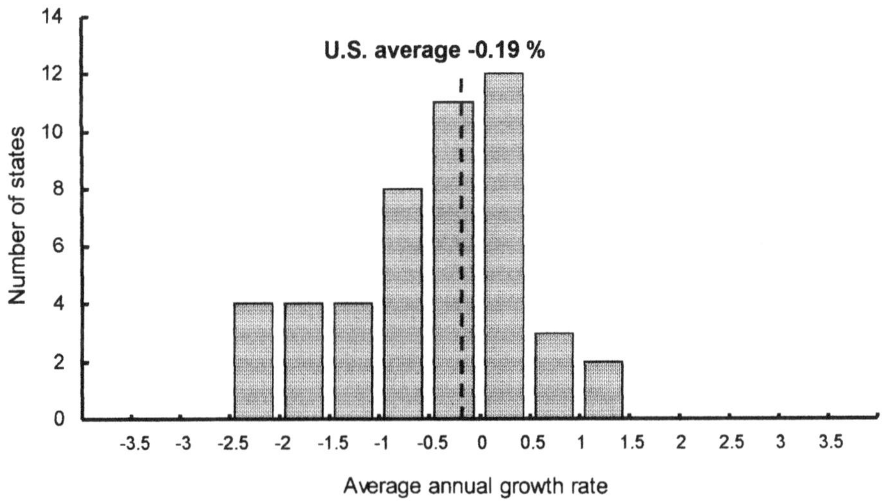

Figure 3.4. Distributions of growth rates in quantity indices of aggregate inputs by states.

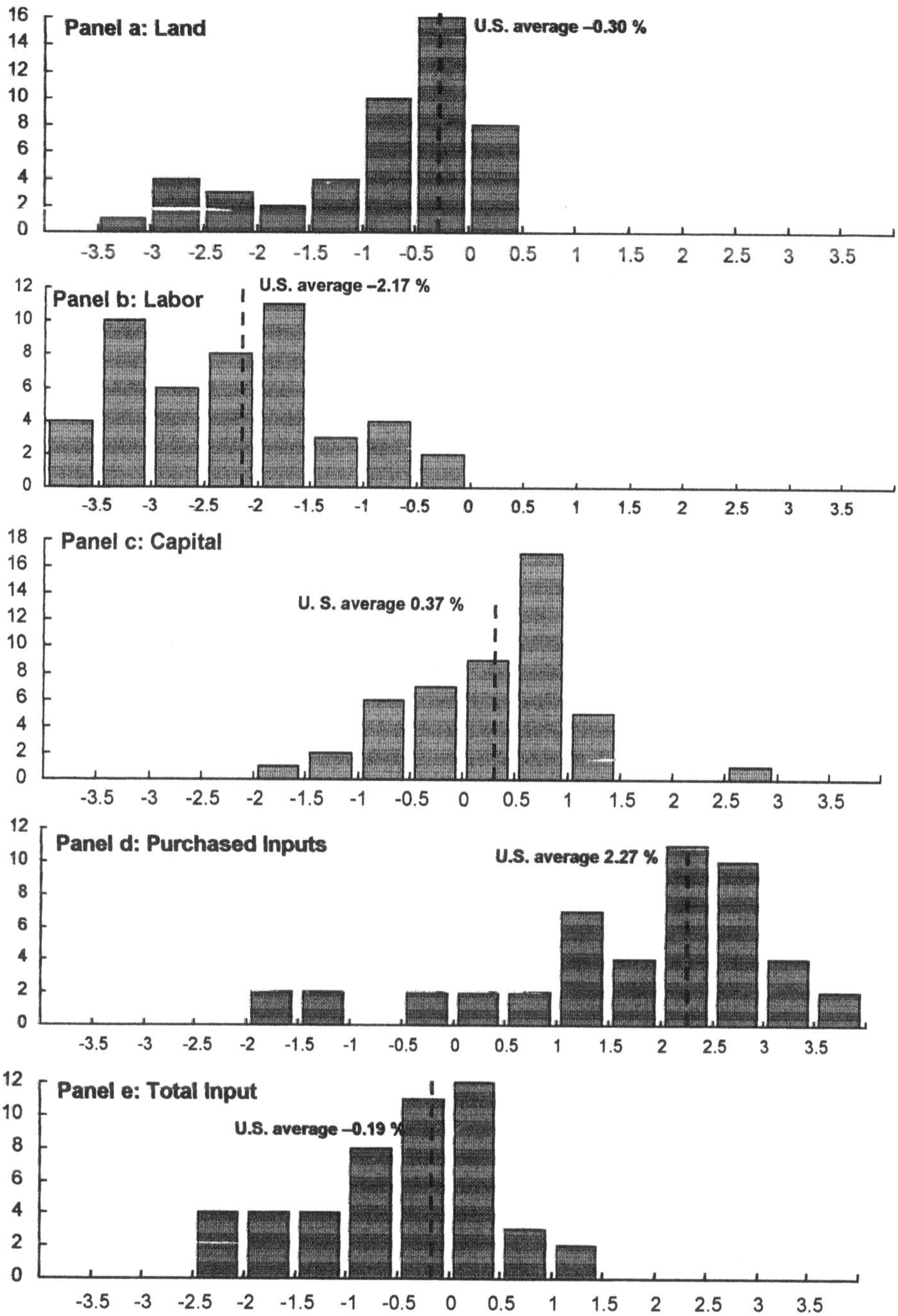

Figure 3.5. Distributions of growth rates in quantity indices for input sub-aggregates.

Northeast 1 region) during 1949-91. Florida had the highest annual average growth rate of capital.

Output Growth Patterns

Figure 3.6 shows distributions of state-specific growth rates of aggregate output among the 48 states over the 1949-91 period. Panel a presents linearized distributions of state-specific output growth rates for each USDA production region, and panel b is a frequency distribution of the 48 state-specific output growth rates; most but not all of the states (i.e., 42) increased their total output. The vertical line indicates the growth rate of the national index. Like aggregate input, output grew fastest generally in the Pacific region, and contracted fastest in the Northeast 1 region.

Patterns of change differed among output categories. Figure 3.7 shows frequency distributions of state-specific output growth rates by output category. National agricultural output grew by 1.72 percent per year between 1949 and 1991. Output in Florida grew the fastest. The aggregate input index concealed differences among categories, whereas the aggregate output index also masks important variation within categories, among states. Greenhouse and nursery products is the only category having a relatively concentrated distribution, with a positive growth rate for every state; it is also positively skewed, reflecting, in particular, high growth rates in California and Florida. For every other category of output, the distribution of state-specific growth rates was much flatter, with a much wider range than for aggregate output.

The Changing Location of Production

Production has tended to become more spatially concentrated, with a greater specialization of states in particular outputs, and a greater concentration of output among states (figure 3.8). The Midwestern states produced about 60 percent of the total value of field crop production in 1949; in 1991, 68 percent. The Northeastern states were generally getting out of field crop production, and by 1960 New Hampshire and Rhode Island no longer reported production of any field crops. In 1949 soybeans accounted for only 4.1 percent of the value of U.S. field crop production, but in 1991 it was the second-most valuable field crop and accounted for 22 percent of value of field crop production. Production of soybeans increased predominantly in the Lake States and Corn Belt sub-regions.

Wheat production stayed predominantly in the Northern Plains, with 32 percent of national wheat production in 1949 rising to 42 percent in 1991.

Panel a

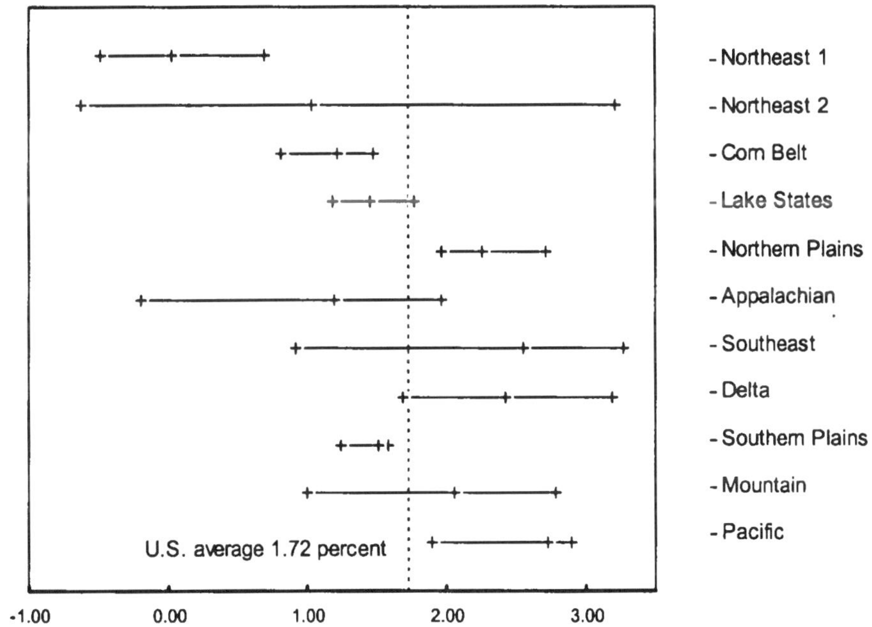

Panel b

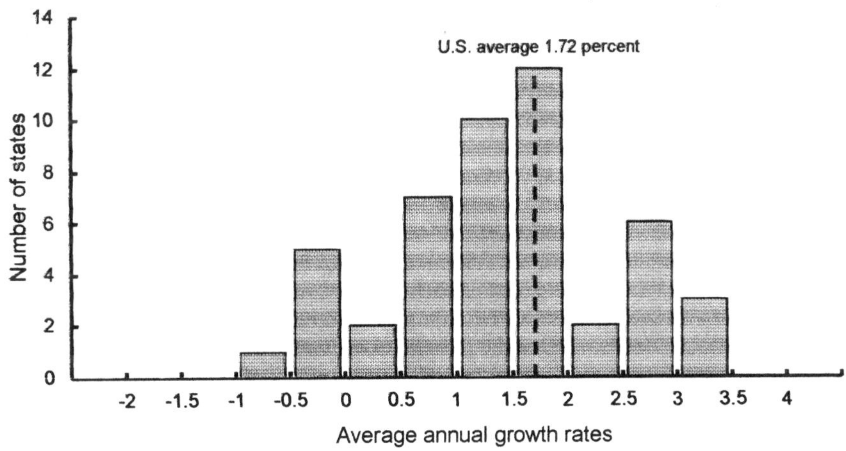

Figure 3.6. Distributions of growth rates in quantity indices of aggregate outputs by states.

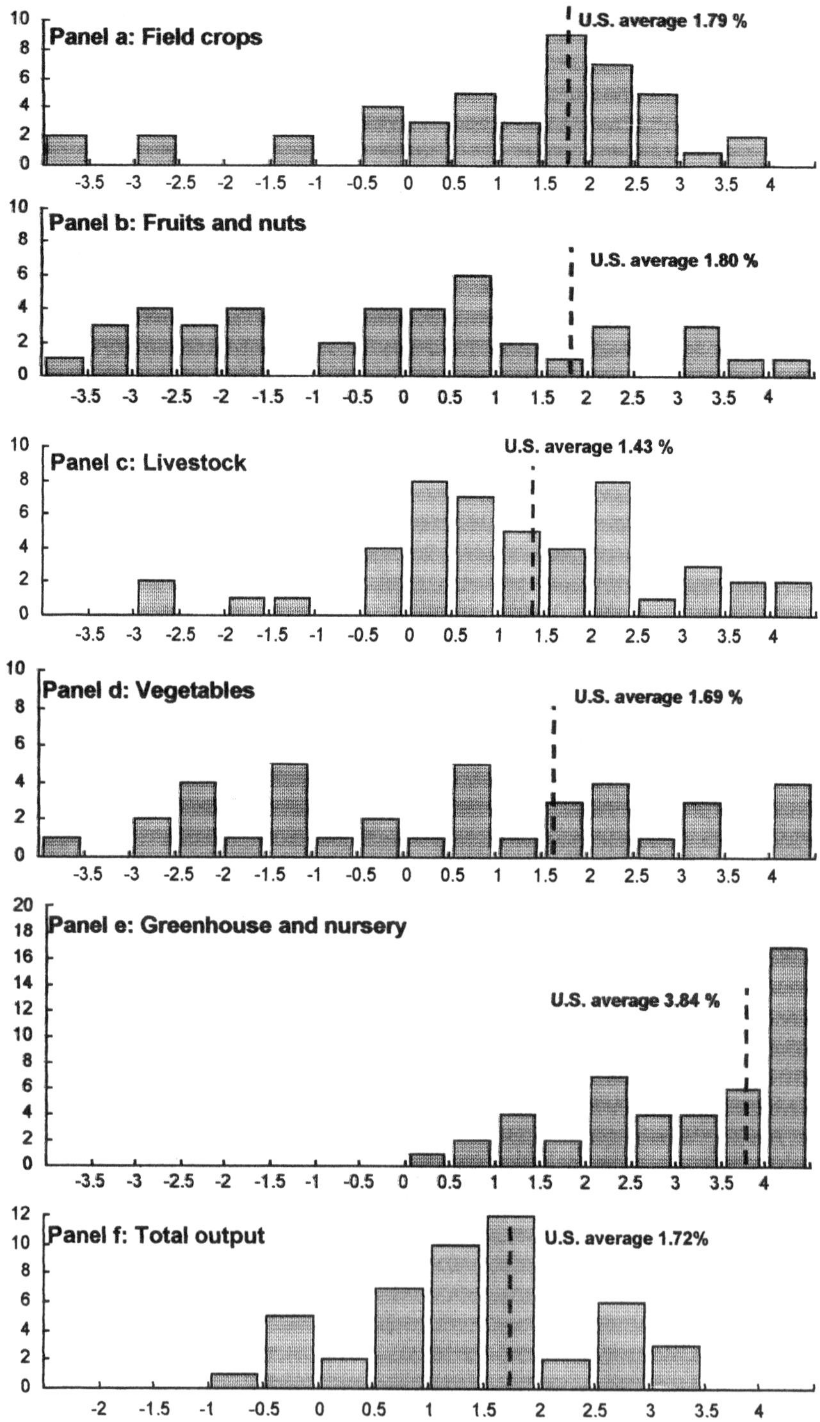

Figure 3.7. Distributions of growth rates in quantity indices for output sub-aggregates.

Figure 3.8a. Changing location of agricultural production, Soybeans.

Panel a: Share of national value of soybean output

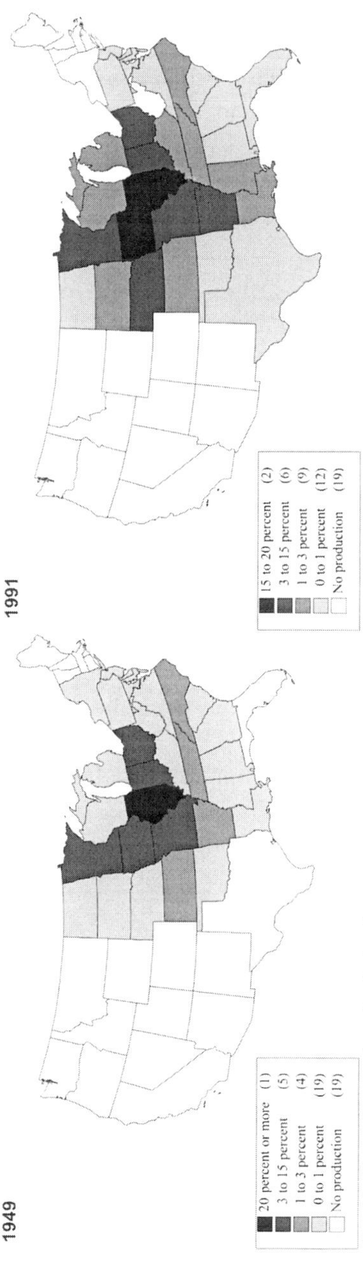

Panel b: Share of state value of agricultural output

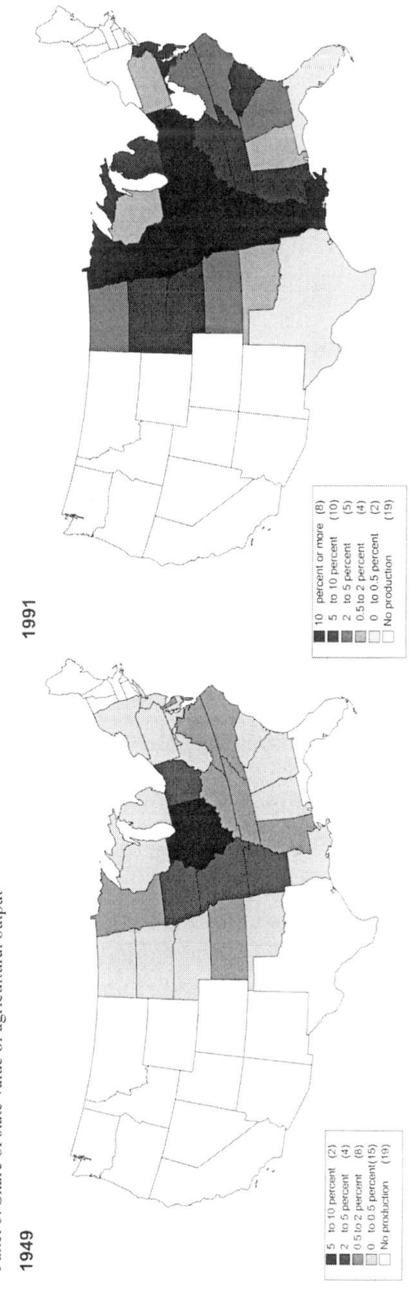

Figure 3.8b. Changing Location of Agricultural Production, Barley.

Panel a: Share of national value of barley output

Figure 3.8c. Changing Location of Agricultural Production, Greenhouse and Nursery Products

Panel a: Share of national value of greenhouse and nursery products output

1949

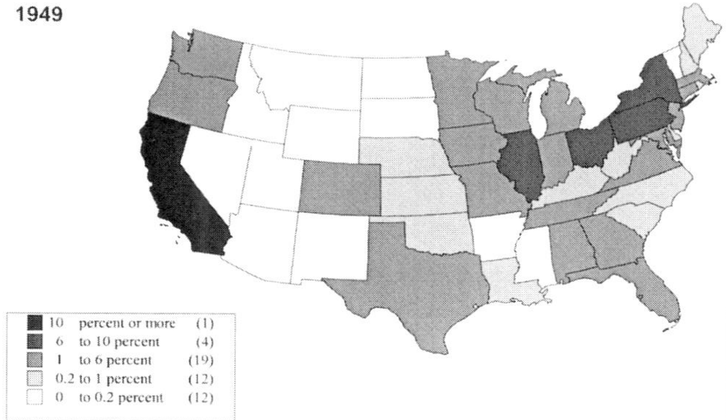

1991

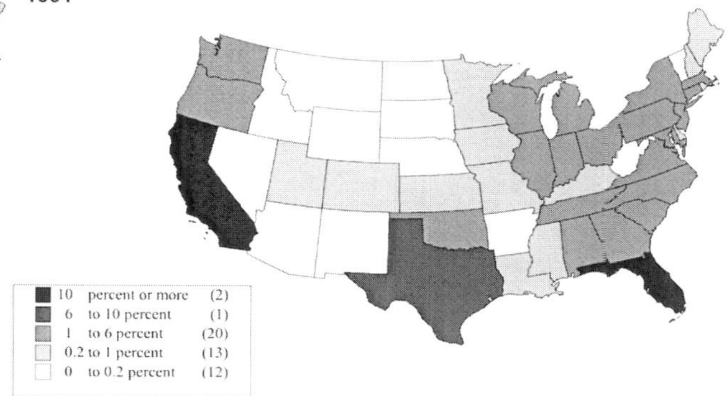

Panel b: Share of state value of agricultural output

1949

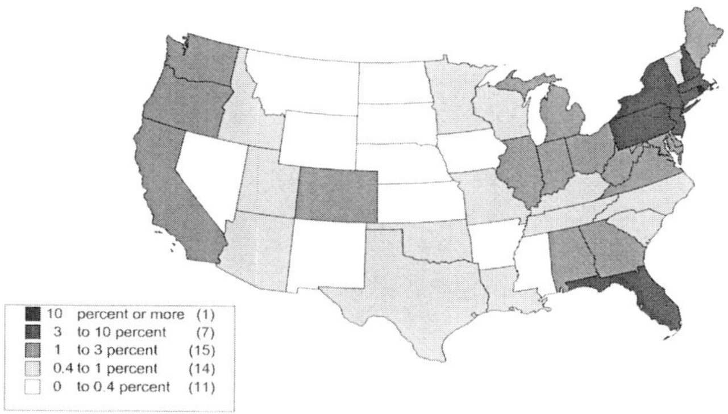

1991

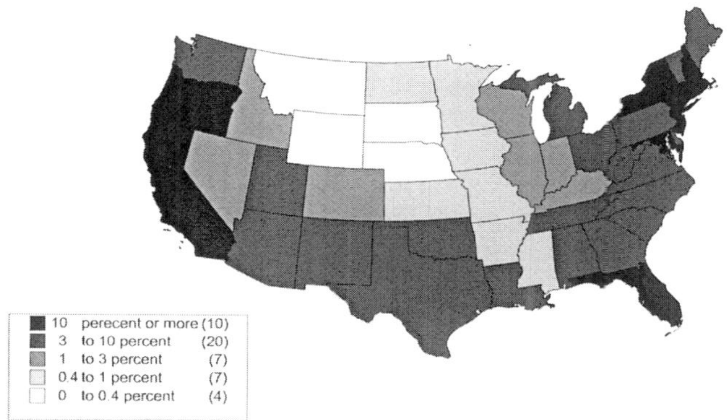

Figure 3.8d. Changing Location of Agricultural Production, Broilers.

Panel a: Share of national value of broiler output

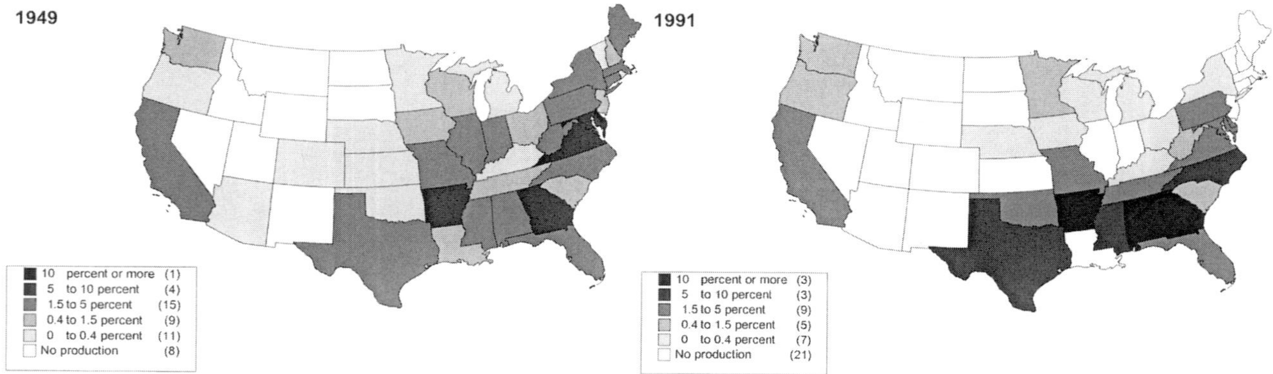

Panel b: Share of state value of agricultural output

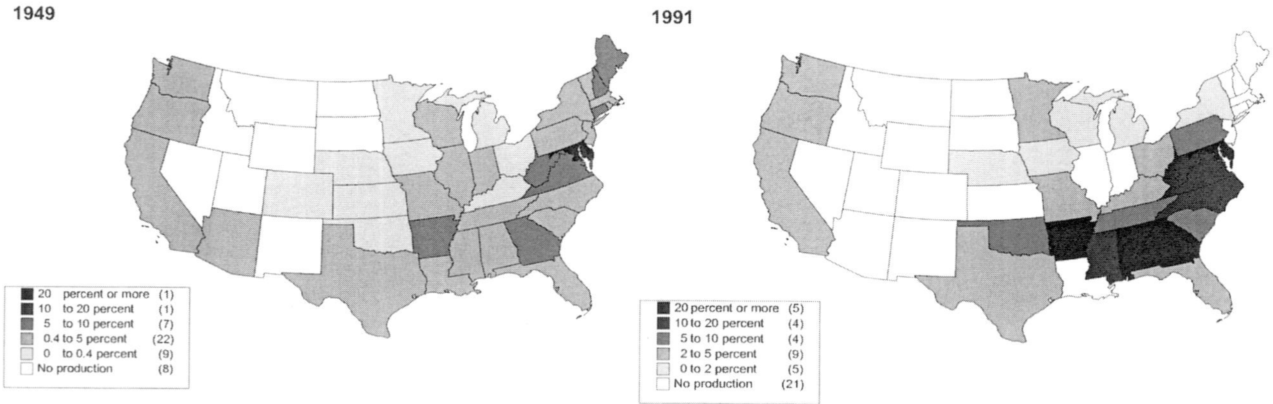

Figure 3.8e. Changing Location of Agricultural Production, Hogs.

Panel a: Share of national value of hog output

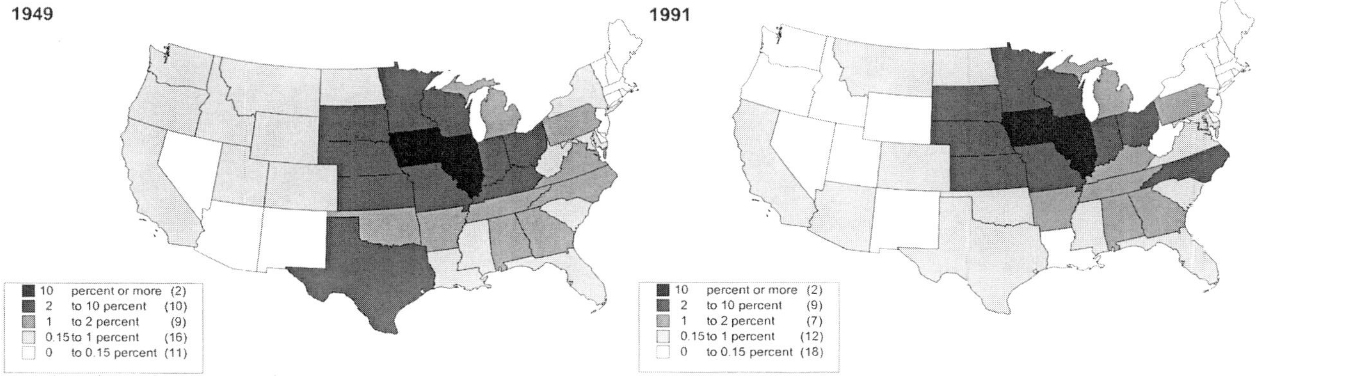

Panel b: Share of state value of agricultural output

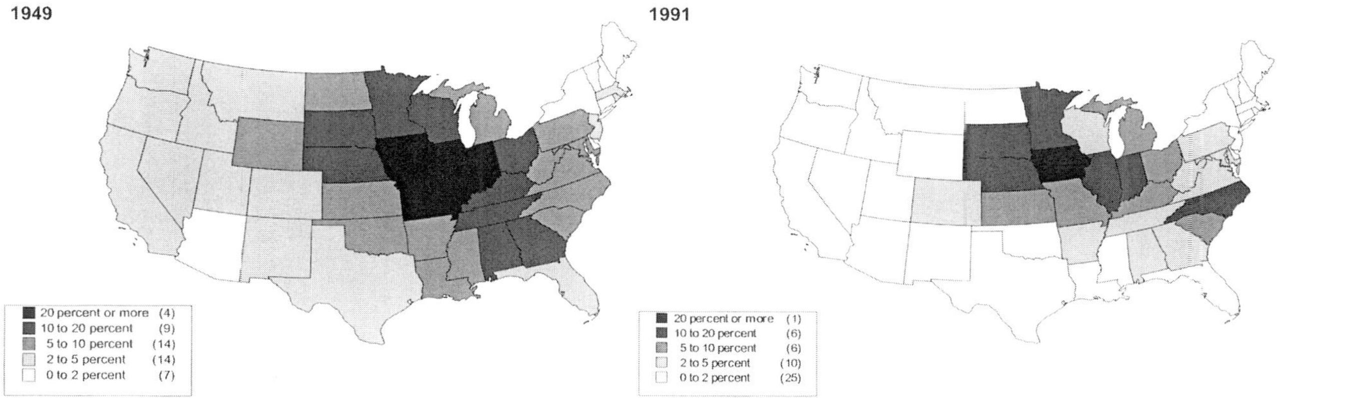

Three regions – Corn Belt, Lake States, and Northern Plains – were responsible for most of the field corn production. In 1991 these three regions accounted for 49 percent, 18 percent, and 20 percent of field corn production respectively. Tobacco production stayed mainly in the Appalachian region, with 74 percent and 81 percent of total production in 1949 and 1991, respectively, a consequence of the farm program for tobacco, that restricted where tobacco could be grown. In 1949, the Southern Plains produced 41 percent of the total cotton crop, but in 1991 the Southern Plains share had fallen to 28 percent and the Delta region produced 30 percent.

Fruits and nuts are mainly produced in the Pacific and Southeast regions; together they accounted for about 69 percent of the value of national production of fruits and nuts in 1949, and 84 percent in 1991. The Pacific and Southeast regions also accounted for the bulk of production of vegetables; the Pacific region accounting for 29 percent of the total value of vegetable production in 1949 and 41 percent in 1991. National output of greenhouse and nursery products in 1991 was about five times as much as in 1949, while national output of the other sub-groups just about doubled. California and Florida accounted for about 36 percent of the national value of greenhouse and nursery products in 1991, compared with 16 percent in 1949.

Besides broilers, every category of livestock production was present in all 48 contiguous states in 1949. Eight states (North Dakota, South Dakota, Idaho, Montana, Nevada, New Mexico, Utah, and Wyoming) did not produce broilers in 1949; by 1991 the number of non-broiler-producing states had increased to 21. By 1991, production of sheep and wool was specialized to 41 states, and turkey production had also become specialized to 28 states.

A Comparison of Input and Output Growth Rates

Higher-than-average rates of output growth in some regions, such as the Pacific, were associated with correspondingly higher-than-average growth rates of input use. As figure 3.9 shows, some states had both inputs and outputs growing, some had both falling, but the majority had output growing against a declining input quantity. The solid 45-degree line in figure 3.9 passes through the observation for the national aggregate. A point above that line indicates a relatively fast output growth rate for the given input growth rate (or a relatively fast reduction in inputs for a given rate of output growth), and a point below the line, the converse. In turn, we can think of the points above the line as reflecting faster-than-average productivity growth. The dashed 45-degree line through the origin passes through points where the rates of input and output growth are equal — zero productivity growth. In a few states, mostly states

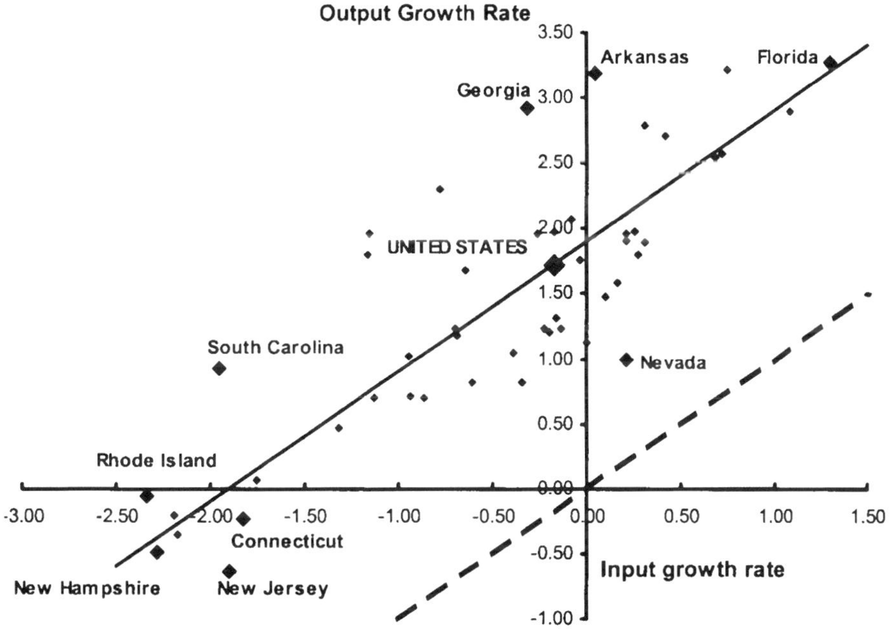

Figure 3.9. Input versus output growth rates, by state.

in the Northeast region, productivity growth reflected a contraction in input use that outweighed declining aggregate output.

Multi-factor productivity (MFP) growth measures calculated using Fisher Ideal indices were obtained by subtracting the input growth rate from the corresponding output growth rate. U.S. agricultural productivity grew, on average, at a rate of 1.90 percent per annum from 1949 to 1991. All 48 states enjoyed some growth in productivity (all states were above the lower 45-degree line in figure 3.9). Output growth was primarily responsible for this productivity growth; input contraction accounted for only 0.19 percent (most states are in the northwest quadrant in figure 3.9). Figure 3.10 shows distributions of state-specific MFP growth rates among the 48 states from 1949 to 1991. Panel a presents linearized distributions of state-specific MFP growth rates for each USDA production region, and panel b the frequency distribution of the 48 state-specific MFP growth rates. The vertical line indicates the growth rate of the national productivity index. The Delta, Southeast, and Appalachian regions recorded the highest regional productivity growth rates; the Corn Belt and Southern Plains regions, the lowest. The Pacific region, with relatively high growth rates of input and output, was below the national average.

Productivity growth was uneven over the period. Figure 3.11 shows the distributions of state-specific MFP growth rates in four sub-periods representing approximately the 1950s, 1960s, 1970s, and 1980s, as well as for the entire period. It can be seen that the shapes of the distributions as well as their central tendency and range are different among the sub-periods, and none of the sub-periods is well represented by the distribution over the entire period. In every sub-period as well as the 43-year period, virtually every state experienced positive productivity growth; the exceptions were Maine and New Hampshire in the 1970s and Massachusetts in the 1980s.

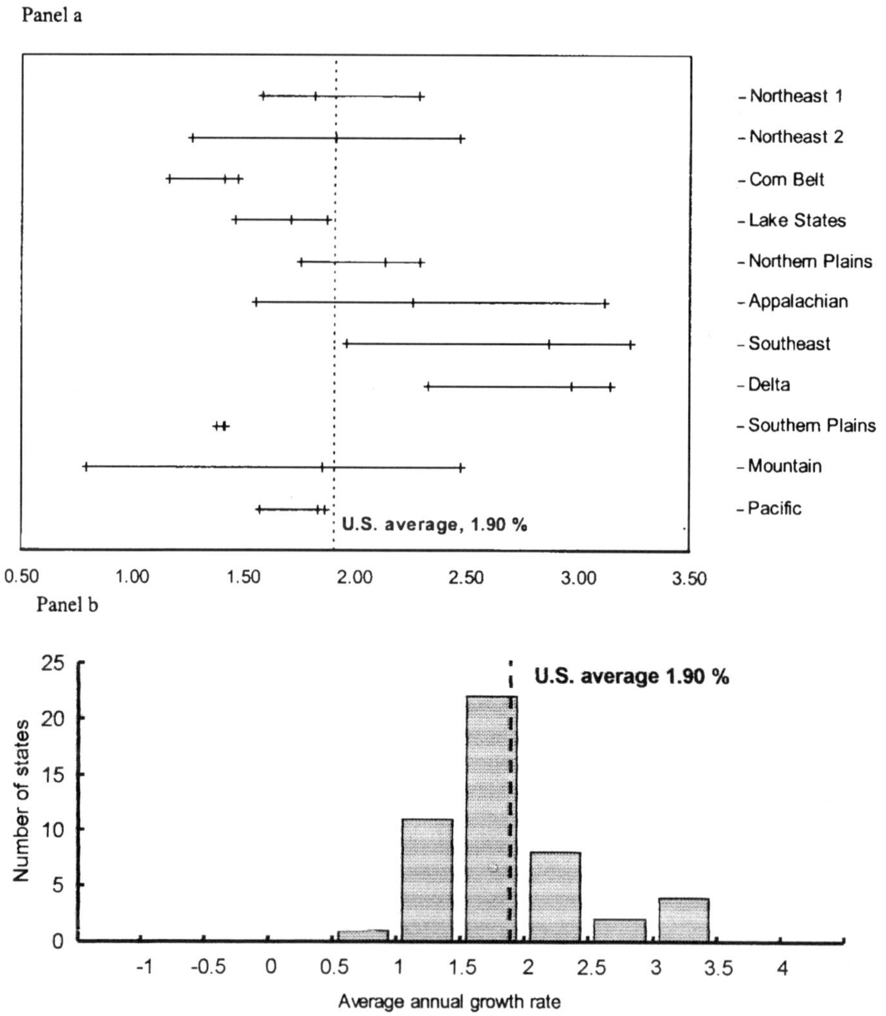

Figure 3.10. Distributions of state-specific and regional MFP growth rates.

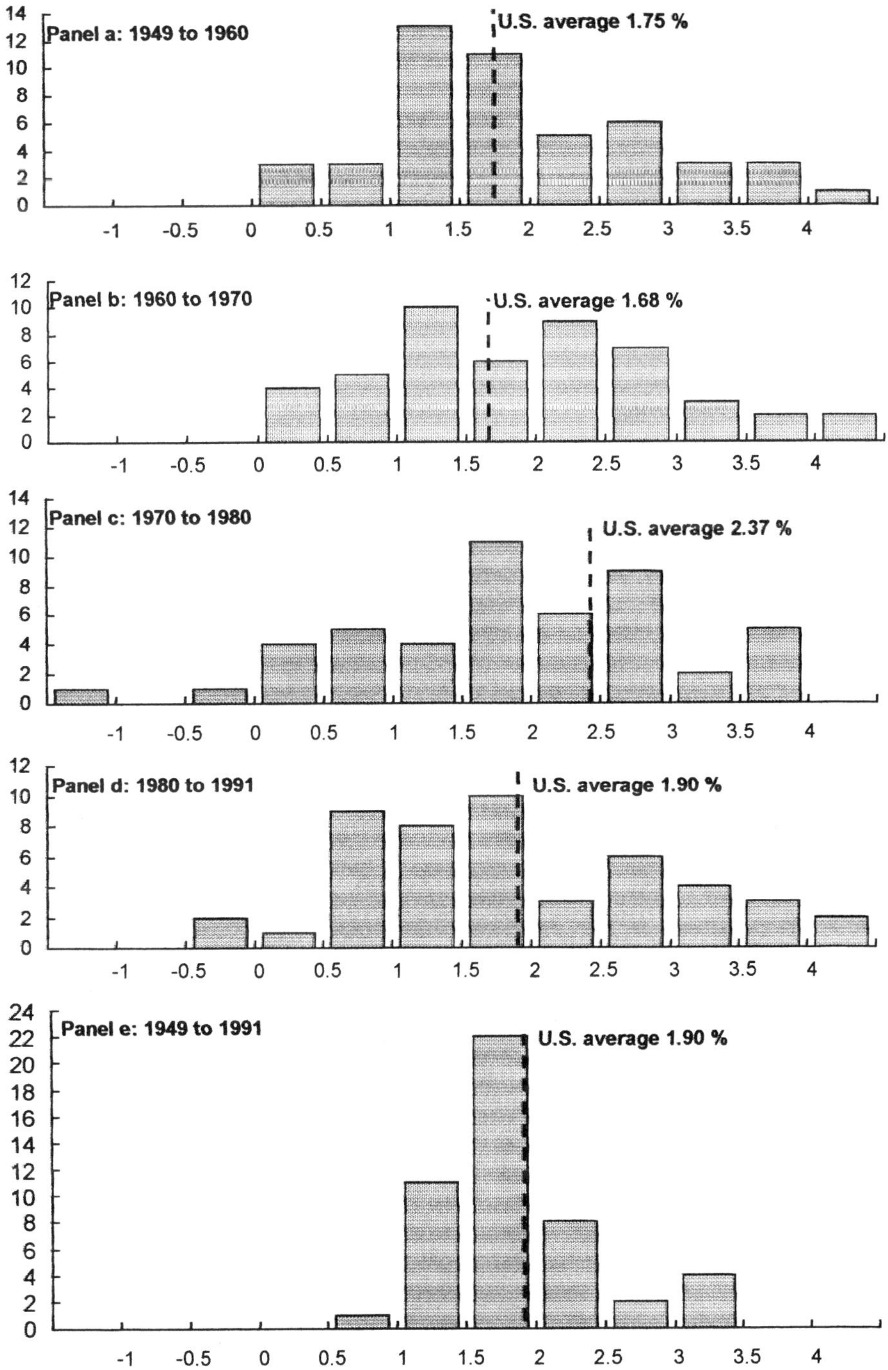

Figure 3.11. Distributions of state-specific MFP growth rates, by sub-period.

Comparison with Ball et al. (1999)

To compare results, we estimated rates of growth in our state-specific and national productivity indices for the 1960-1990 period, as reported by Ball et al. (1999). The average annual growth rates are compared in Appendix table 3A.4. We also computed the state-specific productivity indices in 1990, using a value of 100 in 1960 with the compound growth implied by both our estimates and those of Ball et al. In figure 3.12, panel a, these alternative indices are plotted against each other, state by state. The national average annual growth rates are essentially indistinguishable — 1.99 percent per year versus 2.00 percent for MFP, 1.75 percent versus 1.73 percent for aggregate output, and -0.24 percent versus -0.27 percent for aggregate inputs. These aggregate figures mask substantial differences among categories and states. Unfortunately, among individual input categories, the data presented in Ball et al. allow a direct comparison only for labor. For the United States, labor used in agriculture contracted by 1.83 percent per year by our reckoning, 2.8 percent per annum according to Ball et al.[11]

The state-specific annual growth rates differ quite substantially in some cases, with positive or negative differences of up to 40 percent of the estimates from Ball et al., and little in the way of systematic patterns, apart from the Northeast region where the Ball et al. estimates were generally substantially greater than ours. Comparing the state-by-state estimates, by subtracting our estimate from that of Ball et al. (1999), the mean difference was small (0.07 percent per annum) but some of the differences were quite large (the standard deviation of the differences was 0.37 percent per annum). The simple correlation between the estimates was 0.78.

Possible explanations for these differences include differences in the raw data, and differences in the inputs and outputs included in the definition of agriculture, as well as differences in the treatments of the data, which have been the focus of this paper. It would be desirable to be able to completely reconcile the two sets of estimates (and, indeed, these and other sets of estimates), but to do so is practically impossible given the multitude of differences between the series and the underlying procedures. Rather than attempt a complete reconciliation, here we look at what seem to be prime candidates as sources of significant differences. Specifically, we evaluate the effects of index number formulation and quality adjustments using our own series, and compare the primary elements of growth in output and in major input categories between the two studies.

Ball et al. (1999) used a Törnqvist-Theil index whereas we used a Fisher Ideal index. However, when we computed both indices for each state over the 1960-1990 period using the subset of our data without any zero observations, we found that the differences were negligible (the simple correlation was 0.999),

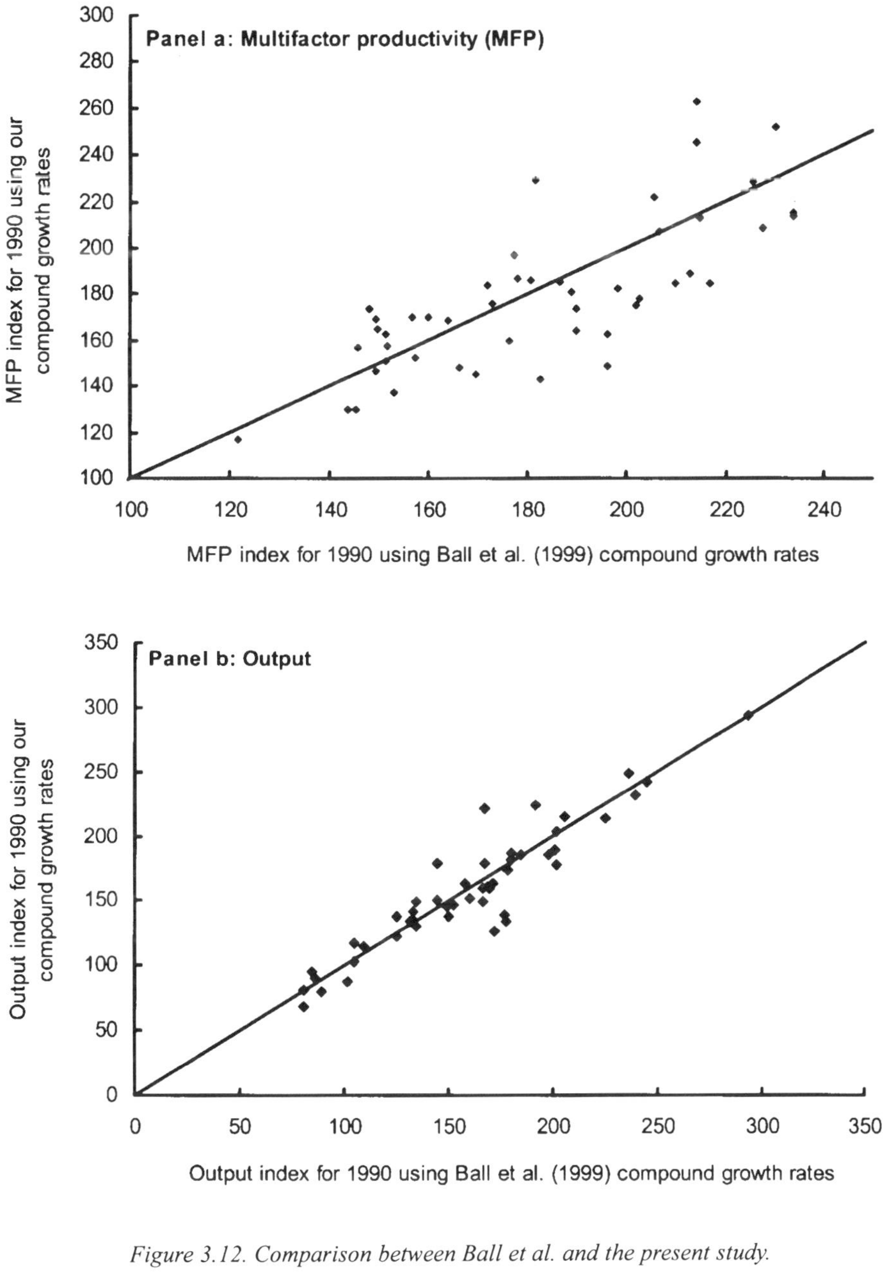

Figure 3.12. Comparison between Ball et al. and the present study.

so the contrast in results is probably attributable to something other than the choice of index.

Another possibility is differences in the definition and measurement of output. In particular, our output measures are based on sales, whereas Ball et al. (1999) use estimates of changes in inventories to adjust sales with a view to more-precisely representing production within a calendar year. Inventory adjustments could make for some substantial differences in measured year-to-year productivity growth, but on-farm inventories are notoriously hard to measure. In any event, over time, we would expect measures of year-to-year inventory adjustments to wash out of the measures of output growth. Hence, we should not expect differences in the treatment of on-farm stocks to account for important differences in estimates of output and productivity growth over the period 1960-1990. Indeed, as can be seen in figure 3.12, panel b, although the measures of output growth differ, there is no systematic difference of a form that could account for the productivity comparison. Apart from differences in the treatment of on-farm stocks, differences in measured output growth could be a result of other differences in the definitions of output (e.g., our inclusion of CRP acreage as an output), different treatments of livestock capital, or different aggregation procedures.

A more likely source of differences in productivity measures, however, is differences in the procedures used to form input aggregates, especially the degree of detail and spatial specificity of the prices used for the disaggregated inputs. To explore that possibility, we would want to compare the detailed input quantities implied by the two studies and how they were calculated. A useful first step is to consider the effects on our own indices resulting from our disaggregated treatment of land and labor (as opposed to simply counting hours and acres). Figure 3.13 shows that the adjustments in land and labor alone had substantial effects on our state-specific measured productivity growth rates. It is not as straightforward to derive alternative estimates that would show the effects of our treatments of capital and purchased inputs.

Finally, in figure 3.14, we compare our state-level estimates of the growth of the main categories of inputs — land, labor, capital, and purchased inputs — over 1960-90 with those of Ball et al. (1999). Unlike the output growth comparison, where the observations were scattered around the 45-degree line, for each input category there is a systematic difference reflected in a general tendency for the observations to lie above or below the line. For both land and labor, the observations lie above the 45-degree line, which reflects a strong tendency of Ball et al. (1999) to find a faster rate of decline in the quantities of land and labor — our adjustment for quality improvements of both land and labor, mitigating to a greater extent the effects of reductions in unadjusted hours and acres. For capital, the pattern is somewhat more mixed both in terms of the direction of change and the comparison with Ball et al. (1999). Although many

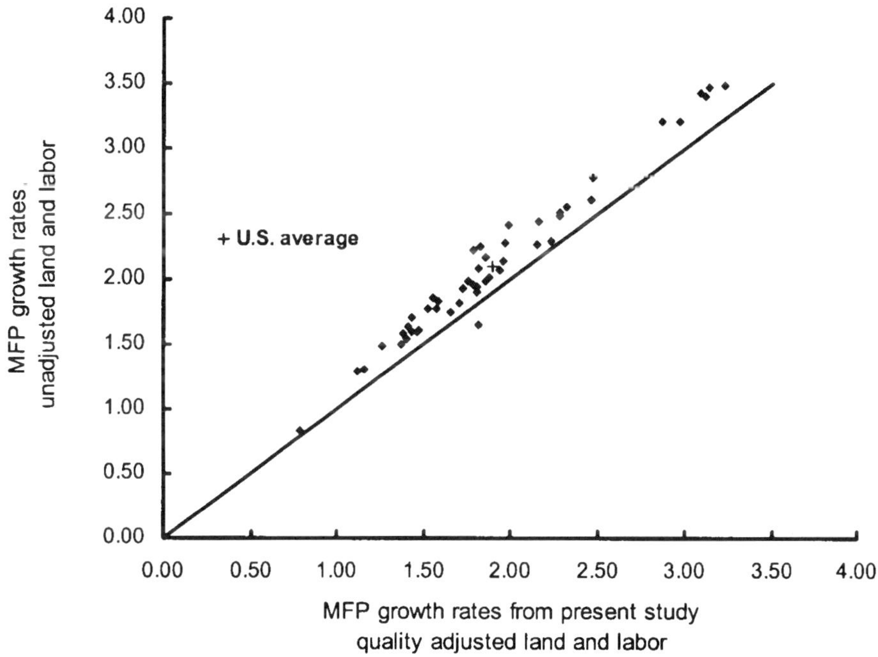

Figure 3.13. Effects of land and labor quality adjustments on MFP indices, 1949-91.

show a decline, a large proportion of states show an increase in the quantity of capital, and, relative to our figures, Ball et al. (1999) tended to report higher rates of growth (or slower rates of decline) in the quantity of capital. That is, they reported an increase in capital use in half the states, compared with one quarter in our estimates. Underpinning this difference could be differences in assumptions about depreciation patterns and service flow patterns, as well as differences in quality adjustments and different measures of capital stocks, especially tractors and combines. The pattern for purchased inputs is different again. Both studies show growth in the use of purchased inputs in most of the states, with a tendency in each instance for our estimate to be greater than that of Ball et al. (1999). It is not obvious to us what might account for these differences.

Conclusion

We have shown the effects of using disaggregated, state-specific data, to obtain improved measures of agricultural productivity growth in the United States. At the national level, on average for the period 1949 to 1991, U.S. agricultural productivity grew by 1.90 percent per annum, reflecting output

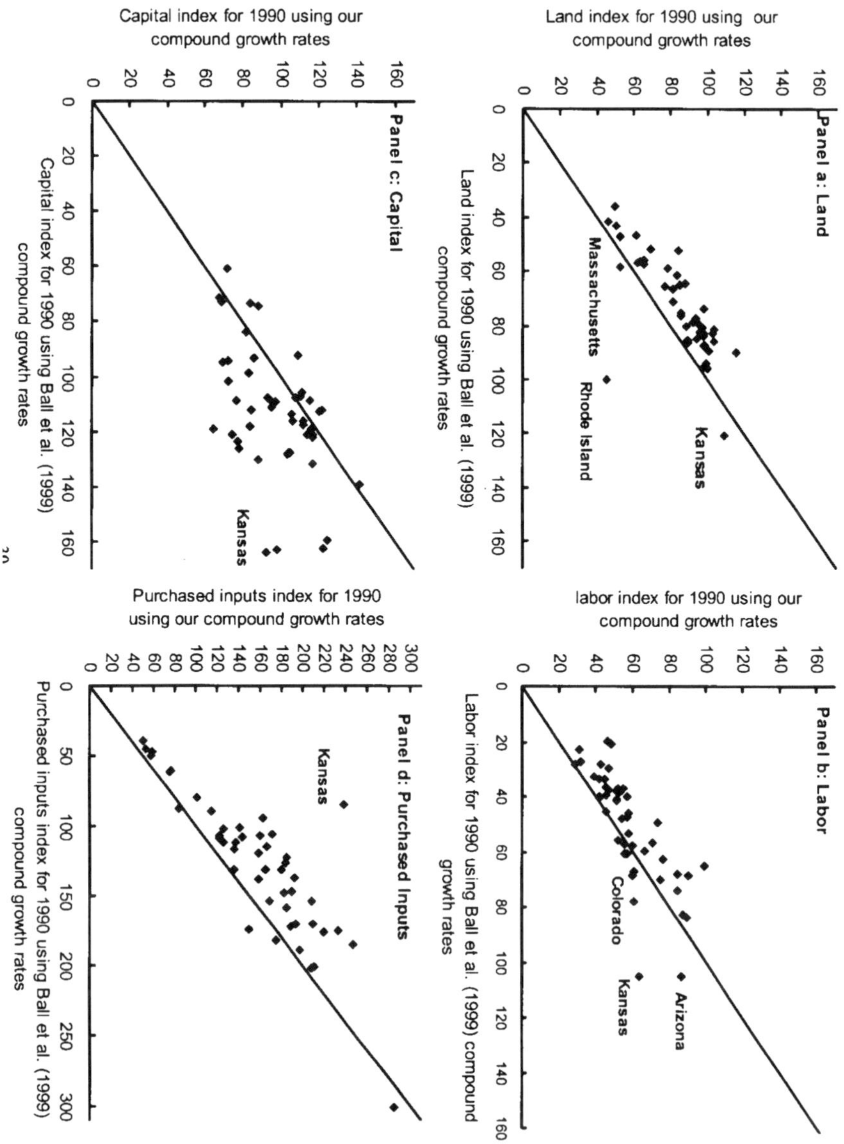

Figure 3.14. Comparison with Ball et al. estimates of input growth, 1960-1991.

growth of 1.71 percent per annum and a reduction in input use of 0.19 percent per annum. These aggregate measures conform in trend with other studies, but we also provide broader coverage to include productivity growth rates for the 48 contiguous states, and for 11 regions.

Our efforts to avoid pre-aggregation bias resulted in generally lower estimates of productivity growth. The adjustments for changes in the composition and quality of inputs generally resulted in increased estimates of input quantities over time: slower rates of decline in some categories and faster rates of growth in others (i.e., generally, more positive growth). The corresponding adjustments on the output side were in the same direction but did not account for as much of the growth in outputs. These aspects varied among states and over time.

Some of the modifications discussed here, such as the use of Divisia indexing procedures, reduce bias in the index numbers. The Fisher Ideal indices implied generally higher estimates of annual average productivity growth rates than did the Laspeyres index for this data set. Some other procedures, such as the adjustment for the composition and quality of input aggregates, also have implications for the interpretation of the measure. For instance, a measure of productivity that does not involve adjustments for changes in labor quality includes the effects on productivity of education and investments in human capital, and these may be desired for some purposes, but will nevertheless involve some index number bias, if the mix of labor quality has changed.

The implications of our procedures may be more important at the more-disaggregated state level. At the national level, our estimates are similar to those from other studies that have used Divisia indices and similar data, but different treatments of the details of measurement of inputs and outputs. At the state level, however, the differences may become more important as the consequences of pre-aggregation vary among states, as attention moves in the direction of accounting for state-level patterns of productivity change, and using the state-specific results to better understand the national picture.

The observation that our measures of agricultural inputs, outputs, and productivity depend on the details of the measures should give pause to those who want to use the measures in their econometric models. In recent months, a number of studies have reported results based on an application of the Ball et al. (1999) data, which have been made available as the official USDA data. The authors of these studies might have taken the "data" on faith as being appropriate for their purpose, presuming no need to consider how they were constructed. But as Griliches (1994, p. 2) stressed in his Presidential Address to the American Economic Association:

> "[We] often misinterpret the available data because of inadequate attention to how they are produced and that same inattention by us to the sources of our data helps explain why progress [in understanding the sources of productivity growth] is so slow.... Great advances have been made in theory and in econometric techniques, but these will be wasted unless they are applied to the right data."

Different data or treatments of data are likely to be required, depending on the purpose at hand. The series used in this study was developed with a view, in particular, to measuring the effects of public agricultural R&D on productivity. It may be an open question whether our series is more or less suitable for that purpose, let alone any other purpose, than the series developed by Ball et al. (1999). At a minimum, however, we have established that the two data sets are different in important ways, and that they are likely to lead to substantially different findings concerning productivity patterns and factor biases of technical change. Similar stories are likely to apply in the analysis of productivity in other industries and other countries. That is, the measures will depend in important ways on the procedural and data details. While the issue of indexing formulas has been addressed, and would seem to be largely resolved, relatively little attention has been paid to the other issues emphasized here: the underlying data, and aggregation over time and across space and form.

NOTES

1. An earlier version of this paper was presented at the NC-208 Conference on "Agricultural Productivity: Data, Methods, and Measures," Waugh Auditorium, USDA-ERS, Washington D.C., March 9-10, 2000. The authors gratefully acknowledge the considerable help they received from Barbara Craig in the construction of the data that underpin this paper. Helpful comments were provided by Bruce Beattie, Paul Cashin, Jim Chalfant, Bruce Gardner, Alan Olmstead, and participants at workshops at UC Berkeley and UC Davis, as well as the NC-208 conference. Financial assistance for this work was provided by the Economic Research Service, United States Department of Agriculture, the Farm Foundation, and the University of California Pacific-Rim Research Program. Senior authorship is not assigned.

2. Also, see Capalbo and Vo (1988). These studies were responding in part to the recommendations made by the American Agricultural Economics Association (AAEA) task force that reviewed USDA productivity statistics Shumway (1988). See also Jorgenson and Gollop (1992).

3. Precedence is not clear. Huffman and Evenson put out a series of five staff papers in the late 1980s, as a draft of their 1993 book, in which they used state-level estimates from Evenson, Landau, and Ballou (1987).

4. In this type of work, a myriad of somewhat arbitrary decisions has to be made in translating available data into the measures of interest. Sometimes these decisions are guided by theory or the purpose to which the estimates are to be put. Sometimes the choice is dictated by practical reality. What cannot be determined a priori is how important such decisions will be individually, let alone taken together, for the estimates. In particular we make no claims here about the general superiority of either series over the other.

5. For a more detailed discussion on the index number procedures presented here, see Diewert (1976), Caves, Christensen, and Diewert (1982), Capalbo and Antle (1988), and Alston, Norton, and Pardey (1995).

6. An alternative, but somewhat more demanding, approach would be to use every observation that could be used (by allowing the vector of quantities to vary from observation to observation), as done by Craig and Pardey (1996a, b).

7. Counting inputs (outputs) of states separately in forming a national index, in our data set we have 2,718 different inputs and 967 different outputs for each year. Note that the indices developed here are not panel measures, but merely disaggregated measures, best suited for looking at changes over time, at different spatial scales of aggregation. Our measures can be used to compare growth rates but not levels of inputs, outputs, or productivity among states.

8. In this paper we use the term "aggregation bias" to describe the index number problem introduced by using prices that fail to correspond with the quantity components of the respective aggregates. For example, state-specific quantities are expected to grow at different rates and state-specific prices to differ from one another. Thus using a single price (such as a national average price for cattle, or even a state-specific composite price for land) will likely yield biased estimates of the resulting quantity aggregate. The use of state-specific, disaggregated quantities and their corresponding prices will reduce this bias.

9. Over the period 1949 to 1991 the number of acres of non-irrigated cropland declined by 0.36 percent per annum, the number of acres of irrigated cropland grew by 1.58 percent per annum, and the number of acres of pasture and rangeland declined by 0.61 percent per annum. The number of acres in the highest-quality land category, irrigated cropland, grew most quickly. Hence, given a heavier weighting on a faster-growing category, the growth rate of quality-adjusted acres is greater than the growth rate of total acreage without adjusting for quality.

10. To illustrate the range among states, Illinois had the equivalent of about 10,000 medium-capacity combines, compared with one medium-capacity combine in Rhode Island in 1991; Texas had the equivalent of about 73,000 numeraire (two-wheel drive, 55 horsepower) tractors compared with 263 numeraire tractors in Rhode Island in 1991. In 1949, Illinois had the equivalent of about 3,000 medium-capacity combines compared to one medium-capacity combine in Rhode Island, and Texas had the equivalent of about 51,000 numeraire tractors compared to 257 numeraire tractors in Rhode Island.

11. In table 1, Ball et al. decomposes the rate of growth in output into its input and productivity parts. They estimate that labor accounted for –0.57 percentage points per year of the 1.73 percent per year growth in output, capital accounted for –0.12 percentage points, and purchased inputs 0.42 percentage points. The comparable figures using our estimates are that labor accounted for -0.62 percentage points of the annual growth in output, capital (here, grouping our capital and land measure to mimic the Ball et al. capital measure) accounted for –0.07 percentage points, and purchased inputs 0.45 percentage points.

References

Acquaye, A. K. A. 2000. "Parametric and Nonparametric Measures of State-Level Productivity Growth and Technical Change in U.S. Agriculture: 1949-1991." Unpublished Ph.D. Dissertation, Department of Agricultural and Resource Economics, University of California, Davis.

Ahearn, M., J. Yee, V. E. Ball, and R. Nehring. 1998. "Agricultural Productivity in the United States." Agricultural Information Bulletin No. 740, Economic Research Service, U.S. Department of Agriculture.

Alston, J. M., G. W. Norton, and P. G. Pardey. 1995. *Science Under Scarcity: Principles and Practice for Agricultural Research Evaluation and Priority Setting*. Ithaca: Cornell University Press.

American Agricultural Economics Association (AAEA). 1980. *Measurement of U.S. Agricultural Productivity: A Review of Current Statistics and Proposals for Change.* ESCS Technical Bulletin No. 1614. Washington, D.C.: U.S. Department of Agriculture.

Ball, V. E. 1985. "Output, Input, and Productivity Measurement in U.S. Agriculture." *American Journal of Economics*, 67(3):475-86.

Ball, V. E. 1994. "Measuring Agricultural Productivity in U.S. Agriculture." Proceedings of the NC-208 symposium at Orlando, Florida, March 1993. In *Evaluating Research and Productivity in an Era of Resource Scarcity*, ed. W. B. Sundquist. Staff paper P94-2, Department of Agricultural and Applied Economics, University of Minnesota, February.

Ball V. E., F. M. Gollop, A. Kelly-Hawke, and G. P. Swinand. 1999. "Patterns of State Productivity Growth in the U.S. Farm Sector: Linking State and Aggregate Models." *American Journal of Agricultural Economics*, 81(1):164-79.

Ball V. E., and R. Nehring. 1998. "Patterns of State Productivity Growth in the U.S. Farm Sector." Economic Research Service staff paper No. 9804, Washington D.C.: U.S. Department of Agriculture.

Ball V. E., J. C. Bureau, R. Nehring, and A. Somwaru. 1997. "Agricultural Productivity Revisited." *American Journal of Agricultural Economics*, 79(4):1045-63.

Barton, G. T., and M. R. Cooper. 1948. "Relation of Agricultural Production to Inputs." *Review of Economics and Statistics,* 30:117-26.

Capalbo, S. M., and J. M. Antle. 1988. *Agricultural Productivity: Measurement and Explanation.* Washington, D.C.: Resources for the Future.

Capalbo, S. M., and T. T. Vo. 1988. "A Review of the Evidence on Agricultural Productivity and Aggregate Technology." Chapter 3 in *Agricultural Productivity: Measurement and Explanation*, ed. S. M. Capalbo and J. M. Antle. Washington, D.C.: Resources for the Future.

Caves, D. W., L. R. Christensen, and W. E. Diewert. 1982. "The Economic Theory of Index Numbers and the Measurement of Input, Output, and Productivity." *Econometrica,* 50(6):1393-414.

Craig, B. J., and P. G. Pardey. 1990a. "Multidimensional Output Indices." Department of Agricultural and Applied Economics Staff Paper P90-63. St. Paul: University of Minnesota, October.

Craig, B. J., and P. G. Pardey. 1990b. "Patterns of Agricultural Development in the United States." Department of Agricultural and Applied Economics Staff Paper P90-72. St. Paul: University of Minnesota, December.

Craig, B. J., and P. G. Pardey. 1996a. "Inputs, Outputs, and Productivity Developments in U.S. Agriculture." Contributed paper presented at the *Global Agricultural Science Policy for the Twenty-First Century* Conference, Melbourne, Australia, August.

Craig, B. J., and P. G. Pardey. 1996b. "Productivity Measurement in the Presence of Quality Change." *American Journal of Agricultural Economics*, 78(5):1349-54.

Diewert, W. E. 1976. "Exact and Superlative Index Numbers." *Journal of Econometrics,* 4(2): 115-45.

Diewert, W. E. 1992. "Fisher Ideal Output, Input, and Productivity Indexes Revisited." *The Journal of Productivity Analysis,* 3(3):211-48.

Evenson, R. E., D. Landau, and D. Ballou. 1987. "Agricultural Productivity Measures for U.S. States 1950-82." In *Evaluating Agricultural Research and Productivity*, ed. W. B. Sundquist. University of Minnesota, Miscellaneous Publication 52-1987.

Griliches, Z. 1960. "Measuring Inputs in Agriculture: A Critical Survey." *Journal of Farm Economics*, 42:1411-27.

Griliches, Z. 1961. "An Appraisal of Long-Term Capital Estimates: Comment." *Output, Input and Productivity Measurement.* A Report of the National Bureau of Economic Research. Princeton: Princeton University Press.

Griliches, Z. 1963. "The Sources of Measured Productivity Growth: Agriculture, 1940-1960." *Journal of Political Economy*, 71:331-46.

Griliches, Z. 1994. "Productivity, R&D, and the Data Constraint." *American Economic Review*, 84(1):1-23.

Hayami, Y., and V. W. Ruttan. 1971. *Agricultural Development: An International Perspective*. Baltimore: The Johns Hopkins University Press.

Huffman, W. E., and R. E. Evenson. 1992. "Contributions of Public and Private Science and Technology to U.S. Agricultural Productivity." *American Journal of Agricultural Economics*, 74(3):752-56.

Huffman, W. E., and R. E. Evenson. 1993. *Science for Agriculture: A Long-Term Perspective*. Ames: Iowa State University Press.

Jorgenson, D. W., and F. M. Gollop. 1992. "Productivity Growth in U.S. Agriculture: A Postwar Perspective." *American Journal of Agricultural Economics*, 74(3):745-50.

Jorgenson, D. W., F. Gollop, and B. Fraumeni. 1987. *Productivity and U.S. Economic Growth*. Cambridge: Harvard University Press.

Jorgenson, D. W., and Z. Griliches. 1967. "The Explanation of Productivity Change." *Review of Economic Studies*, 34:249-83.

Kendrick, J. W. 1961. *Productivity Trends in the United States*. National Bureau of Economic Research, New York. Princeton: Princeton University Press.

Loomis, R. A., and G. T. Barton. 1961. *Productivity of Agriculture*. U.S. Department of Agriculture, Technical Bulletin No. 1238, April.

Pardey, P. G., B. J. Craig, and K. W. Deininger. 1994. "A New Look at State-Level Productivity Growth in U.S. Agriculture." Proceedings of the NC-208 symposium at Orlando, Florida, March 1993. In *Evaluating Research and Productivity in an Era of Resource Scarcity*, ed. W. B. Sundquist. Staff paper P94-2, Department of Agricultural and Applied Economics, University of Minnesota, February.

Shumway, C. R. 1988. "The Statistical Base for Agricultural Productivity Research: A Review and Critique." Chapter 4 in *Agricultural Productivity: Measurement and Explanation*, ed. S. M. Capalbo and J. M. Antle. Washington, D.C.: Resources for the Future.

Star, S. 1974. "Accounting for the Growth of Output." *American Economic Review*, 64(1): 123-35.

APPENDICES

Table 3A.1. Input Cost Shares and Quantity Indices

Year	Land	Labor	Capital	Purchased inputs	All inputs	Land	Labor	Capital	Purchased inputs
	Quantity indices (1949 = 100)					*Cost shares (%)*			
1949	100	100	100	100	100	17.58	40.24	23.53	18.64
1950	99	97	102	99	99	17.37	39.82	23.58	19.24
1951	99	92	107	106	99	17.82	36.94	24.58	20.66
1952	98	88	112	109	99	18.22	35.19	25.85	20.73
1953	98	83	115	111	98	18.26	35.78	25.75	20.22
1954	98	80	115	113	97	18.82	35.63	24.70	20.85
1955	97	78	117	123	99	18.58	36.09	24.14	21.18
1956	96	77	122	122	99	18.57	35.77	25.09	20.57
1957	95	75	120	128	98	18.76	35.75	24.92	20.57
1958	95	74	117	135	98	18.83	35.11	24.79	21.28
1959	94	73	123	146	100	18.31	34.04	26.01	21.64
1960	94	71	125	148	100	17.87	35.03	25.82	21.28
1961	94	69	125	154	100	17.91	35.40	25.18	21.51
1962	93	68	128	163	101	17.67	35.52	24.82	21.99
1963	93	66	132	169	101	17.98	35.34	24.24	22.43
1964	93	64	137	173	102	17.35	35.92	24.50	22.23
1965	92	63	142	175	102	17.87	36.01	24.32	21.80
1966	93	61	141	184	102	18.04	35.70	23.65	22.62
1967	93	58	136	196	101	18.14	35.92	22.82	23.12
1968	93	56	132	205	100	18.70	36.67	22.31	22.32
1969	93	55	125	213	99	18.62	37.33	21.33	22.72
1970	92	55	121	219	99	18.03	37.94	20.63	23.40
1971	92	54	117	222	98	18.37	38.34	20.03	23.26
1972	91	53	116	229	98	18.07	38.88	19.95	23.10
1973	91	54	116	213	96	16.82	36.87	19.34	26.98
1974	90	54	115	194	94	18.56	34.51	18.57	28.36
1975	91	53	115	191	93	20.10	34.00	18.67	27.22
1976	91	52	121	214	96	19.93	33.76	18.77	27.53
1977	92	51	128	215	97	21.76	32.36	19.58	26.31
1978	93	49	131	252	100	20.17	30.92	20.23	28.68
1979	92	48	132	259	101	19.90	28.96	20.70	30.44
1980	92	47	129	239	97	19.55	27.70	21.08	31.67
1981	91	46	127	221	94	19.82	26.91	21.40	31.87
1982	91	44	123	223	92	19.98	27.52	21.14	31.36
1983	90	43	120	221	91	19.75	28.02	20.78	31.46
1984	90	42	122	227	91	19.32	28.09	20.49	32.10
1985	90	42	122	224	91	19.44	29.65	20.90	30.01
1986	90	40	122	233	90	18.40	30.79	21.51	29.30
1987	89	40	121	250	92	16.95	31.48	21.66	29.90
1988	89	40	125	245	92	16.46	31.29	21.72	30.54

continued

Table 3A.1, *continued*

Year	Land	Labor	Capital	Purchased inputs	All inputs	Land	Labor	Capital	Purchased inputs
	Quantity indices (1949 = 100)					*Cost shares (%)*			
1989	89	40	126	244	92	16.59	31.30	21.73	30.39
1990	88	41	120	255	93	16.33	32.06	21.01	30.61
1991	88	40	117	259	92	16.13	32.38	20.57	30.92
	Annual average growth rates (%)								
49-60	-0.56	-3.15	2.04	3.56	0.002	0.15	-1.26	0.84	1.20
60-70	-0.17	-2.60	-0.30	3.94	-0.12	0.09	0.80	2.24	0.95
70-80	-0.08	-1.48	0.59	0.86	-0.18	0.81	-3.15	0.22	3.03
80-91	-0.35	-1.43	-0.88	0.75	-0.44	-1.75	1.42	-0.22	-0.22
49-91	-0.30	-2.17	0.37	2.27	-0.19	-0.21	-0.52	-0.32	1.20

Table 3A.2. Output Value Shares and Quantity Indices

Year	Field crops	Fruits and nuts	Vegetables	Greenhouse nursery products	Livestock	All output	Field crops	Fruits and nuts	Vegetables	Greenhouse nursery products	All crops	Livestock
	Quantity indices (1949 = 100)						*Value shares (%)*					
1949	100	100	100	100	100	100	40.60	3.32	3.36	1.32	48.61	51.35
1950	90	98	108	102	102	97	41.7	3.57	2.87	1.29	49.49	50.45
1951	94	100	97	105	107	101	38.90	2.86	3.11	1.14	46.01	53.87
1952	101	98	102	107	109	105	40.73	3.02	3.78	1.19	48.72	51.08
1953	99	101	109	109	111	106	41.50	3.50	2.73	1.32	49.05	50.70
1954	97	100	108	111	116	107	40.93	3.56	3.29	1.40	49.19	50.50
1955	101	105	111	116	119	110	41.16	3.96	3.27	1.53	49.92	49.69
1956	102	106	118	122	120	112	41.31	4.06	3.71	1.62	50.71	48.85
1957	99	100	116	124	119	110	37.52	3.85	3.67	1.73	46.77	52.69
1958	115	111	118	129	122	119	38.46	3.87	2.77	1.62	46.72	52.66
1959	116	113	117	136	127	122	38.28	3.78	3.44	1.78	47.28	52.06
1960	118	106	121	142	124	121	38.72	4.18	3.43	1.93	48.27	51.03
1961	114	118	130	141	130	123	39.22	3.87	3.10	1.91	48.10	51.22
1962	116	107	124	142	131	124	39.28	3.76	3.19	1.95	48.18	51.12
1963	123	110	127	139	134	129	40.42	4.22	3.29	1.96	49.88	49.36
1964	119	122	119	140	137	129	38.58	4.39	4.49	2.12	49.57	49.62
1965	127	127	132	143	134	131	38.63	3.93	4.09	2.13	48.78	50.34
1966	123	133	135	137	138	132	36.17	3.67	3.77	1.98	45.59	53.53

continued

Table 3A.2, *continued*

Year	Field crops	Fruits and nuts	Veget- ables	Greenhouse nursery prod- ucts	Live- stock	All output	Field crops	Fruits and nuts	Veget- ables	Greenhouse nursery prod- ucts	All crops	Live- stock
	Quantity indices (1949 = 100)						*Value shares (%)*					
1967	132	121	140	136	142	138	36.03	4.04	3.77	2.09	45.93	53.02
1968	138	117	138	135	143	140	35.31	4.40	3.86	2.08	45.66	53.25
1969	138	141	139	135	142	141	33.57	4.18	3.82	2.02	43.58	55.31
1970	132	131	141	136	149	142	33.86	3.85	3.52	1.97	43.19	55.80
1971	151	142	140	137	153	151	35.89	4.20	3.44	1.99	45.53	53.37
1972	155	126	139	141	154	152	39.35	3.93	3.58	1.80	48.65	50.44
1973	165	153	142	157	150	156	44.77	3.57	3.36	1.48	53.18	46.04
1974	146	155	153	166	152	149	47.45	3.54	3.43	1.67	56.09	43.19
1975	172	168	148	178	143	157	46.65	3.62	3.85	1.94	56.06	43.09
1976	171	167	158	205	150	161	45.41	3.75	3.52	2.20	54.89	44.30
1977	189	174	158	214	152	170	43.84	4.51	3.54	2.44	54.34	44.78
1978	193	169	163	236	152	171	42.22	4.77	3.34	2.46	52.78	46.62
1979	216	181	166	253	157	184	43.09	4.57	2.84	2.31	52.81	46.61
1980	189	204	158	277	162	176	43.74	4.49	3.44	2.54	54.22	45.29
1981	229	190	167	283	164	193	43.56	4.42	3.58	2.68	54.24	45.24
1982	228	193	177	301	162	193	43.68	4.36	3.39	2.93	54.36	44.68
1983	154	190	168	330	164	165	37.90	4.43	4.23	3.63	50.19	49.08
1984	212	191	183	367	162	188	40.83	4.36	4.09	3.82	53.10	46.10
1985	228	184	192	366	166	196	41.06	4.73	3.73	4.11	53.63	45.24
1986	205	179	190	396	167	189	32.00	5.40	4.68	5.06	47.14	51.88
1987	200	219	195	436	170	194	34.14	5.82	4.32	5.16	49.44	49.44
1988	164	220	193	442	171	181	33.03	6.09	4.65	5.23	49.00	49.90
1989	200	204	203	469	174	196	34.25	5.35	4.94	5.16	49.72	49.10
1990	219	202	208	491	178	205	33.68	5.47	4.23	5.31	48.69	50.12
1991	212	213	203	502	182	206	33.63	5.98	4.21	5.53	49.34	49.52
	Annual average growth rates (%)											
49-60	1.49	0.55	1.71	3.17	1.98	1.75	-0.43	2.09	0.19	3.45	-0.06	-0.06
60-70	1.11	2.12	1.53	-0.39	1.83	1.57	-1.34	-0.82	0.26	0.21	-1.11	0.89
70-80	3.61	4.41	1.14	7.08	0.80	2.19	2.56	1.54	-0.23	2.54	2.27	-2.09
80-91	1.05	0.38	2.31	5.41	1.09	1.39	-2.39	2.61	1.84	7.07	-0.86	0.81
49-91	1.79	1.80	1.69	3.84	1.43	1.72	-0.45	1.40	0.53	3.41	0.04	-0.09

Table 3A.3. State-Level Annual Average Productivity Growth Rates

Region	1949-60	1960-70	1970-80	1980-91	1949-91
	Average annual growth rate (percentage)				
Northeast 1	**2.93**	**3.24**	**0.18**	**0.88**	**1.81**
Connecticut	2.32	2.02	0.74	1.23	1.59
Massachusetts	2.84	2.91	1.87	-0.25	1.81
Maine	3.52	4.56	-1.52	0.65	1.82
New Hampshire	3.28	3.49	-0.24	0.61	1.79
Rhode Island	2.71	2.58	0.56	3.15	2.28
Vermont	2.67	2.83	0.50	1.27	1.82
Northeast 2	**2.02**	**1.80**	**1.92**	**1.91**	**1.91**
Connecticut	3.85	2.43	0.88	2.56	2.47
Massachusetts	2.42	2.52	1.25	2.40	2.16
Maine	2.30	0.48	0.36	1.77	1.27
New Hampshire	1.59	1.85	2.00	1.72	1.78
Rhode Island	1.92	1.67	2.44	1.74	1.94
Corn Belt	**1.09**	**0.70**	**2.78**	**1.13**	**1.41**
Iowa	0.87	1.01	2.70	1.01	1.38
Illinois	1.09	0.21	2.69	1.92	1.48
Indiana	1.11	0.66	2.84	1.09	1.41
Missouri	1.48	0.56	1.98	0.65	1.16
Ohio	0.79	0.87	3.63	0.59	1.43
Lake States	**1.39**	**1.58**	**2.56**	**1.39**	**1.71**
Michigan	0.61	2.09	3.76	1.23	1.87
Minnesota	1.67	1.45	2.59	1.55	1.81
Wisconsin	1.36	1.36	1.82	1.33	1.46
Northern Plains	**2.70**	**1.20**	**1.34**	**3.14**	**2.13**
Kansas	3.32	0.50	1.07	1.96	1.75
North Dakota	2.00	1.45	0.02	5.18	2.23
Nebraska	2.34	1.38	1.90	3.41	2.29
South Dakota	2.76	1.60	1.10	3.00	2.15
Appalachian	**1.82**	**1.83**	**2.44**	**2.91**	**2.26**
Kentucky	1.02	1.66	1.82	1.78	1.56
North Carolina	2.60	2.43	3.05	4.33	3.12
Tennessee	1.42	1.17	2.94	1.16	1.65
Virginia	1.29	1.36	1.74	3.40	1.97
West Virginia	1.95	0.75	2.29	2.86	1.99
Southeast	**3.05**	**2.91**	**3.18**	**2.37**	**2.87**
Alabama	3.34	2.38	2.62	3.90	3.09
Florida	1.62	2.58	3.88	0.01	1.96
Georgia	4.08	2.85	2.15	3.71	3.23
South Carolina	1.79	3.09	3.07	3.59	2.87

continued

Table 3A.3, *continued*

Region	1949-60	1960-70	1970-80	1980-91	1949-91
	Average annual growth rate (percentage)				
Delta States	**2.82**	**3.97**	**2.33**	**2.80**	**2.97**
Arkansas	2.81	3.98	2.78	3.03	3.14
Louisiana	1.41	4.12	2.14	1.78	2.32
Mississippi	3.70	3.75	1.55	2.82	2.97
Southern Plains	**0.71**	**0.88**	**1.89**	**2.18**	**1.42**
Oklahoma	1.80	0.30	2.77	0.67	1.38
Texas	0.36	1.07	1.57	2.64	1.41
Mountain	**1.60**	**2.18**	**1.83**	**1.83**	**1.86**
Arizona	1.37	1.20	3.58	1.37	1.86
Colorado	1.52	2.00	1.54	1.86	1.73
Idaho	1.83	3.08	2.37	2.66	2.47
Montana	1.48	2.03	0.93	1.64	1.52
New Mexico	1.03	2.16	1.44	2.25	1.71
Nevada	0.39	1.17	1.90	-0.15	0.79
Utah	1.23	2.10	0.02	2.34	1.44
Wyoming	2.03	0.84	0.91	0.66	1.12
Pacific	**1.57**	**2.34**	**2.89**	**0.66**	**1.83**
California	1.71	2.33	2.75	0.57	1.81
Oregon	1.33	1.81	2.49	0.79	1.58
Washington	0.42	2.65	3.65	0.96	1.86
48 States	**1.75**	**1.68**	**2.37**	**1.83**	**1.90**

**Table 3A.4: MFP Growth Rates, 1960-90—
Present Study versus Ball et al. (1999)**

Region	Average annual MFP growth rate 1960-90		Percentage difference	Ranking of states according to MFP growth rate	
	Ball et al. 1999	Present Study		Ball et al. 1999	Present Study
	(percentage per year)		*(percentage)*	*(percentage per year)*	
United States	**2.00**	**1.99**	**0.33**	–	–
New England					
Connecticut	2.25	1.33	40.74	17	40
Maine	1.70	1.31	22.78	32	41
Massachusetts	2.01	1.19	40.63	23	44
New Hampshire	1.76	1.25	29.10	31	43
Rhode Island	2.28	2.00	12.19	16	20
Vermont	1.51	1.41	6.63	35	38
Middle Atlantic					
New Jersey	1.21	0.88	27.11	47	47
New York	1.57	1.77	-12.57	34	28
Pennsylvania	2.08	2.06	0.86	22	16
South Atlantic					
Delaware	2.12	1.98	6.56	21	21
Florida	2.52	2.12	15.68	10	13
Georgia	2.55	2.52	1.03	7	9
Maryland	1.92	2.09	-8.89	26	14
North Carolina	2.78	3.08	-10.69	3	2
South Carolina	2.54	2.99	-17.84	8	3
Virginia	2.47	2.05	16.94	11	17
West Virginia	2.14	1.84	14.25	19	26
East North Central					
Illinois	1.83	1.88	-2.79	29	23
Indiana	2.34	1.87	20.09	15	24
Michigan	2.74	2.45	10.74	4	10
Ohio	2.58	2.05	20.63	6	18
Wisconsin	1.39	1.51	-8.67	38	36
West North Central					
Iowa	1.38	1.62	-17.49	39	33
Kansas	1.38	1.37	0.52	40	39
Minnesota	1.81	2.03	-12.08	30	19
Missouri	1.42	1.05	25.99	37	45
Nebraska	1.91	2.26	-18.41	27	12
North Dakota	2.42	2.43	-0.56	12	11
South Dakota	1.97	2.07	-5.13	25	15

continued

Table 3A.4, *continued*

Region	Average annual MFP growth rate 1960-90			Ranking of states according to MFP growth rate	
	Ball et al. 1999	Present Study	Percentage difference	Ball et al. 1999	Present Study
	(percentage per year)		*(percentage)*	*(percentage per year)*	
East South Central					
Alabama	2.40	2.66	-10.87	13	6
Kentucky	2.25	1.62	28.00	18	34
Mississippi	2.83	2.53	10.47	1	8
Tennessee	2.14	1.65	23.06	20	32
West South Central					
Arkansas	2.54	3.22	-26.93	9	1
Louisiana	2.83	2.56	9.56	2	7
Oklahoma	1.26	1.50	-18.72	45	37
Texas	1.35	1.67	-24.03	41	31
Mountain					
Arizona	1.50	1.77	-18.08	36	27
Colorado	1.31	1.84	-40.37	44	25
Idaho	1.99	2.76	-38.52	24	4
Montana	1.34	1.27	5.24	42	42
Nevada	1.25	0.88	29.43	46	46
New Mexico	1.65	1.74	-5.42	33	3
Utah	1.89	1.57	16.76	28	35
Wyoming	0.65	0.51	21.41	48	48
Pacific					
California	2.35	1.93	17.75	14	22
Oregon	1.34	1.75	-30.93	43	29
Washington	2.71	2.75	-1.31	5	5

Table 3A.5: Comparison of Output Growth Rates with and without CRP Output

	Average Annual Growth Rate, 1949—91		Average Annual Growth Rate, 1986—91	
	with CRP	without CRP	with CRP	without CRP
	(percentage per year)			
United States	**1.715**	**1.715**	**1.684**	**1.691**
New England				
Connecticut	-0.236	-0.236	-0.882	-0.882
Maine	0.065	0.064	-2.520	-2.519
Massachusetts	-0.357	-0.357	-0.475	-0.474
New Hampshire	-0.488	-0.488	-0.486	-0.486
Rhode Island	-0.052	-0.052	-1.609	-1.609
Vermont	0.696	0.696	0.423	0.423
Middle Atlantic				
New Jersey	-0.633	-0.633	-0.876	-0.876
New York	0.465	0.465	0.298	0.299
Pennsylvania	1.237	1.237	-0.504	-0.502
South Atlantic				
Delaware	3.212	3.212	4.646	4.646
Florida	3.267	3.267	0.643	0.645
Georgia	2.915	2.916	4.545	4.558
Maryland	2.076	2.076	1.115	1.115
North Carolina	1.963	1.963	5.902	5.904
South Carolina	0.922	0.923	4.900	4.918
Virginia	1.025	1.025	2.880	2.884
West Virginia	-0.203	-0.203	4.895	4.895
East North Central				
Illinois	1.311	1.310	-0.603	-0.603
Indiana	1.211	1.211	-0.744	-0.743
Michigan	1.188	1.187	1.285	1.279
Ohio	0.823	0.823	-1.717	-1.719
Wisconsin	1.239	1.239	0.431	0.427
West North Central				
Iowa	1.477	1.477	0.163	0.165
Kansas	1.964	1.961	-0.360	-0.372
Minnesota	1.767	1.767	0.880	0.889
Missouri	0.818	0.817	-1.227	-1.220
Nebraska	2.706	2.705	1.636	1.637
North Dakota	1.967	1.964	-0.264	-0.281
South Dakota	1.980	1.979	2.215	2.213

continued

Table 3A.5, *continued*

	Average Annual Growth Rate, 1949—91		Average Annual Growth Rate, 1986—91	
	with CRP	without CRP	with CRP	without CRP
	(percentage per year)			
East South Central				
Alabama	2.308	2.309	5.592	5.619
Kentucky	0.692	0.692	2.499	2.513
Mississippi	1.804	1.805	5.260	5.294
Tennessee	0.714	0.715	-0.333	-0.313
West South Central				
Arkansas	3.186	3.186	5.364	5.367
Louisiana	1.681	1.681	1.923	1.923
Oklahoma	1.239	1.238	0.476	0.479
Texas	1.582	1.581	4.496	4.498
Mountain				
Arizona	2.571	2.571	2.004	2.004
Colorado	1.983	1.987	1.384	1.488
Idaho	2.783	2.783	2.788	2.817
Montana	1.799	1.791	2.906	2.856
Nevada	0.998	0.999	-1.114	-1.111
New Mexico	1.917	1.922	2.902	2.992
Utah	1.046	1.047	1.717	1.739
Wyoming	1.122	1.121	1.998	1.991
Pacific				
California	2.892	2.892	3.399	3.401
Oregon	1.891	1.892	2.616	2.646
Washington	2.544	2.544	3.001	3.013

4

Transitive Multilateral Comparisons of Agricultural Output, Input, and Productivity: A Nonparametric Approach

D. S. Prasada Rao, Christopher J. O'Donnell, and V. Eldon Ball

Abstract

EKS Index number methods are examined for multilateral comparisons of input, output, and productivity. The EKS index number procedure was introduced by Caves, Christensen, and Diewert (1982a,b) and is now routinely applied in multilateral comparisons. The EKS method is currently used by Ball et al. (1999, 2000) in the construction of transitive multilateral output and input index numbers. These numbers, in turn, are used in measuring TFP growth in different states. One of the problems associated with using the EKS index is that it considers all binary comparisons to be equally reliable. However, in the context of interstate comparisons, some binary comparisons are intrinsically more reliable than others. The reliability issue arises because of the different output and input structures in different states. This study explores the use of Minimum Spanning Trees (MST) introduced by Hill (1999) in constructing output and productivity comparisons, and also describes a method for generalizing EKS indices. This approach, first proposed by Rao and Timmer (2000), assigns weights to different binary comparisons. The empirical part of this chapter focuses on the implementation of these two methods using inter-state data. The chapter also highlights the problems associated with the application of the MST approach. Empirical results reported and analytical arguments suggest that the weighted EKS method is likely to provide an improved method for constructing transitive multilateral index numbers of output and productivity.

Introduction

A number of studies have examined the sources of growth in the U.S. farm sector (Ball, 1985; Jorgenson, Gollop, and Fraumeni, 1987; Jorgenson and

Gollop, 1992; Ball et al., 1997). Though their methods differ in important ways, these studies share a common conclusion. Productivity growth was the single most important source of economic growth in post-war agriculture.

More recently, Ball et al. (1999, 2000) focus on agricultural productivity growth at both sector and state levels. A model that accounts for interstate transactions in farm goods links sector-wide and state-specific measures of total factor productivity growth. An interesting conclusion is that the smooth, persistently positive trend typically observed for farm sector productivity growth masks considerable variation across states and regions. The results also indicate that farm sector productivity growth is wholly a function of productivity trends in the individual states. Interstate shifts in production activity and resource reallocations have had little impact.

In this chapter, we provide estimates of the relative levels of productivity for the states. The U.S. Department of Agriculture's Economic Research Service (ERS) has developed a set of state farm production accounts. Salient features of the state accounts are well documented in Ball et al. (1999, 2000). Consequently, our focus will be on methodological issues that arise in the context of making transitive multilateral comparisons. In particular, we will examine critically the widely used Eltetö-Köves-Szulc (EKS) index number procedure in the context of interstate productivity comparisons and describe alternative procedures based on the concept of minimum spanning trees and a weighted or generalised EKS method.

The outline of the chapter is as follows. Section 2 briefly sets the notation used in the paper. A few relevant concepts including transitivity and base invariance are explained. This section also briefly reviews various index number procedures commonly used to construct measures of output, input, and productivity. Section 3 describes the EKS method for multilateral comparisons. We also discuss some of the problems associated with the use of the EKS procedure. Section 4 examines the possibility of constructing chained comparisons between states using the Minimum Spanning Tree (MST) method. Section 5 presents an alternative strategy for constructing generalised or weighted EKS indices that take into account the reliability of various pair-wise comparisons. Results from the empirical application of the method are presented in Section 6. The chapter concludes with a few comments in Section 7.

Notation and Preliminaries

We begin with some basic assumptions and notational conventions. We assume that there are M firms or states to be compared. We further assume that there are N outputs produced. Let p_{ij} and q_{ij} denote the price and quantity, respectively, of commodity i produced in state j. When the comparison is across states, it is possible that in any given state a number of the N goods are not produced (alternatively, inputs are not utilized). That is, some of the q_{ij}'s are zero. All that can be guaranteed in practice is that $q_{ij} \geq 0$ for all i and j, that for each good i there exists at least one state j such that $q_{ij} > 0$, and that for each state j there is at least one commodity i for which $q_{ij} > 0$. As a consequence, the binary comparisons are usually restricted to the goods common to the two states in the comparison. This restriction implies that if the number of goods in common is low then the output or input comparison between the two states will be less meaningful.

We now turn to a brief discussion of the desired properties of index numbers and list some of the commonly used index number formulae. Let I_{jk} denote a comparison of state k relative to state j. The matrix of all the pair-wise comparisons can be written as:

$$I_{MxM} = \begin{bmatrix} I_{11} & I_{12} & . & . & I_{1M} \\ I_{21} & I_{22} & . & . & I_{2M} \\ . & . & . & . & . \\ . & . & . & . & . \\ I_{M1} & . & . & . & I_{MM} \end{bmatrix} \tag{2.1}$$

We note that $I_{jj} = 1$ for all j and, if the index satisfies the state reversal test, $I_{jk} \times I_{kj} = 1$ for all j and k. It follows that $I_{kj} = 1 / I_{jk}$.

Several index number formulae can be used to construct the matrix in equation (2.1). Since the main concern of this paper is multilateral comparisons, we focus on those procedures that satisfy transitivity and other properties relevant for multilateral comparisons. These are discussed below.

Transitivity

An index number formula I_{jk} is said to satisfy the transitivity property if and only if for all choices of j, k, and λ (j,k,λ = 1,2,..,M), the index satisfies

$$I_{jk} = I_{j\lambda} \times I_{\lambda k} \tag{2.2}$$

Equation (2.2) requires that when the index number formula is used to make a direct comparison I_{jk}, it should result in the same numerical value as when it is used to make an indirect comparison between j and k through a link state λ. Note that the transitivity property ensures internal consistency of the index numbers in the matrix given in equation (2.1). As will be noted below, many of the commonly used index number procedures do not satisfy this requirement. The following result is useful in constructing transitive index numbers (see Section 5).

Result: An index number formula I_{jk} satisfies the transitivity property in (2.2) if and only if there exist M positive real numbers $\lambda_1, \lambda_2, \ldots, \lambda_M$, such that

$$I_{kj} = \frac{\lambda_k}{\lambda_j}$$

for all j and k.

The proof of this result is straightforward (Rao and Banerjee, 1984). The result is quite important since it shows that when the transitivity property is satisfied, all we need to measure are M real numbers $\lambda_1, \lambda_2, \ldots, \lambda_M$, and then all the necessary indices in (2.1) can be calculated using these M numbers, thus reducing the dimension of the problem involved.

Base invariance

An index number formula is said to be base invariant if a comparison between the pair (j,k) is invariant to the order in which the states are listed. That is, multilateral comparisons should be invariant to all possible permutations of the data set. For example, the set of star-comparisons shown in Section 4 would satisfy transitivity but not the property of base invariance.

Characteristicity

Characteristicity is a property that requires that any set of multilateral comparisons satisfying the transitivity property should retain the essential features of the binary comparisons constructed without the transitivity requirement. Since condition (2.2) implies that a transitive comparison between a pair of states j and k is necessarily influenced by the price and quantity data for all the other states, the characteristicity property requires that distortions resulting from adherence to the transitivity property should be kept to a *minimum*. The Eltetö-Köves-Szulc (EKS) method for multilateral comparisons, discussed in Section 3, has its origins in the characteristicity property.

We now briefly describe index number procedures that are commonly used in productivity comparisons. There is considerable literature on the subject, and the present paper is not ideal for a complete discussion of results in this area. The interested reader is referred to a review in Coelli, Rao, and Battese (1998).

Laspeyres index

The Laspeyres output quantity index between a pair of states j and k, denoted by Q_{jk}^{L}, is obtained using the price weights of the base state j. Thus

$$Q_{jk}^{L} = \frac{\sum_{i=1}^{n} p_{ij} q_{ik}}{\sum_{i=1}^{n} p_{ij} q_{ij}}. \tag{2.3}$$

Paasche index

The Paasche output index, denoted by Q_{jk}^{P}, is defined using the prices of state k as weights and is defined by

$$Q_{jk}^{P} = \frac{\sum_{i=1}^{n} p_{ik} q_{ik}}{\sum_{i=1}^{n} p_{ik} q_{ij}}. \tag{2.4}$$

Fisher index

The Fisher index, F_{jk}, is defined as the geometric mean of the Laspeyres and Paasche indices and is given by

$$F_{jk} = \sqrt{Q^{L}{}_{jk} \cdot Q^{P}{}_{jk}}.$$

The Fisher index is frequently used in binary comparisons. Diewert (1976, 1992) has provided several arguments in support of the use of the Fisher index. The Fisher index can be shown to be the only index number formula that satisfies a long list of desirable properties or tests. The Fisher index is also exact for a flexible production or input requirements function. The only other index number formula that possesses comparable properties is the Törnqvist index. Much of the discussion in this paper is based around the Fisher index number formula. However, all the arguments presented and methods discussed are equally applicable to the Törnqvist index.

Turning now to productivity measurement, based on the work of Caves, Christensen, and Diewert (1982a and 1982b), and Diewert (1992) where relationships between the Malmquist productivity index and the standard index number formulae were established, the present paper simply uses

$$\text{TFP index} = \frac{\text{Output index}}{\text{Input index}} \tag{2.5}$$

as the basis for productivity measurement. The Fisher or Törnqvist index formula is applied to compute the respective output and input indices required. Results in the papers listed above show that under some fairly simple conditions the Malmquist productivity index can be shown to be equal to the ratio of Fisher or Törnqvist output and input quantity indices. Caution should be exercised in interpreting the TFP measures derived using this formula since the equivalence of the Malmquist index and the TFP index in (2.5) holds only under some special conditions (see Coelli, Rao, and Battese, 1998). In addition, the equivalence of the Malmquist index to the TFP index in (2.5), established in Caves et al. (1982a, b), requires the assumption of optimal behaviour of all the firms, and therefore all firms are assumed to be technically efficient. In such cases TFP can be interpreted as a measure of technical change.

Eltetö-Köves-Szulc Method for Multilateral Comparisons

The Eltetö-Köves-Szulc method is one of the principal aggregation methods used for international comparisons. The method was proposed independently by two Hungarian statisticians: Eltetö and Köves (1964) and the Polish statistician Szulc (1964). The method is named after all three and is known as the EKS.

Though the EKS method has been discussed in the international comparison literature for over two decades, it was only adopted for use relatively recently in the international comparison work undertaken for 1990, 1993, and 1996 by the OECD and the EUROSTAT. The EKS method for multilateral comparisons of output, input and productivity has gained popularity since the publication of seminal papers by Caves, Christensen, and Diewert (1982a, b).

The EKS method is the transitive multilateral method used by Ball et al. (2000) in deriving agricultural sector input and output indices for interstate comparisons. Since these indices are often used in deriving value aggregates, which are in turn used in parametric estimation of cost and production functions, the EKS method plays a crucial role. In this section we critically examine the EKS indices with a view to outlining possible improvements in methodology for use in future empirical research.

Characteristicity and the EKS Method

The EKS method has its foundations in the property of characteristicity. This property explicitly recognises the fact that multilateral methods for international comparisons satisfying the *transitivity* property have the tendency to distort the binary comparisons involved. For example, if states Florida and California are to be compared, the best or ideal way is to simply compare the price and quantity data for these two states using any of the standard index number formulae (e.g., Fisher or Tornqvist indices). However, if these two states are compared within a multilateral comparison framework where transitivity is required, then comparisons between Florida and California are expected to be consistent with other binary comparisons involving these two states. For example, comparisons between Florida and North Carolina and between North Carolina and California have to be consistent with the Florida-California comparison. That is, the Florida-California comparison is necessarily influenced (or distorted) by the price and quantity data for North Carolina (and for other states included in the comparison).

Transitivity necessarily influences the nature of all binary comparisons, thus leading to a loss in characteristicity. He suggested that such a loss in characteristicity resulting from the imposition of transitivity should be kept to a minimum. The EKS method is an aggregation method, which is exclusively devised with preservation of characteristicity as its primary goal.

EKS Method – Basic Framework

The EKS method may be considered as a generic procedure that can be applied to a matrix of non-transitive binary comparisons, binary comparisons derived using any index number formula that satisfies the "time or country reversal test". It is intuitive that we begin by choosing an appropriate binary index number formula based on statistical and economic-theoretic considerations. It is common to make a choice based on Fisher's tests or on the concepts of exact or superlative indices discussed in several of Diewert's papers. The EKS formulation in its original form used the Fisher formula. The method suggests that a comparison between a pair of states is best achieved by a direct comparison derived using the Fisher index number formula. It is known that binary comparisons derived using the Fisher formula do not satisfy the transitivity property. Therefore, the EKS method sets out to derive a transitive set of comparisons using the Fisher binary comparisons as building blocks such that the resulting comparisons *deviate* as little as possible from binary comparisons. Based on the logarithmic distance, the EKS procedure minimises the sum of squared deviations of logarithms of Fisher indices and the multilateral index numbers. The EKS method provides a formula for multilateral comparisons that satisfies the transitivity property.

For a pair of states j and k (say Florida and California), the EKS method uses the following formula.

$$EKS_{jk} = \prod_{\lambda=1}^{M} \left[F_{j\lambda} . F_{\lambda k} \right]^{1/M} \tag{3.1}$$

where F_{lk} represents a binary comparison between states λ and k using the Fisher formula. It is obvious that once the Fisher indices are known for all pairs of states, the EKS indices can be calculated as the *unweighted geometric average* of all the chained comparisons between j and k derived through a link state λ (λ = 1,2,…,M).

Multilateral index numbers derived using the EKS formula, given in (3.1), possess a number of useful properties.

Property 1: They satisfy transitivity.

Property 2: They can be shown to solve the minimization problem:

$$\text{Min} \sum_{j=1}^{M} \sum_{k=1}^{M} \left(\ln I_{jk} - \ln F_{jk} \right)^2$$

Subject to the restrictions: $I_{jk} = I_{jl} \times I_{lk}$ for all j, k and l.

A simple proof of this property is in Rao and Banerjee (1984). The property relates to the *characteristicity* property of multilateral index numbers. The EKS index numbers deviate the least from the Fisher binary index numbers according to the metric defined in this property.

Property 3: They are base invariant.

Property 4: If the EKS approach is extended to all linked comparisons involving chains of length more than two (j and k linked through more than one state), the resulting index is identical to the EKS index in (3.1).

The formula in (3.1) is the method originally proposed in the works of Eltetö and Köves (1964) and Szulc (1964). However, the EKS formulation can be applied to indices derived using any other index number formula, such as the Törnqvist index, that satisfies the state reversal test.

A simple interpretation of the EKS index: Consider a comparison between Florida and California. The EKS method suggests that comparison between these two states can be derived using different link states and the Fisher index number formula.

Figure 4.1 illustrates the basic structure of EKS comparisons. The method suggests that indirect comparisons between Florida and California be derived

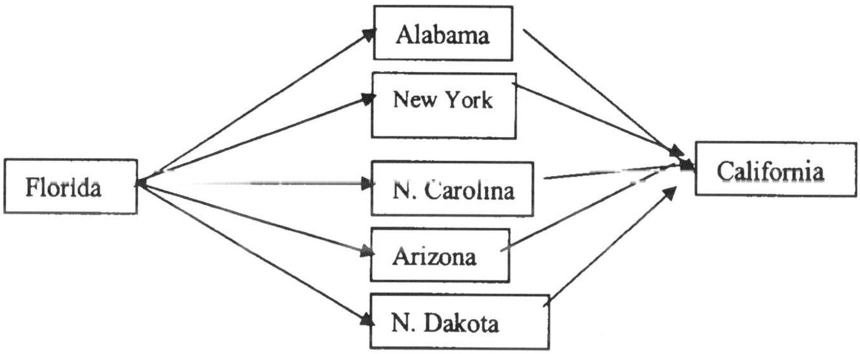

Figure 4.1. EKS Comparisons

using each of the states as a link. Comparisons involved in each of the links are derived using the Fisher formula. Since the Fisher index formula does not satisfy the transitivity property, each indirect comparison results in a different numerical value for the comparison between Florida and California. There are M (equal to the number of states in the comparison) such indirect comparisons between these two states. The EKS method suggests that an unweighted geometric average be taken to yield a single comparison between Florida and California. Comparisons derived using this approach between different pairs of states then satisfy transitivity.

There are a few criticisms of the EKS method. The first criticism stems from its adherence to the principle of characteristicity. One of the basic premises underlying the characteristicity requirement is that a direct comparison between two states provides the best possible outcome, and a multilateral comparison should preserve this outcome to the greatest extent possible. However, in empirical situations it is known that when comparisons between states with very different output vectors are involved, a direct comparison is usually not very reliable. This problem is reflected in large differences in the bounding Laspeyres and Paasche indices. In such cases maintaining characteristicity is not a virtue to be pursued at all costs.

The second criticism relates to the unweighted nature of the average used in combining all the indirect comparisons (shown in Figure 4.1). The method does not distinguish between the quality and reliability of different linked comparisons even though certain indirect comparisons are likely to be more meaningful.

Recent research (Selvanathan and Rao [1994] and Hill [1999]) has established that it is feasible and desirable to define the relevant indices using either a weighted averaging process or chained indices (based on the concept of a minimum spanning tree proposed in Hill (1999)). Both of these options are superior to the regular EKS index number procedure, and both are pursued in the next two sections of this paper.

The final point relates to the fact that EKS indices minimise distance between the Fisher binary indices and transitive multilateral index numbers where distance is measured using the sum of squared differences in logarithms of the indices involved (logarithmic distance between binary and multilateral indices). If distance is measured using a different metric instead of the logarithmic distance, then the resulting indices could be quite different from the EKS index numbers. For example, use of simple Euclidean distance would result in quite a different set of multilateral index numbers. The use of logarithmic distance is consistent with the multiplicative nature of the price index numbers, which in turn imply additive index numbers in logarithmic form.

Chained Comparisons Based on Minimum Spanning Trees

A major development, closely related to the EKS system, is the idea that it is possible to obtain transitive indices through a system of chained comparisons. In the case of temporal comparisons a natural chronological order exists so that an index number for a comparison between two periods, t and t+s, can be obtained through the following chained comparison:

$$I_{t,t+s} = I_{t,t+1} \, I_{t+1,t+2} \cdots I_{t+s-1,t+s}$$

In order to make this operational, it is necessary to choose an appropriate index formula such as the Fisher or Törnqvist index.

However, in the case of multilateral spatial comparisons involving several states, no such natural ordering exists. It is possible to arbitrarily select a state and make comparisons between all pairs of states using the pre-designated state as the link state. This approach leads to comparisons of the following type.

Each link in the "star" comparisons in figure 4.2, in which Florida is the star of the comparisons, may be computed using the Fisher formula or the Törnqvist formula. It is easy to check that the set of all pair-wise comparisons derived using the star-system would satisfy the transitivity requirement. But the comparisons derived are not invariant to the choice of a particular state to be the "star" of the comparisons, violating the base invariance condition. It is, however, possible to consider alternative sets of multilateral comparisons, each based on a different star country, and average the resulting comparisons. In fact, the EKS indices described in Section 3 coincide with a geometric average of comparisons derived using different "star" specifications.

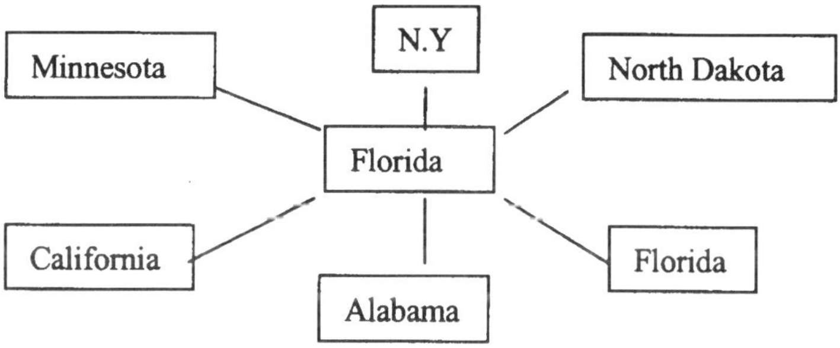

Figure 4.2. Star Comparisons

Hill (1999) proposed a method of linking or chaining based on the "minimum spanning tree" concept, a concept used in graph theory. A few pertinent definitions are given below.

Graph Theory Concepts

Suppose we consider the case where M states are involved in a multilateral comparison. Then consider M vertices (points) used in defining a "graph". We place an edge (branch) between two vertices if we choose to make a direct comparison between the two states.

Graph: A graph defined on a set of M vertices is given by a collection of edges (or direct links) between pairs of vertices (states).

A simple example of a graph is one in which every pair of states has an edge, a case in which each state is directly compared with every other state in the comparison set. Such a graph, known as the "total graph", leads to a matrix of binary comparisons. A total graph forms the basis for the EKS formula.

Connected Graph: A graph is said to be *connected* if there is a path between every pair of vertices or states.

Given a connected graph, it is feasible to make a comparison between any two states through a chain of linked comparisons in the graph.

Spanning Tree: A graph on a set of M vertices is said to be a spanning tree if it is a connected graph with no circuits.

Given a spanning tree, there exists one and only one chain to make a comparison between any pair of vertices. If there were two distinct paths, they can be used to form a circuit, in which case the graph in question is no longer a tree. Figure 4.3 gives examples of two spanning trees defined on five vertices.

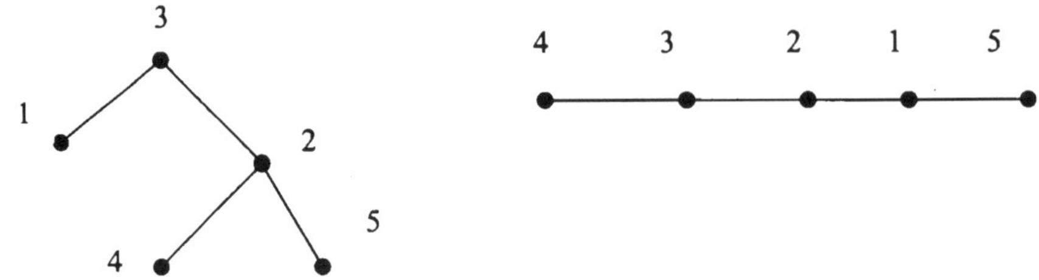

Figure 4.3. Spanning trees with 5 vertices.

It is evident that for a given set of M states or vertices, the spanning trees are not uniquely determined. In fact there are M^{M-2} distinct possible spanning trees defined on M vertices.

If the spanning tree concept is to be used for the construction of chained comparisons, it is necessary to select a single spanning tree from the very large collection of possible spanning trees. Ideally the spanning tree selected would have some desirable properties. Hill (1999) proposes using the minimum spanning tree concept, which is uniquely defined (under very mild conditions) on the basis of a measure of distance defined between pairs of states. The whole concept of the minimum spanning tree is based on the idea that it is possible to assign some sort of weights, for each pair of states, reflecting the reliability of the comparison. Reliability is measured in the form of a distance function, which is inversely related to reliability. Once a distance measure is identified, the minimum spanning tree can be determined using the observed data.

Distance Function

A measure of reliability suggested by Hill (1999) is the well-known Paasche-Laspeyres spread. The Laspeyres and Paasche (quantity or price) index numbers are usually considered to be two extremes, and, for a given comparison, the spread between these two indices reflects the uncertainty associated with the use of an index number formula. The distance function, for a pair of states j and k, is defined as the absolute value of the logarithmic difference in the Laspeyres and Paasche quantity (price) index numbers. It is thus defined as:

$$d(j,k) = \left| \ln\left(\frac{P_{jk}^{L}}{P_{jk}^{P}}\right) \right| = \left| \ln\left(\frac{Q_{jk}^{L}}{Q_{jk}^{P}}\right) \right|$$

where P_{jk} and Q_{jk} represent the price and quantity index numbers respectively. Superscripts P and L refer to the Paasche and Laspeyres indices. The function d(j,k) measures the distance between the two states, and is therefore the reciprocal of a measure of reliability.

The distance, defined through the Paasche-Laspeyres spread, will be zero if the price or quantity vectors in the two states are proportional to each other. If the price or quantity structures are very similar, the distance will be small. Thus the distance function, d(j,k), is an indirect measure of the similarity in the price and quantity structures of states j and k. In the present context of output or input quantity or price comparisons between various states, comparisons are deemed to be reliable (therefore have a smaller distance) if there is similarity in the structures of outputs produced or inputs used.

The following properties are satisfied by the distance function based on the Paasche-Laspeyres spread.

Property 1: For each state, j, $d(j,j) = 0$;

Property 2: For any pair j and k, $d(j,k) \geq 0$;

Property 3: For any j and k, $d(j,k) = d(k,j)$; and

Property 4: Distance based on price index = Distance based on quantity index

It is easy to establish these properties. The last property holds, since the Paasche-Laspeyres ratio of price indices is exactly equal to the corresponding ratio of quantity indices.

Construction of Minimum Spanning Trees

Once a distance measure is selected, the next step in the determination of the minimum spanning tree (MST) is to compute the distance matrix of order $M \times M$ between all pairs of states. The distance matrix is a non-negative symmetric matrix with zeros along the principal diagonal.

By definition, a spanning tree is made up of (M-1) edges. These edges need to be selected from the set of all possible edges, which are M(M-1)/2 in number. The MST is that spanning tree which has the smallest sum of distances for the (M-1) links that make up a spanning tree. If all the distances are distinct, there exists a unique spanning tree that minimises the sum of the distances associated with the edges in the tree.

Several algorithms are available for the construction of the MST. Most mathematical and operations research packages contain a sub-routine for the determination of MST from a given distance matrix. A short program was written in GAUSS for purposes of this study. A listing of the program is given in the Appendix. Construction of the MST is quite simple and follows the steps listed.

The algorithm begins with any state i (in this paper we choose the first state). State i is then connected to the closest state, j (i.e., chosen so that d(i,j) is a minimum among d(i,k) , k=1,2,...,M). The two states i and j now form a "connected" set of two states, i.e., a tree. The remaining states are referred to as "unconnected" states. The next step in the algorithm is to select (and join) the unconnected state that is closest to the tree obtained in the previous step. Ties for the closest node can be broken arbitrarily. This step is repeated until every state is included in the tree. More details concerning this so-called "greedy" algorithm can be found in Winston (1991). The algorithm results in a spanning tree with the minimum sum of distances, thus leading to the desired MST.

Using the MST for Multilateral Comparisons

Once the minimum-spanning tree, associated with a given matrix of distances, is identified, there exists a unique chain linking any pair of states. If states $i_1, i_2, \ldots, i_p$ form the chain that links states j and k, and if the Fisher index is selected for binary comparisons, then the multilateral index number is given by

$$MST_{jk} = F_{j,i_1} F_{i_1,i_2} \; \ldots\ldots \; F_{i_{p-1},i_p} F_{i_p,k}$$

The resulting index provides an indirect or linked comparison between states j and k. If a chain of states $i_1, i_2, \ldots, i_p$ define the path from j to k in the minimum spanning tree, then the chain index between k and j is defined using the same path in the reverse direction. Since we are working with a minimum spanning tree, there is only one path between any two nodes/vertices.

It is quite easy to show that multilateral comparisons constructed using the MST satisfy the transitivity property, provided the binary index number formula satisfies the state (or country or time) reversal test. It is obvious that the MST method is invariant to the order in which the states are introduced into MST computations and, therefore, the resulting index numbers also satisfy the property of base invariance.

Weighted EKS Method for Multilateral Comparisons

In this section we explore another avenue to generalise, and to improve upon, the multilateral comparisons derived using the EKS method. Since the EKS method adheres to the characteristicity principle discussed in Section 2 and ignores the differential nature of the reliability of various pairwise comparisons, we demonstrate the feasibility of attaching weights to different comparisons in providing a modified and improved EKS method. The approach described in this section also overcomes some of the problems associated with the application of the MST method and its sensitivity to the distance measure.

In order to generalize the EKS method to incorporate weights into the various linked comparisons involved in equation (3.1), it is necessary to look at the EKS method from a different perspective. Suppose we wish to derive a set of index numbers $I_{jk,}$ which are transitive and minimize the log-distance from the Fisher indices. Then we

$$\text{minimize} \quad \sum_j \sum_k (\ln I_{jk} - \ln F_{jk})^2$$

$$\text{subject to} \quad I_{jk} = I_{jl} \cdot I_{lk} \qquad \forall j,k,l$$

Using the result stated in Section 2 on transitive index numbers, the above problem can be restated as one of finding the $Q_1, Q_2, \ldots, Q_M$, which

$$\text{minimize} \quad \sum_j \sum_k (Q_k - Q_j - \ln F_{jk})^2 \tag{5.1}$$

Then the required quantity index I_{jk} is defined as the ratio $\exp(\hat{Q}_k)/\exp(\hat{Q}_j)$ where the (^) show that these are solutions to the minimization problem. After some simple algebraic manipulation, it can be shown that the EKS index is related to the solution above as:

$$EKS_{jk} = \frac{\exp(\hat{Q}_k)}{\exp(\hat{Q}_j)} = \exp(\hat{Q}_k - \hat{Q}_j)$$

In this function, $\exp(\hat{Q}_k)$ can be interpreted as a volume measure of agricultural output in state k. Ratios of $\hat{Q}_k$'s provide output/input quantity index numbers.

It is evident that the $\hat{Q}$'s are the ordinary least-squares estimators of the Q's (so the $\hat{Q}_i$'s are the best linear unbiased estimators) in the following model specification

$$\begin{aligned} &\ln F_{jk} = Q_k - Q_j + u_{jk} \\ &\text{with} \quad E(u_{jk}) = 0 \quad \text{and} \quad Var(u_{jk}) = \sigma^2 \end{aligned} \tag{5.2}$$

Given the model specification in (5.2), it is possible to discriminate between different pairs of states using some indicators of reliability. This discrimination can be achieved using the following model:

$$\ln F_{jk} = Q_k - Q_j + u_{jk}$$

$$\text{with} \quad E(u_{jk}) = 0 \quad \text{and} \quad Var(u_{jk}) = \frac{\sigma^2}{w_{jk}} \tag{5.3}$$

where w_{jk} is a measure of reliability. If w_{jk} is large, we consider a particular Fisher index, F_{jk}, to be reliable. Modified EKS indices can be obtained by applying generalized least squares to (5.3) or ordinary least squares to

$$\sqrt{w_{jk}} \ln F_{jk} = \sqrt{w_{jk}} Q_k - \sqrt{w_{jk}} Q_j + u^*_{jk}$$

$$\text{with} \quad E(u^*_{jk}) = 0 \quad \text{and} \quad var(u^*_{jk}) = \sigma^2 \qquad \forall j,k = 1,..M,\ j \neq k$$

The least squares estimates solve the following equations:

$$\begin{bmatrix} 2\sum_{j\neq 1}^{M} w_{1j} & -2w_{12} & . \quad . & -2w_{1M} \\ -2w_{21} & 2\sum_{j\neq 2}^{M} w_{2j} & & -2w_{2M} \\ . & & & . \\ . & & & . \\ -2w_{M1} & -2w_{M2} & . \quad . & 2\sum_{j\neq M}^{M} w_{Mj} \end{bmatrix} \begin{bmatrix} \hat{Q}_1 \\ \hat{Q}_2 \\ . \\ . \\ \hat{Q}_M \end{bmatrix} = \begin{bmatrix} -2\sum_{j\neq 1}^{M} w_{1j} \ln F_{1j} \\ -2\sum_{j\neq 2}^{M} w_{2j} \ln F_{2j} \\ . \\ . \\ -2\sum_{j\neq M}^{M} w_{Mj} \ln F_{Mj} \end{bmatrix} \tag{5.4}$$

The matrix on the left-hand side is singular with rank equal to M-1. Thus, we can only solve for $\hat{Q}_1, \ldots \hat{Q}_M$ after restricting one of the $\hat{Q}$'s to be zero. To set $\hat{Q}_1 = 0$ implies that $Q^*_1 = \exp(\hat{Q}_1) = 1$ and the volume level of the first state is equal to one (i.e., the first state is the numeraire).

It is possible to use standard software packages to obtain the weighted EKS index numbers. The Appendix provides details on how to do the computations using either EXCEL or the econometrics software package SHAZAM.

Some Remarks

The procedure outlined above is relatively new. It was devised as a technique for handling some of the data problems encountered in the International Comparison project. This procedure was used in constructing transitive multilateral comparisons for the manufacturing sector-results reported in the study by Rao and Timmer (2000).

The versatility of the weighted EKS method lies in the fact that the weights can be specified to reflect the reliability associated with a given binary comparison. Two possible specifications of w_{jk} are potentially useful in the case of inter-state comparisons. The first specification sets

$$w_{jk} = \frac{1}{d_{jk}} \qquad \text{for all } j \neq k$$

where d_{jk} is the Paasche-Laspeyres spread described in Section 4. Use of these weights with the procedure outlined in Section 5 can provide an alternative to the comparisons based on the minimum spanning tree.

Another specification of w_{jk} may be related to the coverage ratios associated with binary comparisons. Since the computation of Laspeyres, Paasche, and Fisher indices has to be based on all the commodities that are produced in both the states, the reliability of such a comparison increases with the percentage of output in these states covered by the common list of commodities.

Empirical Application: Inter-State Input, Output, and Productivity Comparisons

The idea of using the minimum spanning tree and the generalised EKS methods as plausible alternatives to the use of the EKS method is explored using data drawn from 48 states of the United States for the benchmark year, 1987. The basic data, in the form of Laspeyres, Paasche, and Fisher matrices for the quantities of outputs and inputs in different states, are drawn from the worksheets of Ball et al. (2000). These matrices are constructed using state level data on prices and quantities. Multilateral index numbers resulting from the application of the generalized EKS method and the Minimum Spanning Tree are available from the authors upon request. Selected summary results are presented in Table 4.1.

Based on the Laspeyres and Paasche index matrices, distance matrices based on the Paasche-Laspeyres spread are computed for inputs and outputs separately, since similar patterns in inputs used could be quite different from those observed for outputs produced (more so when the inputs and outputs are used

Table 4.1. Output, Input, and Productivity Indices
(U.S. All-State Average = 1.000)

		Output Index Numbers			
Nos.	State	Fisher	EKS	MST	Generalised EKS
1	Alabama	1.138	1.138	1.149	1.135
2	Arkansas	1.873	1.873	1.886	1.872
3	Arizona	0.958	0.942	0.946	0.941
4	California	8.756	8.669	8.484	8.680
5	Colorado	1.878	1.916	1.939	1.912
6	Connecticut	0.161	0.159	0.158	0.159
7	Delaware	0.227	0.230	0.230	0.230
8	Florida	2.469	2.443	2.440	2.458
9	Georgia	1.735	1.750	1.755	1.749
10	Iowa	5.796	5.827	5.828	5.826
11	Idaho	1.362	1.362	1.362	1.363
12	Illinois	3.917	3.980	3.866	3.890
13	Indiana	2.614	2.611	2.607	2.611
14	Kansas	3.692	3.706	3.731	3.703
15	Kentucky	1.495	1.506	1.510	1.505
16	Louisiana	0.806	0.794	0.795	0.796
17	Massachusetts	0.158	0.155	0.156	0.155
18	Maryland	0.491	0.501	0.503	0.500
19	Maine	0.213	0.214	0.213	0.214
20	Michigan	1.715	1.712	1.709	0.713
21	Minnesota	3.946	3.926	3.943	3.927
22	Missouri	2.251	2.258	2.271	2.257
23	Mississippi	1.192	1.193	1.201	1.193
24	Montana	0.908	0.896	0.889	0.898
25	North Carolina	1.943	1.946	1.944	1.945
26	North Dakota	1.719	1.715	1.714	1.722
27	Nebraska	4.443	4.508	4.504	4.499
28	New Hampshire	0.061	0.060	0.059	0.060
29	New Jersey	0.287	0.285	0.283	0.285
30	New Mexico	0.642	0.641	0.638	0.639
31	Nevada	0.157	0.156	0.156	0.156
32	New York	1.532	1.533	1.524	1.533
33	Ohio	2.240	2.220	2.220	2.220
34	Oklahoma	1.723	1.725	1.743	1.722
35	Oregon	1.131	1.127	1.131	1.131
36	Pennsylvania	1.945	1.956	1.959	1.958
37	Rhode Island	0.025	0.024	0.024	0.024
38	South Carolina	0.585	0.586	0.586	0.586
39	South Dakota	2.010	2.042	2.030	2.041
40	Tennessee	1.130	1.128	1.140	1.128
41	Texas	4.989	5.004	5.047	4.993
42	Utah	0.391	0.395	0.399	0.394
43	Virginia	0.945	0.938	0.940	0.938
44	Vermont	0.212	0.211	0.205	0.211
45	Washington	1.780	1.781	1.786	1.781
46	Wisconsin	3.337	3.390	3.432	3.394
47	West Virginia	0.180	0.182	0.181	0.181
48	Wyoming	0.411	0.422	0.423	0.422

continued

Table 4.1, *continued*
(U.S. All-State Average = 1.000)

Nos.	State	Input Index Numbers Fisher	EKS	MST	Generalised EKS
1	Alabama	1.088	1.088	1.116	1.093
2	Arkansas	1.579	1.607	1.666	1.613
3	Arizona	1.357	1.332	1.320	1.341
4	California	5.007	4.818	4.949	4.818
5	Colorado	1.702	1.702	1.682	1.699
6	Connecticut	0.241	0.245	0.233	0.246
7	Delaware	0.234	0.234	0.214	0.233
8	Florida	1.977	1.958	2.033	1.955
9	Georgia	1.453	1.457	1.399	1.452
10	Iowa	5.184	5.347	5.176	5.405
11	Idaho	1.001	1.007	1.083	1.005
12	Illinois	3.905	3.946	4.195	3.936
13	Indiana	2.140	2.150	2.305	2.146
14	Kansas	2.900	2.927	2.975	2.930
15	Kentucky	1.485	1.486	1.427	1.491
16	Louisiana	0.720	0.724	0.764	0.721
17	Massachusetts	0.264	0.267	0.254	0.267
18	Maryland	0.662	0.663	0.656	0.664
19	Maine	0.243	0.245	0.235	0.247
20	Michigan	1.415	1.419	1.531	1.418
21	Minnesota	3.509	3.567	3.491	3.573
22	Missouri	2.143	2.143	2.114	2.142
23	Mississippi	0.899	0.903	0.972	0.899
24	Montana	0.885	0.876	0.879	0.876
25	North Carolina	1.969	1.968	2.009	1.969
26	North Dakota	1.368	1.386	1.312	1.387
27	Nebraska	2.987	3.006	3.165	3.005
28	New Hampshire	0.112	0.112	0.107	0.112
29	New Jersey	0.492	0.492	0.517	0.497
30	New Mexico	0.893	0.876	0.915	0.876
31	Nevada	0.176	0.171	0.181	0.170
32	New York	1.353	1.359	1.236	1.365
33	Ohio	2.299	2.296	2.445	2.295
34	Oklahoma	1.492	1.484	1.403	1.488
35	Oregon	1.062	1.053	1.025	1.052
36	Pennsylvania	2.494	2.520	2.513	2.541
37	Rhode Island	0.036	0.036	0.038	0.036
38	South Carolina	0.479	0.477	0.470	0.475
39	South Dakota	1.361	1.387	1.360	1.391
40	Tennessee	1.333	1.342	1.377	1.342
41	Texas	5.645	5.520	5.461	5.466
42	Utah	0.541	0.539	0.535	0.537
43	Virginia	1.157	1.145	1.070	1.145
44	Vermont	0.253	0.254	0.240	0.254
45	Washington	1.474	1.492	1.411	1.496
46	Wisconsin	2.450	2.491	2.511	2.514
47	West Virginia	0.192	0.185	0.179	0.181
48	Wyoming	0.392	0.380	0.388	0.376

continued

Table 4.1, *continued*
(U.S. All-State Average = 1.000)

Nos.	State	Total Factor Productivity Index Numbers Fisher	EKS	MST	Generalised EKS
1	Alabama	1.045	1.045	1.030	1.038
2	Arkansas	1.186	1.166	1.132	1.160
3	Arizona	0.706	0.707	0.717	0.702
4	California	1.749	1.799	1.714	1.802
5	Colorado	1.104	1.126	1.153	1.126
6	Connecticut	0.669	0.649	0.678	0.647
7	Delaware	0.971	0.985	1.074	0.988
8	Florida	1.249	1.248	1.200	1.257
9	Georgia	1.194	1.201	1.254	1.204
10	Iowa	1.118	1.090	1.126	1.078
11	Idaho	1.361	1.352	1.258	1.356
12	Illinois	1.003	0.986	0.922	0.988
13	Indiana	1.222	1.214	1.131	1.216
14	Kansas	1.273	1.266	1.254	1.264
15	Kentucky	1.006	1.013	1.058	1.009
16	Louisiana	1.120	1.096	1.040	1.105
17	Massachusetts	0.597	0.579	0.614	0.580
18	Maryland	0.742	0.755	0.767	0.752
19	Maine	0.879	0.873	0.904	0.867
20	Michigan	1.212	1.206	1.116	1.208
21	Minnesota	1.125	1.101	1.130	1.099
22	Missouri	1.050	1.054	1.074	1.053
23	Mississippi	1.326	1.321	1.236	1.327
24	Montana	1.026	1.023	1.010	1.025
25	North Carolina	0.987	0.989	0.967	0.988
26	North Dakota	1.256	1.237	1.306	1.241
27	Nebraska	1.487	1.500	1.423	1.497
28	New Hampshire	0.543	0.535	0.549	0.535
29	New Jersey	0.584	0.579	0.547	0.573
30	New Mexico	0.719	0.732	0.698	0.729
31	Nevada	0.891	0.913	0.865	0.919
32	New York	1.132	1.127	1.234	1.124
33	Ohio	0.975	0.967	0.908	0.968
34	Oklahoma	1.155	1.163	1.242	1.157
35	Oregon	1.065	1.071	1.103	1.075
36	Pennsylvania	0.780	0.776	0.779	0.771
37	Rhode Island	0.706	0.688	0.637	0.686
38	South Carolina	1.220	1.229	1.246	1.234
39	South Dakota	1.477	1.472	1.493	1.467
40	Tennessee	0.848	0.841	0.828	0.840
41	Texas	0.884	0.906	0.924	0.913
42	Utah	0.724	0.733	0.744	0.734
43	Virginia	0.817	0.819	0.878	0.819
44	Vermont	0.837	0.829	0.855	0.833
45	Washington	1.208	1.194	1.266	1.191
46	Wisconsin	1.362	1.361	1.367	1.350
47	West Virginia	0.934	0.979	1.011	1.000
48	Wyoming	1.048	1.110	1.090	1.121

at a more aggregated level). The distances computed were quite small for most pairs of states. The small magnitudes may be attributable to aggregate level data from which the Laspeyres and Paasche indices are computed. It would be useful to repeat this exercise with Laspeyres and Paasche index numbers calculated using more disaggregated data. We expect that for some pair-wise comparisons there may be no overlap in the output vectors. In such cases where the overlap is very small or non-existent, it is possible to assign an arbitrarily large distance to such pairs so that no direct link is produced in the MST.

It was expected that the MSTs based on input structures and on output structures would be different. Using the GAUSS program listed in the appendix, separate spanning trees for inputs and outputs were derived. These are presented in figures 4.4 and 4.5. As expected, the spanning trees for inputs and outputs appear to be structurally different. In the case of output spanning trees, we find similarities between the links shown in the spanning tree with geographical contiguity. Such a phenomenon reflects the fact that contiguous states share common climatic and agronomic characteristics that influence the composition of output. Since the spanning trees are computed using the distances defined through the Laspeyres-Paasche spread, and because of the low numerical values of distances in the present example, the spanning tree links can be sensitive to even small differences in distances. This sensitivity is partly the reason for the presence of direct links between some states in the spanning tree, which may be difficult to explain. Unless the spanning trees are constructed using more disaggregated data from the state accounts, results from the spanning tree presented here need to be interpreted with a degree of caution.

The MST for input comparisons has resulted in links between states, which differed significantly from geographical contiguity of states. Several links seem to connect states that have very dissimilar output patterns. Such links are consistent with the fact that the distance matrix used in deriving the minimum spanning tree is based on similarity in input use. The present empirical application is based on aggregated inputs. Therefore it is possible that some of the dissimilarities that are present at a very disaggregated level would have been scaled down when measured at the aggregated level.

The MSTs for inputs and outputs were used in constructing multilateral comparisons between all pairs of states, which in turn are used in total factor productivity comparisons. In addition to the indices computed using links identified through the MSTs, the paper also presents results based on the weighted EKS method. As discussed in Section 4, the new method assigns weights to different binary comparisons, where weights are inversely related to the Laspeyres-Paasche spread associated with the binary comparison. These results are obtained using the SHAZAM code presented in the Appendix.

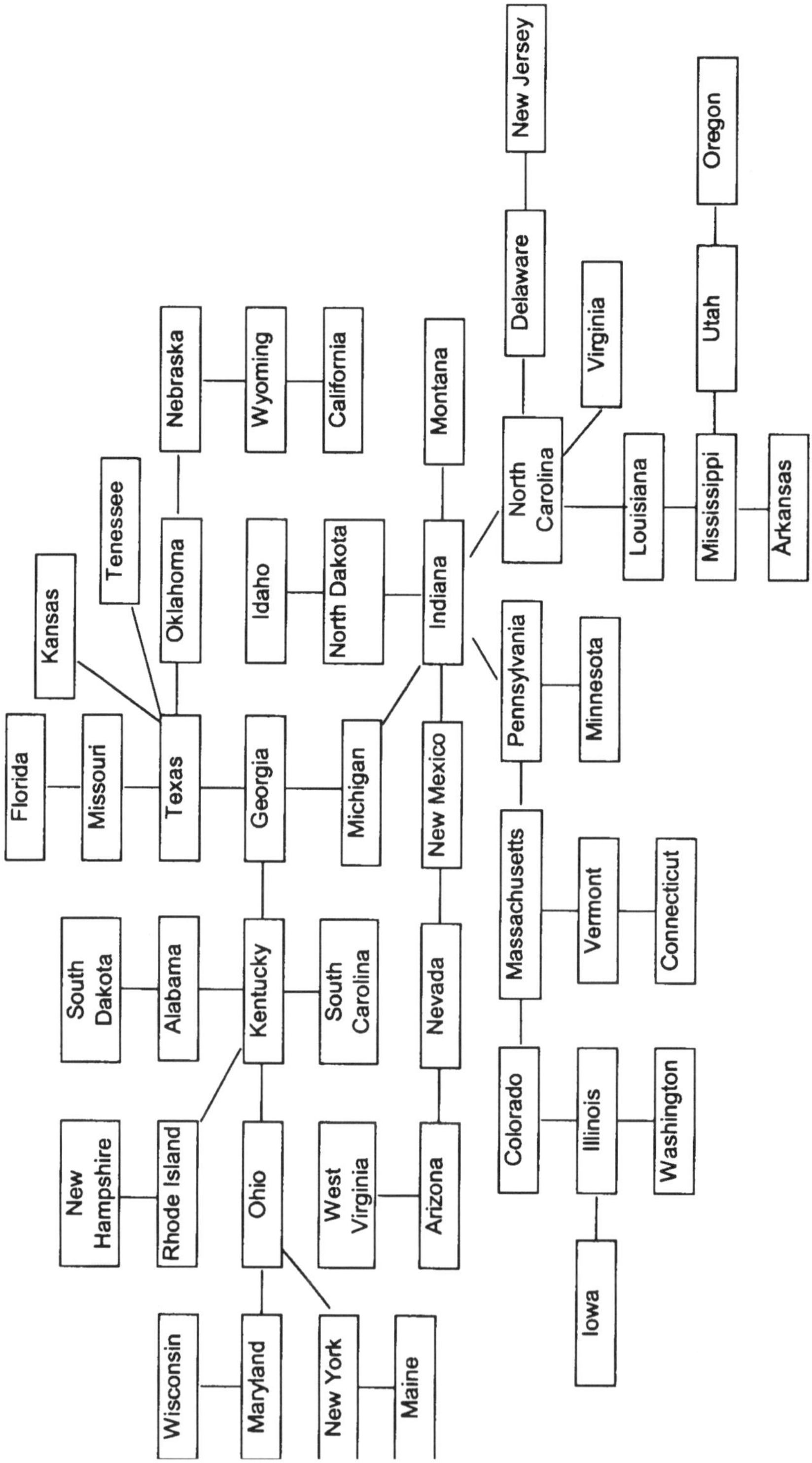

Figure 4.4. Input Minimum Spanning Tree.

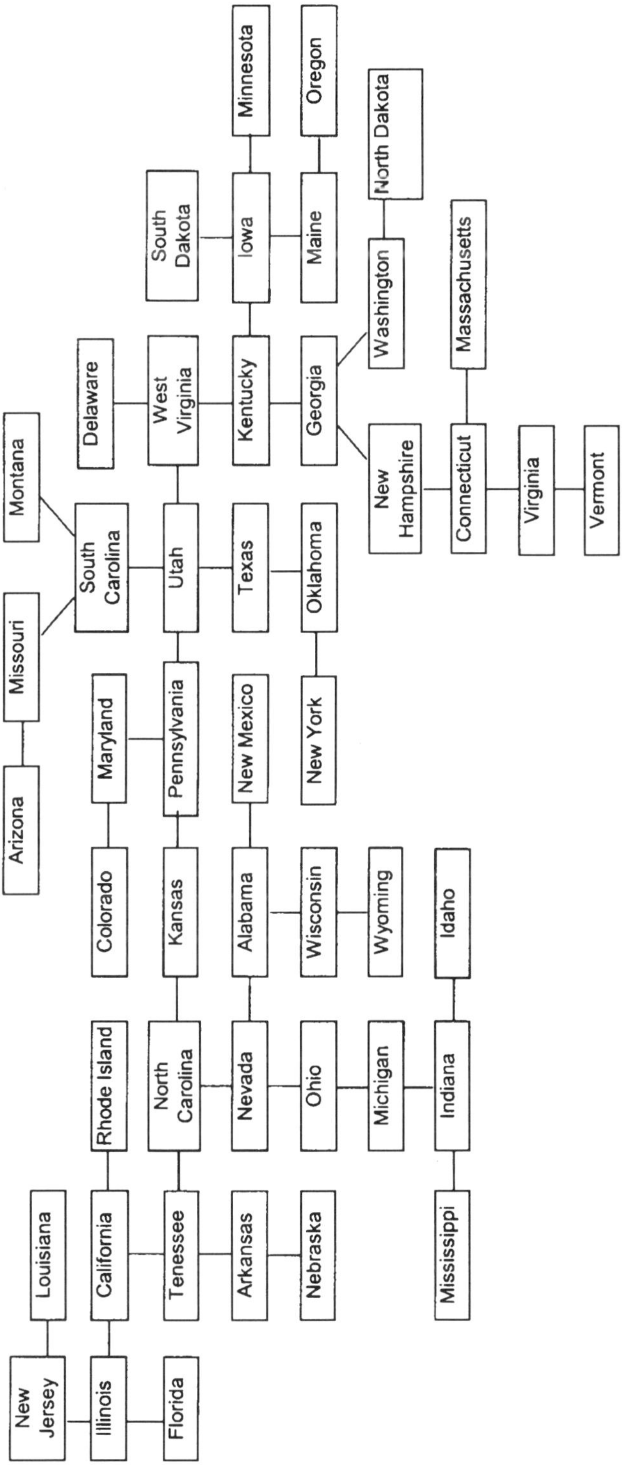

Figure 4.5. Output Minimum Spanning Tree.

Table 4.1 shows input, output, and TFP indices computed using alternative methods. Since EKS, generalised EKS, and MST indices are transitive, it is sufficient to present indices with one of the states selected as the base. In table 4.1, all the indices are shown relative to the U.S. all-state average = 1.0. Since all comparisons are transitive, comparisons for any other pair of states can be derived by simply rebasing the indices to the state that is used as the base.

The results in table 4.1 reveal some interesting features. In general, differences between the EKS and MST indices do not appear to be large. However, there are cases where differences are quite significant, in some cases around 10 percent. Differences appear to be small for pairs of states that exhibit similarity in input and output structures, but more significant for states that tend to have higher Paasche-Laspeyres spreads. Indices based on the MST method are likely to be influenced by the type of spanning tree resulting from the distance matrix based on the Paasche-Laspeyres spread. Since MSTs are sensitive to changes in magnitudes of the elements of the distance matrix, results based on the MST approach should be interpreted and used with caution.

In table 4.1, we also present results derived using the weighted EKS method that is designed to address some of the shortcomings associated with the basic EKS method that is currently followed in the measurement and comparison of inter-state productivity in agriculture. As expected, the generalised EKS indices are fairly close to the EKS indices. In a very few instances, the differences between the EKS and the generalised EKS methods are large. This result is largely as expected and is consistent with the generally small magnitudes of the elements of the distance matrix based on the Paasche-Laspeyres spread. We have also checked the fact that the generalised EKS method proposed here contains the standard EKS method as a special case in which all the binary comparisons are considered to be equally reliable. The new procedure may be viewed as a way of making finer adjustments to the EKS comparisons, taking into account any information concerning the reliability of any given binary comparisons. We feel confident that the generalised EKS method will in due course replace the current practice of computing transitive indices using the EKS procedure.

Conclusion

The main objective of the paper is to bring to the attention of practitioners some of the analytical issues associated with the construction of transitive index numbers using the EKS method, which is quite routinely used in most international and multilateral comparisons. The paper critically examines the intuitive underpinning of the EKS method and its basic philosophy of maintaining a degree of "characteristicity", the principle of preserving the binary comparisons to the greatest extent possible, within the context of constructing transitive index numbers. The principal question addressed in the paper is whether there is a need to preserve binary comparisons even when we have reason to believe that such a comparison may not be reliable. Once this issue is recognised, it follows that the EKS method needs to be modified in such a way that only those binary comparisons that are fairly reliable (e.g., comparisons between states that have similar production structures) are preserved.

The paper described two alternative procedures that provide transitive index number comparisons after explicitly accounting for the reliability of each and every binary comparison. The first procedure considered here is based on the concept of a minimum spanning tree (MST), introduced by Hill (1999), which provides a way of constructing index numbers based on a series of links between different states. Identification of the links is based on the Paasche-Laspeyres spread that was proposed by Hill as a measure of reliability. In the application of this procedure, we found that the distances measured were quite small and that the MST is quite sensitive to the magnitudes of the reliability measures. Results from the application suggest that the MST procedure certainly offers a viable alternative to the EKS method. However, the spanning trees for output and input comparisons have suggested linking of states that may be counter to intuition and considered wisdom on what links may be meaningful. Since the MST is somewhat mechanical in its application, and since it depends upon the type of distance measure used for measuring similarity in the price and quantity structures, further work on the identification of appropriate measures of similarity is necessary before the method can be applied on a regular basis.

The second aggregation procedure discussed in the paper provides a generalisation of the EKS procedure, the generalisation derived using the least squares property associated with the EKS index numbers. The procedure explicitly recognises that each binary comparison has a different degree of reliability attached to it and uses this information in modifying the EKS procedure. The procedure is quite simple to apply; a simple way of computing the weighted EKS indices using either EXCEL or SHAZAM is provided in the paper. Empirical results derived using the new method are very encouraging. Results from the new method do not seem to deviate too much from the EKS indices. The main advantage of this procedure is that it makes use of all the information that is available, in the form of a complete matrix of binary Fisher indices. This complete matrix of indices is in sharp contrast to the MST procedure in which all the binary comparisons that do not figure in the spanning tree are completely discarded. Another advantage of the procedure is that it is more robust to small numerical differences in the distances involved. Small changes in the distance matrix are likely to produce little effect on the final comparisons.

A common feature of both the methods presented is the underlying need to identify a measure of reliability associated with each binary comparison or, alternatively, a measure of the similarity in price and/or quantity structures involved. Future research needs to be focused on this critical step in the process of improving methods for multilateral comparisons of output and productivity.

Given the important role transitive index numbers play in providing various output and input aggregates at constant prices, which are in turn used in more advanced applied econometric analysis of efficiency and productivity, it is imperative that practitioners are aware of the issues associated with the construction of transitive index numbers. The present paper is designed to provide useful insights into the commonly used EKS procedure and to highlight the need to conduct further research in refining the methodology underlying the construction of various index numbers.

NOTES

1. Similarly we let w_{kj} and x_{kj} denote the input price and input quantities of k-th agricultural input (k=1,2,...K) in the j-th state (j=1,2,...M). All the index number formulae discussed in the text refer to output prices and output quantities (defined using p_{ij}'s and q_{ij}'s.). Input price and quantity indices can be defined using the same formulae by simply substituting data on input commodities. Output quantities refer to gross outputs of agricultural commodities.

2. There can be several alternative measures of reliability, and each measure has the potential to result in a different spanning tree. See Diewert (1992) for a discussion on various measures of reliability. In this paper we restrict ourselves to the original specification described in Hill (1999).

3. In the current application of the minimum spanning tree method, the PLS measure is computed separately for outputs and inputs. Since the principal aim is to obtain the best possible links leading to chained output and input index numbers, minimum spanning trees are identified for outputs and inputs separately. In order to make the exposition simple and general, no superscripts (or subscripts) are used to distinguish between outputs and inputs.

4. In the presence of ties, there is a possibility of multiple solutions to the MST problem. However, given that the PLS measure is based on observed price and quantity (output or input) data, the probability of having a tie is zero and, therefore, the event of having a tie (and, therefore, multiple MSTs) has a probability of zero.

5. The authors would have preferred to conduct the empirical work starting from the basic state-level agriculture sector accounts. However, the present work is based on matrices of Laspeyres and Paasche indices computed using data aggregated to fairly broadly defined groups. In fact, in some cases where overlap was minimal, the Laspeyres and Paasche indices were computed using some links. Such detailed work is currently in its planning stages. The present paper is designed to highlight the need to look beyond the standard EKS procedure.

6. The problem, inherent in the MST algorithm, is now well recognised in the literature. Some attempts (see Aten et al., 2001) have been made to replace the PLS measure of reliability with more robust measures of similarity of price structures based on statistical measures of similarity. The sensitivity of MST still remains a problem. The generalised or weighted EKS indices described in Section 4 offer a less sensitive yet viable alternative to the MST approach.

7. In most studies on U.S. agriculture productivity comparisons, Alabama (the first state in alphabetical order) seems to be used as the base. In order to reflect the truly transitive nature of the comparisons, we present our results, which are expressed relative to the all state average. Inter-state relativities remain unchanged.

REFERENCES

Aten, B., R. Summers, and A. H. Heston. 2001. "Price structures, the quality factor, and chaining." Paper presented at the Joint OECD-World Bank Seminar on PPPs, 30 Jan-2 Feb.

Ball, V. E., J. C. Bureau, R. Nehring, and A. Somwaru. 1997. "Agricultural Productivity Revisited." *American Journal of Agricultural Economics*, 79(4):1045-63.

Ball, V. E., F. M. Gallop, A. Kelly-Hawke, and G. P. Swinand. 1999. "State Productivity Growth in the U.S. Farm Sector." *American Journal of Agricultural Economics*, 81(1): 164-79.

Ball, V. E., J. P. Butault, and R. Nehring. 2000. "United States Agriculture, 1960-1996: A Multilateral Comparison of Total Factor Productivity." Paper presented at *The Workshop on Agricultural Productivity: Data, Methods, and Measures*, March 9-10, Washington, D.C.

Caves, D. W., L. R. Christensen, and W. E. Diewert. 1982a. "Multilateral Comparison of Output, Input and Productivity Using Superlative Index Numbers." *Economic Journal*, 92; 73-86.

Caves, D. W., L. R. Christensen, and W. E. Diewert. 1982b. "The Economic Theory of Index Numbers and the Measurement of Input, Output and Productivity." *Econometrica*, 50(6): 1393-1414.

Coelli, T. J., D. S. Prasada Rao, and G. E. Battese. 1998. *An Introduction to Efficiency and Productivity Analysis*. Kluwer Academic Publishers.

Diewert, W. E. 1976. "Exact and Superlative Index Numbers." *Journal of Econometrics*, 4: 115-45.

Diewert, W. E. 1992. "Fisher Ideal Output, Input and Productivity Indexes Revisited." *The Journal of Productivity Analysis*, 3(3): 211-48.

Eltetö, O., and P. Köves. 1964. "On an Index Computation Problem in International Comparisons." (in Hungarian). *Statisztikai Szelme*, 42: 507-18.

Gini, C. 1931. "On the Circular Test of Index Numbers." *International Review of Statistics*, 9(2): 2-35.

Hill, R. J. 1999. "International Comparisons using Spanning Trees." In *International and Interarea Comparisons of Income, Output and Prices*, ed. A. Heston and R. E. Lipsey. Chicago: University of Chicago Press, pp. 109-20.

Jorgensen, D., and F. M. Gollop. 1992. "Productivity Growth in U.S. Agriculture: A Postwar Perspective." *American Journal of Agricultural Economics*, 74: 745-50.

Rao, D. S. Prasada. 1993. *Inter-country Comparisons of Agricultural Output and Productivity*. FAO Economic and Social Development Paper, No. 112, Rome.

Rao, D. S. Prasada, and K. S. Banerjee. 1984. "A Multilateral Index Number System Based on the Factorial Approach." *Statistische Hefte*, 27: 297-313.

Rao, D. S. Prasada, and M. P. Timmer. 2000. "A Framework For Multilateral Comparisons of Manufacturing Sector Comparisons. " In *The ICOP Project: Issues, Methods And Empirical Results*. Groningen Growth and Development Centre Research Memorandum Series.

Selvanathan, E. A., and D. S. Prasada Rao. 1994. *Index Numbers: A Stochastic Approach*. New York: Macmillan.

Szulc, B. 1964. “Index Numbers of Multilateral Regional Comparisons.” (in Polish). *Przeglad Statysticzny,* 3:239-54.

Winston, W. L. (1991), *Operations Research: Applications and Algorithms*, Boston: PWS-Kent.

A GAUSS Program for Obtaining A Minimum Spanning Tree

(data is contained in the matrix x)

```
maxval = maxc(maxc(x));
k = rows(x);
for kk(1,k,1);
   x[kk,kk] = maxval+1;
endfor;

@ TREE @
m = zeros(k,k);
r = x[1,1]*ones(k,k);
c = zeros(k,1);
u = zeros(k,1);
one = ones(k,1);

@ Start node @
c[1,1] = 1;
u = one - c;

for iter(1,k-1,1);
  r = x[1,1]*ones(k,k);
  for i(1,k,1);
     ci = c[i,.];
     if ci.>0;
        for j(1,k,1);
           uj = u[j,.];
```

```
            if uj.>0; r[i,j] = x[i,j]; endif;
          endfor;
        endif;
    endfor;
    minval = minc(minc(r));
    for i(1,k,1),
        for j(1,k,1);
            rij = r[i,j];
            if rij.<=minval; rowind = i; colind = j ; endif;
        endfor;
    endfor;
    m[rowind,colind] = 1;
    c = (sumc(m')+sumc(m)).>0;
    u = one - c;
endfor;
d=m.*x;
dist = sumc(sumc(d));

@ REPORT RESULTS @
" ";"    Incidence Matrix";m;" ";
" ";"       From          To          ";
  "       State         State         Distance";" ";
for i(1,k,1);
   for j(1,k,1);
      mij = m[i,j];
      if mij.>0; i~j~d[i,j]; endif;
   endfor;
endfor;
" ";" ";"    Total Distance";" ";
dist;
end;
```

EXCEL Steps for Obtaining Weighted EKS Index Numbers

Step 1. Compute the Fisher binary matrix

$$F = \begin{bmatrix} F_{11} & F_{12} & . & . & F_{1M} \\ F_{21} & F_{22} & & & F_{2M} \\ . & & & & . \\ . & & & & . \\ F_{M1} & . & . & . & F_{MM} \end{bmatrix}$$

Note that $F_{ii} = 1$ for all i and that $F_{ij} = 1/F_{ji}$.

Step 2. Compute the weight matrix for all binary comparisons

$$w = \begin{bmatrix} w_{11} & w_{12} & . & . & w_{1M} \\ w_{21} & w_{22} & & & w_{2M} \\ . & & & & . \\ . & & & & . \\ w_{M1} & . & . & . & w_{MM} \end{bmatrix}$$

Step 3. Compute the matrix on the left-hand side of equation (5.4),denoted by P

$$P = \begin{bmatrix} \sum_{j \neq 1}^{M} w_{1j} & -w_{12} & . & . & -w_{1M} \\ -w_{21} & \sum_{j \neq 2}^{M} w_{2j} & & & -w_{2M} \\ . & & & & . \\ . & & & & . \\ -w_{M1} & -w_{M2} & . & . & \sum_{j \neq M}^{M} w_{Mj} \end{bmatrix}$$

Step 4. Drop the first row and column, and denote the resulting matrix as P^*. Compute P^{*-1}, the inverse of P^*.

Step 5. Compute the vector on the right-hand side of (5.4).

$$q = \begin{bmatrix} -\sum_{j\neq 1}^{M} w_{1j} \ln F_{1j} \\ -\sum_{j\neq 2}^{M} w_{2j} \ln F_{2j} \\ \cdot \\ \cdot \\ -\sum_{j\neq M}^{M} w_{Mj} \ln F_{Mj} \end{bmatrix}$$

Step 6. Drop the first element of q and denote it by q^*.

Step 7. The solution for $\hat{Q}_2, \hat{Q}_3, \ldots \hat{Q}_M$, given that $\hat{Q}_1 = 0$, can be computed using P^{*-1} and q^* as

$$\begin{bmatrix} \hat{Q}_1 \\ \hat{Q}_2 \\ \cdot \\ \cdot \\ \hat{Q}_M \end{bmatrix} = P^{*-1} \cdot q^*$$

Step 8. The new Q*s , or modified Q*s, based on weighted EKS method are given by

$$Q*_j = \exp(\hat{Q}_j)$$
$$\text{with} \quad Q*_1 = \exp(Q*_j) = 1$$

A SHAZAM Program for Obtaining Weighted EKS Index Numbers

(the variable F is the column vector of length 2304 created by appending the 48 rows of the Fisher index matrix to each other; the variable w is the column vector created by appending the rows of the distance matrix to each other; the variables I and J are associated row and column indices)

```
smpl 1 2304
read (eksoutput.txt) i j f w
genr lnf = log(f)
do # = 1,48
  genr d# = 0
endo
do # = 1,48
  if (i.eq.#) d# = d#+1
  if (j.eq.#) d# = d#-1
endo
* standard EKS
ols lnf d2-d48 / noconstant coef = beta1
end
genr lnfstar = lnf*sqrt(w)
do # = 1,48
  genr dstar# = d#*sqrt(w)
endo
* weighted EKS
ols lnfstar dstar2-dstar48 / noconstant coef = beta2
end
smpl 1 47
genr eks1 = 1/exp(beta1)
genr eks2 = 1/exp(beta2)
set output
print eks1 eks2
```

5

Productivity Versus Urban Sprawl: Spatial Variations in Land Values

Charles B. Moss, Gregory Livanis, Vince Breneman, and Richard F. Nehring

Abstract

How farmland is valued has major implications for the measurement of agricultural productivity and for the financial viability of the farm sector. This study examines the relative effects of productivity and urban growth on farmland values using spatial data. The results suggest that the land stock in the current productivity accounts includes urban pressures not related to productivity.

Introduction

The importance of real estate to the net worth of domestic agriculture is well documented. In 1950, real estate represented 62 percent of all agricultural asset values while real estate debt accounted for 48 percent. By 1981, real estate accounted for 80 percent of all agricultural assets, while real estate debt peaked at 58 percent of all farm debt in 1986. Given this significance of farm real estate, two policy issues related to farmland valuation need to be addressed. First, as the dominant asset in the agricultural portfolio, how farmland is valued has major implications for the measurement of agricultural productivity. Second, the valuation of farmland has important implications for the financial viability of the farm sector. This study examines the relative effects of productivity and urban growth on farmland values using spatial data.

The spatial dimension of farmland prices is particularly important for productivity measurement. Spatial differences in farmland prices may result from differences in productivity that arise from differences in land quality. Land quality can involved a multitude of physical factors such as soil composition, drainage, and so forth. In addition, land quality may be affected by its proximity to urban centers. Specifically, transportation advantages imply that land closer to cities may have a competitive advantage in the production of vegetable, fruit, and horticultural crops. These crops are typically high-value crops with a significant premium on freshness. Naturally, this competitive advantage should be manifested through land values. Unfortunately, land close to urban centers is also affected by demand for housing and other urban uses. As a result, the value of farmland may imperfectly reflect differences in productivity.

The discussion that follows addresses the effect of farmland values on productivity measures in agriculture. A spatial formulation of land values reflecting differences in productivity and urban pressures is presented. An empirical model of productivity versus urban sprawl is followed by a discussion of the empirical results. Finally, the study concludes with a discussion of the implications of this analysis.

EFFECT OF FARMLAND VALUES ON PRODUCTIVITY MEASUREMENT

Productivity measurement analyzes whether the quantity of outputs that can be obtained from a given quantity of inputs has changed over time. In general, if the quantity of outputs that can be obtained from a fixed quantity of inputs has increased, then agriculture has become more productive. Unfortunately, this intuitively appealing definition can be subject to several significant pitfalls. Specifically, changes in the relative input or output bundles raise the possibility of substitution along a particular isoquant or isoproduct surface rather than a shift in technology.

One approach used to measure agricultural productivity is the index number approach (Ball et al., 2000a). State level input and output indices can be computed using the Fisher quantity index:

$$Q_F^{jk}(p^j, p^k, x^j, x^k) = \left[\frac{p^k x^j}{p^k x^k}\frac{p^j x^j}{p^j x^k}\right]^{\frac{1}{2}} \tag{1}$$

where p^j denotes the price of either the output or input price faced by state j and x^j is the corresponding quantity. Total factor productivity (TFP) is then measured as the ratio of outputs to inputs.

While the basic formulation in (1) is based on input quantities, the use of values has several desirable attributes. The use of values instead of quantities allows for the use of service flows to measure the use of intermediate and long-term assets. In addition, the use of values allows for hedonic adjustment of input quantity. The measurement of productivity gains requires all input qualities to be homogeneous. The reason for this requirement can be seen most readily in labor. Assume that the quality of labor used over time responds to state and national investments in education. Then increased human capital should lead to an increase in output, but the increase in output would not represent gains in productivity. While labor inputs are treated in a slightly different manner, the same adjustment must be made to the machinery and land inputs.

Hedonic adjustments in land values are required for two reasons. First, in many regions of the United States, farmland has been converted into other uses. If the farm sector maintains the same overall quantity of output given this conversion (all other factors held constant), then productivity gains have occurred. However, the relative gain in productivity is a function of the quality of farmland removed from production. If relatively more productive land is removed from production, the gain in productivity should be relatively greater than if low productivity land is removed. Second, in comparing changes in productivity across states, the productivity of the land must be taken into account. If all other factors are constant, the value of farmland can be used as a measure of its relative productivity.

A Spatial Formulation of Land Values

Enke (1942) develops six spatial formulations for markets. The first follows the simple Marshallian formulation in which buyers and sellers are located in close proximity to one another. The second case corresponds to the simple von Thunen central-place model. Under this scenario, consumers are concentrated in a central place, while production occurs various distances away. In a related scenario, Enke allows for both producers and consumers to be distributed broadly across the plane, but requires all goods to pass through the central point. In general, this scenario yields the same spatial patterns observed under the pure von Thunen formulation. Enke's other formulations assume that production and consumption are distributed throughout the plane, but allow for direct interaction between producers and consumers.

Each of these scenarios can be found in various markets. Several agricultural markets such as hay and feeder cattle can be thought of as local markets. On the other hand, declining rural populations makes the central location of consumption an increasingly viable model. Similarly, for commodities such as wheat for bread, widely dispersed economies of scale have given rise to central markets for wheat in the manufacturing of bread.

Miyao (1977) constructs a model of land values in the von Thunen plane such that production occurs in concentric rings around the central place. In each ring, land is valued such that the rent to land absorbs all the profit from production. In this model, land values decline the farther land is located from the central point. However, while Miyao's formulation focuses on the development of homogeneous rings and the comparative statics of those rings, our focus is on the rental values of farmland.

As a starting point, consider farmland values as the sum of economic rents from farming plus the value of future conversion to urban use:

$$V(t)=E\left[\int_t^{\infty} e^{-rs}R_{AG}(s)ds\right]+E\left[\int_t^{\infty} e^{-rs}R_U(s)ds\right] \tag{2}$$

where $R_{AG}(s)$ is the return to farmland in period s, $R_U(s)$ is the return to urbanization in period s, r is the discount rate, and $E[.]$ is the expectation function. Breaking (2) down into parts,

$$E\left[\int_t^{\infty} e^{-rs}R_{AG}(s)\,ds\right] \tag{2a}$$

is the expected return on farmland from agricultural activities. This expectation is based on a variety of random variables including agricultural prices, weather, etc. In addition, for the purpose of this study, the expected value of agricultural production is a function of urbanization. We assume the existence of a random variable $z(t)$ that is equal to zero before urbanization and one after urbanization. The expectation function in (2a) can then be rewritten as

$$\int_t^{\infty} e^{-rs}R_{AG}(s)\,z(s)\,f(z,s)\,ds \tag{2a'}$$

where $f(z,s)$ is the probability that urbanization occurs in period s. By extension, the second half of (2) represents the value of conversion of farmland into urban use. Taken together, the two halves of the equation form a switching function by which the value of farmland is determined by the return in agriculture up to the point of urbanization plus the value of farmland for urbanization. In this formulation, we assume that the urbanization event itself is random.

If we assume that urbanization follows a Poisson arrival process, the probability of farmland remaining in agriculture is:

$$f(t,\theta)=\theta e^{-\theta t} \tag{3}$$

(Feller 1950, pp. 44-48). To derive the probability that urbanization occurs in period t, we assume that no urbanization has occurred until period t-δ. θ is a parameter related to the expected time to urbanization. Specifically, the expected arrival time is 1/θ. The probability that urbanization occurs in period t, given that no urbanization has occurred in period t-δ, assuming that the two events are independent, becomes

$$g(t,\theta)=f(t-\delta,\theta)(1-f(t,\theta))=\theta e^{-\theta(t-\delta)}\left(1-\theta e^{-\theta t}\right) \tag{4}$$

Taking the limit as δ approaches zero then yields

$$g(t,\theta)=\theta e^{-\theta t}-\theta^2 e^{-2\theta t} \tag{5}$$

Assuming that the rates of return to agriculture and return to urbanization are constant over time, and substituting equations (3) and (5) into equation (2), yields

$$V(t)=\frac{\theta e^{-t(r+\theta)}}{r+\theta}R_{AG}+\frac{e^{-t(r+2\theta)}\theta\left(\theta(r+\theta)-(r+2\theta)e^{-t\theta}\right)}{(r+\theta)(r+2\theta)}R_U \tag{6}$$

Evaluating this expression at t=0 yields

$$V(0)=\frac{\theta}{r+\theta}R_{AG}+\frac{\theta\left(r+2\theta-r\theta-\theta^2\right)}{(r+\theta)(r+2\theta)}R_U \tag{7}$$

Breaking this expression down into parts, the first part is the discounted value of agricultural returns. As in the standard farmland pricing formula, the value of farmland is an increasing function of the return to agriculture and a decreasing function of the discount rate. In addition, the value of farmland is now a function of the θ parameter in the exponential distribution. Note that since the expected value of the exponential distribution is 1/θ, the discounted value of agricultural returns is inversely related to the expected length of time to urbanization. Dividing equation (7) through by its left-hand side implies that the longer the expected time to urbanization the larger the percent of farmland value contributed by agricultural returns. The second term on the right-hand side of equation (7) is the discounted expected time of development.

Taking the variation around equation (6) at t=0 and holding the discount rate constant yields:

$$V'(0)=\frac{\theta}{r+\theta}dR_{AG}+\frac{\theta\left(r+2\theta-r\theta-\theta^2\right)}{(r+\theta)(r+2\theta)}dR_U+ \tag{8}$$

$$\left[-\frac{2\theta+r}{(r+\theta)^2(r+2\theta)^2}R_{AG}+\frac{r^2(1-2\theta)+2r^2(2-3\theta)+2r(2-3\theta)-2\theta^4}{(r+\theta)^2(r+2\theta)^2}R_U\right]d\theta$$

The results from equation (8) can then be used to form a first-order Taylor series expansion of farmland values:

$$
\begin{aligned}
V &= \alpha_0 + \alpha_1\left(R_{AG} - \bar{R}_{AG}\right) + \alpha_2\left(R_U - \bar{R}_U\right) + \alpha_3\left(\theta - \bar{\theta}\right) + O\left(\Delta^2\right) \\
&= \tilde{\alpha}_0 + \tilde{\alpha}_1 R_{AG} + \tilde{\alpha}_2 R_U + \tilde{\alpha}_3\theta + O\left(\Delta^2\right)
\end{aligned}
\tag{9}
$$

where the as are estimated parameters and $O(D^2)$ is a second-order approximation error.

Thus, equation (9) relates cross-sectional changes in farmland values to changes in agricultural returns, changes in the value of urbanization, and changes in the probability of urbanization. The next step is to construct a model of farmland values that is consistent with the von Thunen effect on agricultural prices. We begin by maximizing profit:

$$
\begin{aligned}
&\max_{y,x,D} p'y - w'x - rD \\
&st \quad f(y,x,A,I) = 0 \\
&(I - I_0) + (A - A_0)V = D - D_0
\end{aligned}
\tag{10}
$$

where p is a vector of output prices, y is a vector of outputs, w is a vector of input prices, x is a vector of inputs, r is the interest rate on farm debt, D is the level of farm debt, $f(.)$ is a multiproduct production function, A is the acres of farmland, I is the level of intermediate assets, V is the value of farmland, and the subscript zeros denote initial levels. Given that we are typically interested in the decisions at the farm level, we assume that the input and the output prices are exogenous. However, given that we are interested in the spatial variation of farmland prices, we expand the formulation in equation (10) as

$$
\begin{aligned}
&\max_{y,x,D} (p - \tau(\delta))'y - w'x - rD \\
&st \quad f(y,x,A,I) = 0 \\
&(I - I_0) + (A - A_0)V = D - D_0
\end{aligned}
\tag{11}
$$

where t(d) is the transportation cost associated with each commodity and d is some measure of distance. Forming the Lagrange multiplier of equation (11) yields

$$
L = (p - \tau(\delta))'y - w'x - rD - \lambda_1(f(y,x,A,I)) + \lambda_2(D - D_0 - (I - I_0) - (A - A_0)V)
\tag{12}
$$

From this formulation, we want to develop the marginal value of each unit of output given the transportation cost and the marginal value of farmland. Focusing on the marginal value of each output first,

$$\frac{\partial L}{\partial y_i} = (p_i - \tau_i(\delta)) - \lambda_1 \frac{\partial f(y,x,A,I)}{\partial y_i} = 0 \tag{13}$$

This equation yields the standard relationship that the marginal rate of transformation between two products equals the inverse of their price ratios. Note that increases in the transportation cost for each commodity imply a relative reduction in the output of that commodity. Equating the shadow value of production across all outputs yields

$$\lambda_1 = \frac{(p_1 - \tau_1(\delta))}{\frac{\partial f(y,x,A,I)}{\partial y_1}} = \Lambda = \frac{(p_n - \tau_n(\delta))}{\frac{\partial f(y,x,A,I)}{\partial y_n}} \tag{14}$$

Differentiating the shadow value with respect to distance then yields

$$\frac{\partial \lambda_1}{\partial \delta} = -\frac{\frac{\partial \tau_i(\delta)}{\partial \delta}}{\frac{\partial f(y,x,A,I)}{\partial y_i}} \leq 0 \tag{15}$$

as long as the transportation cost is an increasing function of distance.

Turning to the value of farmland to complete the model, we first take the partial equation (12) with respect to debt to yield

$$\frac{\partial L}{\partial D} = -r + \lambda_2 = 0 \Rightarrow \lambda_2 = r \tag{16}$$

Next, taking the first order condition with respect to land values and substituting the result of equation (16) yields

$$-\lambda_1 \frac{\partial f(y,x,A,I)}{\partial A} - rV = 0 \tag{17}$$

Using the implicit market condition on V yields

$$V = \frac{-\lambda_1 \dfrac{\partial f(y,x,A,I)}{\partial A}}{r} \tag{18}$$

Noting that the partial of the multiproduct production function with respect to land is negative, equation (18) is the same value as found in equation (7) if conversion to urban use never occurs. In particular, we are interested in specifying the return to agricultural activities in equation (2) as

$$R_{AG} = -\lambda_1 \frac{\partial f(y,x,A,I)}{\partial A} \tag{19}$$

Merging the results of equation (14), we have

$$R_{AG} = -\frac{(p_i - \tau_i(\delta))}{\dfrac{\partial f(y,x,A,I)}{\partial y_i}} \frac{\partial f(y,x,A,I)}{\partial A} = (p_i - \tau_i(\delta))\frac{dy_i}{dA} \tag{20}$$

where the last derivative is evaluated at the optimal point of production.

Given the results from equation (15) and assuming the quality of farmland is constant, we conclude that the return to farmland is a decreasing function of the transportation cost and distance to the market. In addition, the value of farmland is an increasing function of the relative productivity of farmland. Specifically,

$$\frac{dy_i}{dA} = \frac{\dfrac{\partial f(y,x,A,I)}{\partial A}}{\dfrac{\partial f(y,x,A,I)}{\partial y_i}} \tag{21}$$

The solution depicted in equation (21) assumes that all crops are produced continuously throughout the region. Obviously, the formulation in equation (10) could be changed to guarantee that only nonnegative quantities of crops could be chosen. This restriction would transform the problem into a Kuhn-Tucker optimization problem. The critical point is that not all crops would meet the marginal value condition in equation (14). The result would be the absence of low-value crops close to the urban centers.

The return to farmland in agriculture in (20) and (21) assumes that all land is of constant quality. This assumption generates the symmetric arrangements in von Thunen's theory. However, the quality of farmland is not homogeneous.

Past studies such as Miranowski and Hammes (1984) emphasize the effect of differing land quality on land prices. In addition, climate has a direct effect on land values by affecting the feasible crops. For example, soil quality in New Jersey may be much higher on average than in Hillsborough County, Florida. However, the climate in New Jersey rules out winter tomatoes. The combination of soil quality and climate tends to distort the concentric rings of the von Thunen surface. In addition, economies of scale may affect transportation costs disadvantages. Again, the economies of scale in Ruskin, Florida, may offset the transportation advantages close to an urban center (although Ruskin is close to Tampa and St. Petersburg, significant markets in their own right). Thus, the return to farming presented in (20) and (21) should be modified to incorporate other hedonic effects.

Empirical Model of Productivity Versus Urban Sprawl

The empirical model of farmland values given the existence of a von Thunen effect on both agricultural output and urbanization is a simultaneous three-equation system based on equations (9), (20), and (21) above. Equation (9) is used to model the value of farmland as the sum of the value of land in production agriculture, plus the value of land at the point of urbanization.

$$V_c = a_0 + a_1 R_{AG,c} + a_2 H_c + a_3 DP_c + a_4 DL_c + \varepsilon_{1c} \tag{22}$$

where V_c is the value of farmland in county c, $R_{AG,c}$ is the gross revenue per acre for agriculture in county c, H_c is the housing value in county c, DP_c is the population growth rate in county c, DL_c is the rate of farmland loss in county c, and ε_{1c} is the error term. Within this formulation, we assume that the gross revenue to farmland and the housing value are endogenously determined.

The value of land in production agriculture is equal to the gross return on each acre of farmland. This gross value is computed as the sum of the share of farmland in each crop times the gross revenue for each crop.

$$R_{AG,c} = \sum w_{ic} (p_{ic} q_{ic}) \tag{23}$$

where w_{ic} is the acreage share of crop i in county c, p_{ic} is the price of the output for crop i in county c, and q_{ic} is the quantity of crop i produced in county c. Following the hedonic pricing literature (Miranowski and Hammes, 1984; Ball et al., 2000b) for farmland, the gross revenue for agriculture in each county is modeled as a function of land characteristics:

$$R_{AG,c} = b_0 + b_1 Ac_c + \sum_{j=2}^{m} b_j x_c + \varepsilon_{2c} \tag{24}$$

where Ac_c is a measure of accessibility for county c, and the variables $x_{i,c}$ i=2,...m are a set of soil characteristics.

Finally, the house value in county c is modeled as a function of the accessibility variable, income, and the tax rate.

$$H_C = c_0 + c_1 Y_c + c_2 Ac_c + c_e T_c + \varepsilon_{3c} \tag{25}$$

where H_c is the house value in county c, Y_c is the income in county c, T_c is the real estate tax in county c, and ε_c is an error term.

The comparison between the impact of urban pressure in a productivity sense versus the effect of urban sprawl is captured by the coefficients on accessibility in equations (24) and (25). If increased accessibility causes a change in the relative crop mix, or changes in price for a particular crop, this effect will be manifest through the b_1 coefficient in equation (24). However, if the impact of accessibility comes only through urbanization, it will be captured in the c_2 coefficient in equation (25).

The data used to estimate these three equations (22, 24, and 25) come from a variety of sources. V_c (the market value of land and buildings per acre) was taken from the USDA 1997 *Census of Agriculture*. Also taken from the *Census* were the crop levels for each county and yield per acre for each crop used to compute the gross agricultural margin in equation (23). The prices used for each crop are taken from the state average prices at harvest from The USDA publication *Agricultural Prices*. The median house values and incomes were taken from the general *Census* as well as population levels. The USDA also estimated the accessibility coefficient.

The list of hedonic factors used in equation (24) includes the cation exchange capacity (meg/100g) (x_2), the pH of the soil (x_3), an index of the texture of the ground (x_4), an index of organic matter (x_5), the percent of three-inch rocks in the ground (x_6), the percent of clay content (x_7), an index of k-factor erodibility (x_8), an index of t-factor erosion tolerance (x_9), the salinity of the ground (mmhos/cm) (x_{10}), the water-holding capacity (inches/inch) (x_{11}), the water-table depth (in inches) (x_{12}), the permeability of the ground (in inches) (x_{13}), an index of the drainage of the ground (x_{14}), the bulk density (in grams/ccm) (x_{15}), the percent of calcium carbonate in the soil (x_{16}), the percent of gypsum in the ground (x_{17}), a score of ponding (x_{18}), the percent of ten-inch rocks (x_{19}), a score of shrink and swell (x_{20}), the percent of the slope of the soil (x_{21}), the sodium adsorption ratio (x_{22}), the soil depth in inches (x_{23}), an index of wind erosion (x_{24}), and the percent of irrigated acres (irrigated acres/cropland in acres) (x_{25}). In addition, we appended nine dummy variables for ERS regions: x_{26} is the dummy for the Lake states region; x_{27} is the dummy variable for the Corn Belt, x_{28} is the dummy variable for the Northern Plains, x_{29} is the dummy variable for Appalachia, x_{30} is the dummy variable for the Southeast, x_{31} is the dummy variable for the Delta States, x_{32} is the dummy variable for the Southern Plains, x_{33} is the dummy variable for the Mountain States, and x_{34} is the dummy variable for the Pacific region. Finally, we include one interaction term for the interaction between pH and percent of land irrigated (x_{35}).

Cross sectional data yielded 2,965 counties. The difference in the actual number of counties can be explained by the fact that the states of Alaska and Hawaii were excluded and data are not reported in the *Census* if reporting violates confidentiality (usually less than 4 observations). Equations (22), (24) and (25) were estimated iterated three-stage least squares, although the simultaneity of the model is mitigated by the block recursive nature of the system.

Empirical Results

Table 5.1 presents the estimated parameters for equations (22), (24), and (25). Table 5.2 provides a comparison of farmland prices across states based on the estimated model. Columns (a) and (b) of table 5.2 present the estimated values of farmland for each state, holding the urbanization pressure in equation (25) constant. Column (a) is the estimated land value while column (b) is the estimated land value for each state divided by the estimated land value for Alabama. Columns (c) and (d) of table 5.2 depict the land value for each state holding all hedonic factors constant except for the von Thunen effect. Again, column (c) presents the estimated land value for each state while column (d) presents the land value for each state relative to Alabama. Finally, column (e) presents the observed market value for each state while column (f) presents the observed market value for each state relative to Alabama.

The results in table 5.1 suggest that the return to agriculture, house values, and changes in acreage are statistically significant in determining the value of farmland across counties. Further, consistent with the hypothesized von Thunen effect, accessibility affects the gross margin to agriculture. In addition, the median house price increases as accessibility increases. Thus, farmland prices are affected by increased population pressures both through changes in the gross margin and by urban sprawl.

The effect of changes in acres in table 5.1 is counterintuitive. We would expect that, as urban pressure removes farmland in a county, the price of farmland would increase. One explanation is that the system is slightly mis-specified. Specifically, instead of changes in population and changes in acreage, we

Table 5.1. Results from the Regression of the Three-Stages Least-Squares Model

Parameter	Estimate	Standard Error
	Farm Land Value	
Constant (a_0)	-1,264.170***	66.220
Gross Ag. Margin (a_1)	1.414***	0.102
Median House Value (a_2)	0.041***	0.001
Population Growth Rate (a_3)	151.272	177.893
Change in Acres (a_4)	-494.194***	115.546
	Gross Margin from Agriculture	
Constant (b_0)	334.384***	124.462
Accessibility (b_1)	0.550***	0.030
Cation Exchange (b_2)	0.108	2.178
PH (b_3)	-27.252	18.250
Soil Texture (b_4)	-11.666	13.746
Organic Matter (b_5)	13.478*	7.438
Rocks > 3 Inches (b_6)	-9.197**	3.873
Percent Clay (b_7)	2.552	2.397
k-factor Erodability (b_8)	263.091	260.350
t-factor Erodability (b_9)	14.818	11.324
Salinity (b_{10})	94.290***	20.244
Water Holding Cap. (b_{11})	-5.009	3.568
Water Table Depth (b_{12})	-11.160	10.982
Permeability (b_{13})	25.779***	7.714
Drainage (b_{14})	-40.029***	11.242
Bulk Density (b_{15})	221.841***	71.214
Calcium Carbonate (b_{16})	7.798*	4.205
Gypsum (b_{17})	76.519	61.199
Ponding (b_{18})	-413.441**	206.016
Rocks > 10 Inches (b_{19})	7.502	10.376
Shrink/Swell (b_{20})	-52.911	45.046
Slope (b_{21})	4.441***	1.794
Sodium absorption (b_{22})	-56.555***	11.281
Soil (b_{23})	3.817**	1.821

continued

Table 5.1, *continued*

Parameter	Estimate	Standard Error
Wind Erosion (b_{24})	-1.698	1.967
Percent Irrigation (b_{25})	8.660***	2.107
Lake States (b_{26})	-321.954***	58.685
Corn Belt (b_{27})	-131.089***	51.099
Northern Plains (b_{28})	-20.196	53.629
Appalachia (b_{29})	64.216	44.324
Southeast (b_{30})	135.234***	51.183
Delta States (b_{31})	-45.835	54.804
Southern Plains (b_{32})	-193.584***	61.254
Mountain States (b_{33})	-95.511*	55.113
Pacific (b_{34})	634.069***	62.204
Interaction pH (b_{35})	-1.146***	0.344
	Median House Value	
Constant (c_0)	-8,895.380***	1,472.400
Median income (c_1)	2.246***	0.063
Accessibility (c_2)	11.555***	1.164
Property tax (c_3)	833.380***	22.951

*** Denotes statistical significance at .01 level of confidence, **Denotes statistical significance at .05 level of confidence, *Denotes statistical significance at .10 level of confidence.

could have specified changes in population density over time. Defining population density as the ratio of population to acreage, the results are somewhat closer to expectations.

Conclusions

Two alternative reasons for increases in farmland values in urban areas have been suggested in the literature. The first explanation grows out of the von Thunen plane through the concept of transportation costs. Given that transportation is costly, higher-valued outputs that have higher transportation costs are produced close to urban areas. The second explanation involves a similar concept, but emphasizes the conversion of farmland into residential housing. In this study, we examined the possibility of each of these effects. The results indicate that both effects can be observed in the United States.

Table 5.2. Hedonically Adjusted Land Values Based on Agronomic and Von Thunen Effects

	Hedonic and von Thunen Valuation		Von Thunen Valuation		Market Value of Farmland	
	Value ($/Acre)	Normalized Value	Value ($/Acre)	Normalized Value	Value ($/Acre)	Normalized Value
States	(a)	(b)	(c)	(d)	(e)	(f)
			Northeast			
CT	1,814.759	1.082	2,004.369	1.133	5,917.772	4.087
DE	1,978.408	1.179	1,708.386	0.966	2,648.603	1.829
MA	2,042.471	1.217	2,033.184	1.149	5,097.978	3.521
MD	1,892.090	1.128	1,844.137	1.042	3,154.621	2.179
ME	1,583.005	0.943	1,548.851	0.875	1,196.100	0.826
NH	1,774.219	1.057	1,650.584	0.933	2,268.421	1.567
NJ	2,361.841	1.408	2,268.020	1.282	6,616.483	4.570
NY	1,556.160	0.927	1,656.924	0.937	1,284.179	0.887
PA	1,628.248	0.970	1,752.088	0.990	2,378.607	1.643
RI	1,949.034	1.162	2,000.825	1.131	5,884.741	4.064
VT	1,490.808	0.888	1,559.457	0.881	1,516.792	1.048
			Lake States			
MI	1,438.462	0.857	1,218.664	0.689	1,672.159	1.155
MN	1,241.318	0.740	1,093.586	0.618	1,165.498	0.805
WI	1,224.523	0.730	1,138.038	0.643	1,245.427	0.860
			Corn Belt			
IA	1,350.933	0.805	1,364.611	0.771	1,698.648	1.173
IL	1,367.298	0.815	1,430.060	0.808	2,132.501	1.473
IN	1,628.317	0.970	1,462.116	0.826	2,072.685	1.432
MO	1,357.026	0.809	1,370.937	0.775	1,067.502	0.737
OH	1,575.077	0.939	1,509.789	0.853	2,045.431	1.413
			Northern Plains			
KS	1,299.043	0.774	1,498.255	0.847	578.340	0.399
ND	1,523.838	0.908	1,480.195	0.837	401.463	0.277
NE	1,498.857	0.893	1,489.448	0.842	650.169	0.449
SD	1,407.809	0.839	1,479.684	0.836	350.961	0.242
			Appalachian			
KY	1,682.629	1.003	1,684.366	0.952	1,450.527	1.002
TN	1,660.627	0.990	1,691.985	0.956	1,808.077	1.249
VA	1,514.038	0.902	1,697.495	0.959	1,924.098	1.329
WV	1,550.598	0.924	1,663.503	0.940	1,093.690	0.755
NC	1,934.711	1.153	1,725.200	0.975	2,089.517	1.443
			Southeast			
AL	1,678.000	1.000	1,769.186	1.000	1,447.878	1.000
FL	2,534.205	1.510	1,835.336	1.037	2,256.799	1.559
GA	1,712.236	1.020	1,770.462	1.001	1,506.777	1.041
SC	1,747.639	1.042	1,797.295	1.016	1,495.796	1.033

continued

Table 5.2, *continued*

	Hedonic and von Thunen Valuation		Von Thunen Valuation		Market Value of Farmland	
	Value ($/Acre)	Normalized Value	Value ($/Acre)	Normalized Value	Value ($/Acre)	Normalized Value
States	(a)	(b)	(c)	(d)	(e)	(f)
			Delta States			
AR	1,392.234	0.830	1,484.690	0.839	1,156.732	0.799
LA	1,644.454	0.980	1,508.922	0.853	1,198.032	0.827
MS	1,479.410	0.882	1,489.954	0.842	1,051.402	0.726
			Southern Plains			
OK	1,238.567	0.738	1,267.411	0.716	609.591	0.421
TX	1,225.309	0.730	1,275.282	0.721	622.696	0.430
			Mountain States			
AZ	1,275.166	0.760	1,391.893	0.787	360.722	0.249
CO	1,263.321	0.753	1,395.766	0.789	616.600	0.426
ID	1,248.167	0.744	1,385.954	0.783	1,029.301	0.711
MT	1,182.574	0.705	1,369.422	0.774	294.059	0.203
NM	1,492.374	0.889	1,373.630	0.776	198.665	0.137
NV	1,441.798	0.859	1,372.533	0.776	405.688	0.280
UT	1,568.029	0.934	1,522.657	0.861	577.235	0.399
WY	1,459.306	0.870	1,505.375	0.851	222.270	0.154
			Pacific			
CA	2,608.848	1.555	2,550.736	1.442	2,633.263	1.819
OR	2,076.764	1.238	2,418.139	1.367	959.777	0.663
WA	2,262.695	1.348	2,429.760	1.373	1,208.852	0.835

The decomposition of the two effects becomes important in the modeling of productivity. Specifically, the estimation of productivity measures requires incorporation of differing productivity between similar inputs. In the case of farmland, if the level of outputs remains constant despite losses in farmland, then productivity has increased. This relationship must be adjusted for difference in farmland quality. Loss of more productive farmland implies relatively larger gains in productivity than loss of less productive farmland.

Past attempts to measure this discrepancy in productive capacity have relied primarily on difference in overall farmland values without explicitly adjusting for each pressure. The results of this study suggest that this assumption misstates the cross-section productivity measures in some cases. Specifically, Connecticut's farmland values are 4.087 times those of Alabama. However, the difference in productivity incorporating hedonic factors and the von Thunen effect on agricultural gross revenue is 1.082 times Alabama. This discrepancy compares with similar results from California and Florida whose market values are 1.819 and 1.559 times Alabama respectively, but who have productivity-adjusted land values of 1.555 and 1.510 times Alabama, respectively. Thus, the unadjusted values for land overstate the productivity in Connecticut through time. These results suggest that the land stock in the current productivity accounts includes urban pressures, not related to productivity. In order to remove these pressures, the land stock must be multiplied by the ratio of hedonically adjusted prices divided by the market price for farmland presented in table 5.2.

While the empirical results of this study improve the estimates of farmland values as applied to productivity measurement, we can suggest several alternatives for improving these estimates. First, the data could be extended using the 1987 and 1992 *Census of Agriculture*. Repeated observations would improve the estimates of the decomposition. Second, the number of hedonic (agronomic) factors could be reduced and potentially augmented with pedo-climatic factors. Finally, the estimated values could be used to adjust for relative von Thunen effects over time within a given state.

NOTES

1. The assumption that urbanization is a random event is used here as a convenience to derive a tractable econometric model. A more rigorous development involves the formulation of an optimal control model with a discontinuous or "Bang-Bang" control (Kamien and Schwartz, 1981, pp. 186-91). In this formulation, the farmer maximizes the present value

$$\max_{\delta_i} E\left[\sum_{t=0}^{\infty} \beta^t \left(\{1-\delta_t\} R_{AG,t} + \delta_t R_{u,t}\right)\right]$$

where d_t is the decision to convert farmland and b is the discount rate. For explanatory purposes, the Bellman formulation (Kamien and Schwartz, 1981, pp. 238-49) becomes

$$\max_{\delta_0}\left[E\left(V_{AG,0}\right), E\left(R_{U,0}\right)\right] = \max_{\delta_0}\left[E\left(R_{AG,0} + \max_{\delta_1}\left\{V_{AG,1}, R_{U,1}\right\}\right), E\left(R_{U,0}\right)\right]$$

$$= \max_{\delta_0}\left[E\left(R_{AG,0} + \max_{\delta_1}\left\{\beta R_{AG,1} + \max_{\delta_2}\left\langle V_{AG,2}, R_{U,2}\right\rangle, R_{U,1}\right\}\right), R_{U,0}\right]$$

This equation can be reformulated so that agricultural asset values follow a mixed Wiener and Poisson process similar to the portfolio process developed by Merton (Malliaris and Brock, 1982, pp. 228-30) in which the Wiener process represented the variation in returns to agriculture and the Poisson process depicted the jump process in urbanization. Specifically, the Bellman formulation can be rewritten as

$$\max_{\delta_0} E\left[(1-\delta_0)\left(R_{AG,0}+\beta V^*\left\langle R_{AG,t},R_{U,t},\delta_t^*,t\right\rangle\right)+\delta_0 R_{U,0}\right]$$

$$dR_{AG,t} = \alpha_{11}dt + \alpha_{12}dz_1 + \alpha_{13}d\xi_1$$

$$dR_{U,t} = \alpha_{21}dt + \alpha_{22}dz_2 + \alpha_{23}d\xi_2$$

where $V^*\left\langle R_{AG,t},R_{U,t},\delta_t^*,t\right\rangle$ is the value of farmland remaining in agriculture for an additional year given that the optimal future conversion (δ_t^*) occurs; dz_1 and dz_2 are normal Brownian motion deviations for returns to agriculture and returns to urbanization, respectively; and dx_1 and dx_2 are Poisson deviations for the return to agriculture and the return to urbanization, respectively. In this application, we assume that the future returns to agriculture are known with certainty, so the only variation is in the urbanization process. Finally, since conversion from agriculture to urban uses implies irreversibility, this valuation would be modified using the investment rules Dixit-Pindyck modification discussed in Purvis et al., (1995).

2. Specifically, we used the estimated model to predict land values in each county in the sample. These land values were then aggregated to the state level using the relative share of land in each county.

3. As in the computation of column a, we use the U.S. average value for all the hedonic (agronomic) factors and allow the accessibility to vary only in the von Thunen effect.

References

Ball, V. E., J.-P. Butault, and R. Nehring. 2000a. "U.S. Agriculture, 1960-96: A Multilateral Comparison of Total Factor Productivity." ERS Staff Paper, United States Department of Agriculture, AGES 00-03, October.

Ball, V. E., J.C. Bureau, J.-P. Butault, and R. Nehring. 2000b. "Levels of Farm Sector Productivity: An International Comparison." *Journal of Productivity Analysis*,15:287-311.

Enke, S. 1994. "Space and Value." *Quarterly Journal of Economics*, 56(4)(August):627-37.

Feller, W. 1950. *An Introduction to Probability Theory and Its Applications*, vol. 1, third edition. New York: John Wiley & Sons.

Kamien, M. I., and N. L. Schwartz. 1981. *Dynamic Optimization: The Calculus of Variations and Optimal Control in Economics and Management*, New York: North-Holland.

Malliaris, A. G., and W. A. Brock. 1982. *Stochastic Methods in Economics and Finance*. New York: North-Holland.

Merton, R. C. 1971. "Optimum Consumption and Portfolio Rules in a Continuous-Time Model." *Journal of Financial and Quantitative Analysis*, 3:373-413.

Miranowski, J. A., and B. D. Hammes. 1984. "Implicit Prices of Soil Characteristics for Farmland in Iowa." *American Journal of Agricultural Economics*, 745-49.

Miyao, T. 1977. "Some Dynamic and Comparative Static Properties of a Spatial Model of Production." *Review of Economic Studies*, 44(2)(June):321-27.

Purvis, A., W. G. Boggess, C. B. Moss, and J. Holt.1995. "Technology Adoption Decisions Under Irreversibility and Uncertainty: An Ex Ante Approach." *American Journal of Agricultural Economics*, 77:243-50.

U.S. Department of Agriculture. (Various years). *Agricultural Prices*, Washington, D.C.

U.S. Department of Agriculture. 1997. *Census of Agriculture*, Washington, D.C.

U.S. Department of Commerce. 1987, 1992. *Census of Agriculture*, Washington, D.C.

Part II.
Productivity, Efficiency, and the Role of R&D and Infrastructure

6

Parametric Estimation of Technical and Allocative Efficiency in U.S. Agriculture

C. J. O'Donnell[1]

ABSTRACT

Bayesian methodology is used to impose inequality constraints (including monotonicity and concavity) on a system of equations derived from a translog shadow cost frontier. The estimated parameters are used to estimate levels of technical, allocative, and cost efficiency in forty-eight states. Results are reported mainly in terms of means and standard deviations of estimated marginal probability density functions. Estimated probability distributions are also provided for some efficiency rankings. Florida is found to be the most technically and cost-efficient state, and South Dakota is found to be the most allocatively efficient. Point estimates of levels of technical, allocative, and cost efficiency average 64%, 88%, and 56% respectively across all states.

INTRODUCTION

The parametric approach to measuring technical and allocative efficiency typically involves the estimation of systems of equations comprising a cost frontier and/or a set of first-order derivative equations. Examples include Atkinson and Cornwell (1994), Ferrier and Lovell (1990), Greene (1980) and Kumbhakar (1991). Unfortunately, functions of the estimated parameters of these systems are often inconsistent with important curvature conditions implied by microeconomic theory – estimated matrices of substitution effects typically have sampling (or Bayesian posterior) distributions that are inconsistent with concavity of the underlying cost function, and estimated input demands and cost shares have distributions that are often inconsistent with monotonicity. This paper shows how Bayesian methodology can be used to estimate a cost frontier in such a way that estimated posterior distributions of functions of the unknown parameters are locally consistent with these properties.

The specific methodology used in this paper has previously been used by O'Donnell, Shumway, and Ball (1999) to impose local curvature constraints on a set of conditional input-demand frontiers, and by Griffiths, O'Donnell, and Tan Cruz (2000) to impose similar constraints on a system of equations comprising a translog cost function and associated factor-share equations. Both studies assume the firm is allocatively efficient, and both estimate equations that are linear in the parameters. In contrast, this paper relaxes the assumption of allocative efficiency and, because all allocative inefficiency effects are treated as fixed, it estimates equations that are nonlinear in the parameters. The use of Bayesian methodology to impose local curvature constraints on a nonlinear system is not new – Griffiths and Chotikapanich (1997) have used the approach to impose inequality constraints on the parameters of a nonlinear consumer expenditure system. This paper applies the methodology in a production economics context.

The empirical application involves the use of U.S. agricultural data to estimate an extended version of the translog cost frontier and associated share equations of Atkinson and Cornwell (the extensions simply allow for multiple outputs and Hicks neutral technical change). The use of Bayesian methodology makes it possible to report the results of the empirical analysis in quite powerful and novel ways. For example, I estimate there is a 0.89 probability that South Dakota over-utilizes materials inputs relative to labor, and the probability is 0.17 that Washington is the third most technically efficient of the forty-eight contiguous states. Bayesians routinely make finite sample inferences of this type – a feature of the Bayesian approach is that it summarizes information about unknown functions of parameters in terms of a density function that describes how likely (in a probabilistic sense) different values of the functions are. This capacity to obtain and report probability distributions for efficiency measures and/or rankings is one feature that distinguishes the parametric Bayesian approach from other parametric and non-parametric approaches to the estimation of efficiency.

The structure of the paper is as follows. Section 2 presents a multiple-output version of the translog cost and share equations of Atkinson and Cornwell. This section also describes and motivates a number of constraints that must be satisfied by the parameters of the model, including homogeneity, symmetry, monotonicity, and concavity. Section 3 places the economic model in a stochastic framework and describes the way in which Bayesian methodology can be used to ensure that the parametric constraints are satisfied. Practical implementation of the methodology requires the use of Markov Chain Monte Carlo (MCMC) numerical integration techniques, and a particular MCMC algorithm is described. Section 4 describes the data – the model is estimated using annual state-level data on two outputs and four inputs covering the period 1960 to 1996. Section 5 presents the results of the empirical analysis. The focus in this section is on characteristics of estimated probability density functions of input demand elasticities and measures of technical, allocative, and cost efficiency. Probability distributions for efficiency rankings are also reported for selected states. The paper is concluded in Section 6.

Economic Model

I begin by assuming the firm employs an $N \times 1$ non-negative real-valued input vector $\mathbf{m} = (m_1, \ldots, m_N)'$ to produce an $M \times 1$ non-negative real-valued output vector $\mathbf{q} = (q_1, \ldots, q_M)'$. The input price vector is denoted $\mathbf{w} = (w_1, .., w_N)'$ and the production possibilities set $T(\mathbf{m}, \mathbf{q})$ is assumed to satisfy the standard properties listed by Chambers (1988, p.252). The input-output combination $(\bar{\mathbf{m}}, \bar{\mathbf{q}})$ is defined as technically efficient if and only if there is no $(\mathbf{m}, \mathbf{q}) \in T$ such that $\mathbf{m} \leq \bar{\mathbf{m}}$ and $\mathbf{q} > \bar{\mathbf{q}}$, or $\mathbf{m} < \bar{\mathbf{m}}$ and $\mathbf{q} \geq \bar{\mathbf{q}}$ (i.e., if there is no way to produce more output with the same inputs or to produce the same output with fewer inputs). A cost-minimizing firm is defined as allocatively efficient if and only if it operates on its least cost expansion path (i.e., if it produces output $\mathbf{q}$ using input combinations that minimize the cost of producing $\mathbf{q}$). In the remainder of this section I describe a set of estimable cost and cost-share equations that are founded on these assumptions and definitions.

Shadow and Observed Costs and Cost-Shares

A system of shadow cost and cost-share equations that allows for both technical and allocative inefficiency is (Atkinson and Cornwell, 1994, p.233; O'Donnell, 1996):

$$c_{it}^* = c^*(\mathbf{w}_{it}^*, \mathbf{q}_{it})(1/b_i) \tag{1}$$

and

$$s_{nit}^* = s_n^*(\mathbf{w}_{it}^*, \mathbf{q}_{it}) \tag{2}$$

where c_{it}^* represents the (unobserved) shadow cost of production of firm i ($i = 1, \ldots, I$) in time period t ($t = 1, \ldots, T$), s_{nit}^* is the n-th shadow cost-share ($n = 1, \ldots, N$), $0 < b_i \leq 1$ is a parameter reflecting the extent to which actual input usage differs from minimal (i.e., technically efficient) input usage, and $\mathbf{w}_{it}^* = (w_{1it}^*, w_{2it}^*, \ldots, w_{Nit}^*)'$ is a vector of shadow input prices. The shadow prices are defined as $w_{nit}^* \equiv a_{ni} w_{nit}$ where a_{ni} is a parameter representing optimization (i.e., allocation) errors. Earlier assumptions on the production possibilities set T are enough to ensure that the shadow cost function possesses a standard set of properties (Chambers, pp. 51-52): it is non-negative for all $\mathbf{w}^*_{it}$, $\mathbf{q}_{it} > \mathbf{0}$; non-decreasing in $\mathbf{w}^*_{it}$ and $\mathbf{q}_{it}$ (monotonicity); concave and continuous in $\mathbf{w}^*_{it}$; and positively linearly homogeneous in $\mathbf{w}^*_{it}$. Finally, the i-th firm is technically efficient if and only if $b_i = 1$, and is allocatively efficient if and only if $a_{ni} = a_{ji}$ for all n and j.

This treatment of technical and allocative efficiency is consistent with the models of Kumbhakar (1988, 1992), Atkinson and Cornwell (1994), Toda (1976), Atkinson and Halvorsen (1984, 1990) and Lovell and Sickles (1983). In much of this literature it is argued that technical inefficiency is caused by unobserved variations in the quality of inputs (especially management) leading to variations in flows of factor services, while allocative inefficiency occurs when firms minimize costs subject to constraints over and above the usual technological constraints (these additional constraints may reflect government regulations, or dynamic considerations in the choice of some inputs). In this paper, all inefficiency effects are treated as fixed parameters (rather than random variables), partly to ensure the cost function is consistent with expected utility maximizing behaviour, and partly to avoid arbitrary and restrictive assumptions about the distributions of error terms and functional forms. For more details on cost minimization in the presence of random efficiency effects, see Pope and Chavas (1994) and O'Donnell (1996). For more details on the statistical advantages and disadvantages of fixed versus random effects models, see Bauer (1990) and Atkinson and Cornwell.

The relationships between shadow and observed cost-shares and log-costs are (Atkinson and Cornwell, p.235; O'Donnell, (1996) p.11):

$$s_{nit} = \frac{s^*_{nit} a^{-1}_{ni}}{\sum_{n=1}^{N} s^*_{nit} a^{-1}_{ni}} \tag{3}$$

and

$$\ln(c_{it}) = \ln(c^*_{it}) + \ln\left(\sum_{n=1}^{N} s^*_{nit} a^{-1}_{ni}\right) \tag{4}$$

where c_{it} is observed cost and s_{nit} is the n-th observed cost-share. It is apparent from equations (1) to (4) that observed costs and cost-shares can be regarded as standard cost and cost-share equations evaluated at shadow prices and multiplied or divided by a share-weighted average of inverse allocative inefficiencies. Moreover, shadow costs and shares are identical to observed costs and shares when the firm is allocatively efficient (i.e., if $a_{ni} = a_{ji}$ for all n and j).

Functional Form

I follow Atkinson and Cornwell and assume a translog functional form for the shadow cost function. Under this assumption

$$s^*_{nit} = \beta_n + \sum_{k=1}^{N} \beta_{nk} \ln(a_{ki} w_{kit}) + \sum_{m=1}^{M} \phi_{nm} \ln(q_{mit}) \tag{5}$$

and

$$\ln(c^*_{it}) = [\alpha - \ln(b_i)] + [\delta_1 t + \delta_2 t^2] + \sum_{n=1}^{N} \beta_n \ln(a_{ni} w_{nit}) + \sum_{m=1}^{M} \gamma_m \ln(q_{mit})$$

$$+ 0.5 \sum_{n=1}^{N} \sum_{k=1}^{N} \beta_{nk} \ln(a_{ni} w_{nit}) \ln(a_{ki} w_{kit}) + 0.5 \sum_{m=1}^{M} \sum_{j=1}^{M} \gamma_{mj} \ln(q_{mit}) \ln(q_{jit})$$

$$+ \sum_{n=1}^{N} \sum_{m=1}^{M} \phi_{nm} \ln(a_{ni} w_{nit}) \ln(q_{mit}) \tag{6}$$

where α, δ_1, δ_2 and the β_n, γ_m, β_{nk}, γ_{mj} and ϕ_{nm} are parameters to be estimated. Note that the first term in square brackets on the right-hand-side of (6) incorporates the technical inefficiency parameters, b_i, that vary across firms but not time periods. The second term in square brackets is included to allow for Hicks neutral technical change.

Parametric constraints

Economic theory (and just plain common sense) dictates that the parameters of equations (5) and (6) satisfy a number of simple constraints. In particular, technical regress is ruled out by the parametric inequality constraint

$$\delta_1 + 2\delta_2 t \leq 0, \tag{7}$$

while linear homogeneity of the cost function can be maintained by imposing the constraints

$$\sum_{n=1}^{N} \beta_n = 1, \qquad \sum_{n=1}^{N} \beta_{nk} = 0 \; \forall \, k, \qquad \text{and} \qquad \sum_{n=1}^{N} \phi_{nm} = 0 \; \forall \, m. \tag{8}$$

Constraints that ensure that the estimated cost function exhibits constant returns to scale (CRS) are

$$\sum_{m=1}^{M} \gamma_m = 1, \qquad \sum_{m=1}^{M} \gamma_{mj} = 0 \;\forall\, j, \qquad \text{and} \qquad \sum_{m=1}^{M} \phi_{nm} = 0 \;\forall\, n. \tag{9}$$

The CRS assumption has been maintained in empirical studies of agricultural technologies by Ball (1988), Shumway and Lim (1993), and O'Donnell and Woodland (1995), among others. It is a useful assumption in the present context because it is sufficient to ensure that measures of input-technical inefficiency (i.e., over-utilization of inputs given output and the input mix) and output-technical inefficiency (i.e., failure to produce maximum output given inputs) are equivalent. The assumption is maintained for this reason, and because it is consistent with the assumption of scale efficiency and the observation that different states produce different levels of output.

The constraint given by equation (7) is an inequality constraint involving both the parameters and the data. Two more of these types of constraints arise if the shadow cost function is to satisfy the curvature properties implied by economic theory. Specifically, monotonicity of the shadow cost function will be satisfied if the shadow cost shares given by (5) are non-negative, and concavity of the cost function in shadow prices will be satisfied if the matrix of shadow substitution effects is negative semi-definite. The elements of the shadow substitution matrix $\boldsymbol{S}_{it} = [\zeta_{nkit}]$ are

$$\zeta_{nkit} = \begin{cases} (\beta_{nm} + s^{*2}_{nit} - s^{*}_{nit}) \times \dfrac{c^{*}_{it}}{w^{*}_{nit}} & \text{for } n = k, \\ (\beta_{nk} + s^{*}_{nit} s^{*}_{kit}) \times \dfrac{c^{*}_{it}}{w^{*}_{nit} w^{*}_{kit}} & \text{otherwise.} \end{cases} \tag{10}$$

Associated with the matrix of substitution effects is the matrix of elasticities of shadow input demands with respect to shadow prices. This elasticity matrix has elements

$$\eta_{nkit} = \begin{cases} \dfrac{\beta_{nm}}{s^{*}_{nit}} + s^{*}_{nit} - 1 & \text{for } n = k, \\ \dfrac{\beta_{nk}}{s^{*}_{nit}} + s^{*}_{kit} & \text{otherwise.} \end{cases} \tag{11}$$

Finally, the parameters of the economic model must satisfy constraints that ensure that technical and allocative efficiency effects are identified. Since the cost function is homogeneous of degree one in $\mathbf{w}^{*}_{it}$, it is also homogeneous of degree one in $\mathbf{a}_i = (a_{1i}, \ldots, a_{Ni})'$. Accordingly, I follow Atkinson and Cornwell

(p. 237) and arbitrarily normalize on the N-th input, setting $a_{Ni} = 1$ for all i. I also normalize b_i to one for the most efficient firm (state). More will be said about b_i and the measurement of technical efficiency in Section 5 below.

The Econometric Model and Estimation Method

The first part of this section develops the econometric model, which turns out to be a member of the class of non-linear seemingly unrelated regression models. The rest of the section is concerned with Bayesian methodology for estimating the parameters of the model subject to the equality and inequality constraints described above.

Econometric Model

A first step in building the econometric model is to substitute the shadow functions (5) and (6) into the share equations (3) and the log-cost function (4) to yield a set of N estimating equations of the form:

$$s_{nit} = f_n(\mathbf{x}_{it}, \boldsymbol{\beta}, \mathbf{a}_i) + \varepsilon_{nit} \qquad n = 1, \ldots, N-1, \tag{12}$$

and

$$\ln(c_{it}) = f(\mathbf{x}_{it}, \mu_i, \boldsymbol{\beta}, \mathbf{a}_i) + \varepsilon_{it} \tag{13}$$

where the ε_{nit} and ε_{it} are disturbance terms with zero mean vector and constant covariance matrix Σ; $\mathbf{x}_{it}$ is a data vector containing observed quantities, prices, and a time trend; $\mu_i \equiv [\alpha - \ln(b_i)]$ is a scalar; $\mathbf{a}_i$ is the $N \times 1$ vector of allocative inefficiency parameters introduced above; and $\boldsymbol{\beta}$ is a $K \times 1$ vector containing all other unknown parameters in equations (5) and (6).

A second step is to substitute the equality constraints (8) and (9) into equations (12) and (13). This can be done by writing (8) and (9) in the compact form

$$\boldsymbol{R}\boldsymbol{\beta} = \mathbf{r} \tag{14}$$

where $\boldsymbol{R}$ is $J \times K$ and $\mathbf{r}$ is $J \times 1$. By partitioning $\boldsymbol{R} \equiv [\boldsymbol{R}_1 \;\; \boldsymbol{R}_2]$, where $\boldsymbol{R}_1$ is $J \times J$ and $\boldsymbol{R}_2$ is $J \times (K - J)$, and by also partitioning $\boldsymbol{\beta}$ conformably as $\boldsymbol{\beta} \equiv [\boldsymbol{\lambda}' \;\; \boldsymbol{\gamma}']'$, the equality constraints (14) can be substituted into (12) and (13) by simply replacing $\boldsymbol{\lambda}$ with

$$\boldsymbol{\lambda} = \boldsymbol{R}_1^{-1} \ (\mathbf{r} - \boldsymbol{R}_2\boldsymbol{\gamma}). \tag{15}$$

Finally, it is convenient to stack these equality-constrained versions of (12) and (13), first by firm, then by time period, then by equation, to obtain

$$\mathbf{y} = h(\boldsymbol{X}, \boldsymbol{\theta}) + \boldsymbol{\varepsilon} \tag{16}$$

where $\mathbf{y} = (s_{111}, s_{121}, \ldots, s_{1M1}, s_{112}, s_{122}, \ldots, s_{1IT}, s_{211}, \ldots, \ln(c_{IT}))'$, $\boldsymbol{X}$ and $\boldsymbol{\varepsilon}$ are similarly defined, and $\boldsymbol{\theta} = (\mu_1, \ldots, \mu_I, \boldsymbol{\gamma}', \mathbf{a}_1', \ldots, \mathbf{a}_I')'$.

Bayesian Estimation

Estimation of the parameter vector $\boldsymbol{\theta}$ is conducted within a Bayesian framework. The Bayesian approach overcomes a well-known problem with the maximum likelihood (ML) approach, namely that inequality (e.g., concavity) information concerning the parameters cannot be adequately utilized. For a nice introduction to Bayesian econometrics see Griffiths, Hill, and Judge (1993, Chapters 24 and 25). Also see Griffiths (1988), who gives a nice exposition of the way in which Bayesian econometrics can be used to impose simple inequality constraints. For further discussion of the relative advantages of the Bayesian and sampling theory approaches to inference, see Poirier (1995).

Bayesian estimation essentially involves four steps: i) identification of the likelihood function that summarizes information (about the parameters) contained in the sample; ii) specification of a prior probability density function (pdf) that summarizes all non-sample information; iii) derivation of the joint posterior pdf for the parameters through the application of Bayes's Theorem; and iv) isolation of various characteristics of marginal posterior pdfs (e.g., means and standard deviations) through the evaluation of certain integrals (in practice, many of these integrals are analytically intractable and must be evaluated numerically).

Assuming $\boldsymbol{\varepsilon}$ is multivariate normal, the likelihood function is (Judge et al., 1985, p. 478):

$$L(\mathbf{y} \mid \boldsymbol{\theta}, \boldsymbol{\Sigma}) \propto |\boldsymbol{\Sigma}|^{-IT/2} \exp[-{}_{1/2}(\mathbf{y} - h(\boldsymbol{X}, \boldsymbol{\theta}))'(\boldsymbol{\Sigma}^{-1} \otimes \boldsymbol{I}_{IT})(\mathbf{y} - h(\boldsymbol{X}, \boldsymbol{\theta}))]$$

$$= |\boldsymbol{\Sigma}|^{-IT/2} \exp[-{}_{1/2}\, tr(\boldsymbol{G}\boldsymbol{\Sigma}^{-1})] \tag{17}$$

where $\boldsymbol{G}$ is an $N \times N$ matrix with (n, k)-th element $g_{nk} = (\mathbf{y}_n - h_n(\boldsymbol{X}, \boldsymbol{\theta}))'(\mathbf{y}_k - h_k(\boldsymbol{X}, \boldsymbol{\theta}))$, where $\mathbf{y}_n$ and $h_n(\boldsymbol{X}, \boldsymbol{\theta})$ are obvious sub-vectors of $\mathbf{y}$ and $h(\boldsymbol{X}, \boldsymbol{\theta})$.

In this paper $\boldsymbol{\Sigma}$ and $\boldsymbol{\theta}$ are assumed to be a priori independent. A conventional non-informative prior is used for $\boldsymbol{\Sigma}$, namely $p(\boldsymbol{\Sigma}) \propto |\boldsymbol{\Sigma}|^{-(N+1)/2}$ (see Judge et al., p. 478). A uniform inequality-restricted prior is used for $\boldsymbol{\theta}$, meaning all feasible values of $\boldsymbol{\theta}$ are regarded as equally likely. In this paper, feasible values of $\boldsymbol{\theta}$ are values where technical progress, monotonicity, and concavity constraints hold when evaluated at state average prices and quantities (i.e., at 48 data points). Thus, the joint prior for $\boldsymbol{\theta}$ and $\boldsymbol{\Sigma}$ is

$$p(\boldsymbol{\theta}, \boldsymbol{\Sigma}) \propto |\boldsymbol{\Sigma}|^{-(N+1)/2} I(\boldsymbol{\theta}) \tag{18}$$

where $I(\boldsymbol{\theta})$ is an indicator function that takes the value one if $\boldsymbol{\theta}$ is feasible and zero otherwise.

Bayes's Theorem is used to combine the prior (18) with the likelihood function (17) to yield the joint posterior pdf

$$p(\boldsymbol{\theta}, \boldsymbol{\Sigma} \mid \mathbf{y}) \propto L(\mathbf{y} \mid \boldsymbol{\theta}, \boldsymbol{\Sigma})\, p(\boldsymbol{\theta}, \boldsymbol{\Sigma})$$

$$= |\boldsymbol{\Sigma}|^{-(N+IT+1)/2} \exp[- ½\, tr(\boldsymbol{G\Sigma}^{-1})]\, I(\boldsymbol{\theta}). \tag{19}$$

Because our main interest lies in $\boldsymbol{\theta}$, the covariance matrix $\mathbf{S}$ is regarded as a matrix of "nuisance" parameters and is integrated out of (18) using the properties of the inverted Wishart distribution. This procedure yields the marginal posterior pdf for $\boldsymbol{\theta}$ (see Griffiths and Chotikapanich, p. 333):

$$p(\boldsymbol{\theta} \mid \mathbf{y}) \propto \int p(\boldsymbol{\theta}, \boldsymbol{\Sigma} \mid \mathbf{y})\, d\boldsymbol{\Sigma}$$

$$\propto |\mathbf{G}|^{-IT/2} I(\boldsymbol{\theta}). \tag{20}$$

This posterior pdf is the source of all inferences to be made about $\boldsymbol{\theta}$ (and functions of $\boldsymbol{\theta}$). Drawing these inferences involves the evaluation of certain integrals. For example, to obtain a point estimate of $\boldsymbol{\theta}$, we would typically evaluate the integral

$$E\{\boldsymbol{\theta} \mid \mathbf{y}\} = \int \boldsymbol{\theta}\, p(\boldsymbol{\theta} \mid \mathbf{y})\, d\boldsymbol{\theta}. \tag{21}$$

Markov Chain Monte Carlo methods

Historically, the evaluation of integrals of the type given by (21) has been analytically intractable in all but the simplest of cases. Where analytical integration is impossible, Markov Chain Monte Carlo (MCMC) simulation methods can be used to evaluate integrals numerically. A useful introduction to MCMC methods is provided by Casella and George (1992) and Chib and Greenberg (1996). MCMC methods provide a way of producing a sample of observations $\boldsymbol{\theta}^1, \boldsymbol{\theta}^2, .., \boldsymbol{\theta}^{s+M}$ with the property that, for large s (the "burn-in"), the sequence $\boldsymbol{\theta}^{s+1}, \boldsymbol{\theta}^{s+2}, .., \boldsymbol{\theta}^{s+M}$ can be regarded as a sample of observations from the joint pdf $p(\boldsymbol{\theta} \mid \mathbf{y})$. Thus, any integral of the form

$$\int g(\boldsymbol{\theta})\, p(\boldsymbol{\theta} \mid \mathbf{y})\, d\boldsymbol{\theta}. \tag{22}$$

can be simply estimated as

$$\frac{1}{M} \sum_{m=s+1}^{M} g(\boldsymbol{\theta}^m) \tag{23}$$

This paper, following Griffiths and Chotikapanich, uses a random-walk Metropolis-Hastings algorithm to draw observations on $\boldsymbol{\theta}$. This algorithm is considerably more efficient than some alternative MCMC algorithms (e.g., the Gibbs Sampler) when the feasible region is inequality-constrained. The algorithm involves the following steps:

1. Select $\boldsymbol{\theta}^0$, a feasible initial value for $\boldsymbol{\theta}$. Set $m = 0$.

2. Generate $\mathbf{d}$ from $N(\mathbf{0}, V)$, where V is an arbitrarily chosen covariance matrix.

3. Compute a candidate for the next value in the sequence, $\boldsymbol{\theta}^* = \boldsymbol{\theta}^m + \mathbf{d}$. If $\boldsymbol{\theta}^*$ is infeasible, set $\boldsymbol{\theta}^{m+1} = \boldsymbol{\theta}^m$ and go to step 6; otherwise proceed to step 4.

4. Compute $r = \exp[\ln(p(\boldsymbol{\theta}^* \mid \mathbf{y})) - \ln(p(\boldsymbol{\theta}^m \mid \mathbf{y}))]$. If $r \geq 1$, set $\boldsymbol{\theta}^{m+1} = \boldsymbol{\theta}^*$ and go to step 6; otherwise proceed to step 5.

5. Generate a uniform random variable, u, from the unit interval. If $u < r$, set $\boldsymbol{\theta}^{m+1} = \boldsymbol{\theta}^*$; otherwise set $\boldsymbol{\theta}^{m+1} = \boldsymbol{\theta}^m$.

6. Set $m = m+1$ and return to step 2.

This algorithm provides a means of moving around the parameter space in a manner consistent with $p(\boldsymbol{\theta} \mid \mathbf{y})$. In the empirical work the elements of $\boldsymbol{\theta}^0$ were set to their maximum likelihood estimates, except the β_{nk} elements that were set to zero, and the elements of the vectors $\mathbf{a}_i$ that were set to one. In step 2, V was chosen to be a scalar, k, times a partitioned matrix: the partition corresponding to $(\mu_1, ..., \mu_I, \boldsymbol{\gamma}')'$ was the corresponding partition of the ML covariance matrix; the partition corresponding to $(\mathbf{a}_1', ..., \mathbf{a}_I')'$ was 0.01 times an identity matrix; and the remaining partitions were zero matrices. The scalar, k, was set experimentally, to a value that saw $\boldsymbol{\theta}^*$ ultimately chosen as the next value in the sequence approximately half the time. Using this algorithm, a new observation is accepted into the sequence if it is more probable than the previous one; if it is less probable, then it is accepted into the sequence with a probability equal to the ratio of the two densities. Thus, a relatively large proportion of observations in the MCMC sequence are drawn from regions of the parameter space where the posterior density is relatively high, and a relatively low proportion are drawn from regions with low probability. No observations are drawn from regions with zero probability – in this paper, no observations are drawn from regions of the parameter space where monotonicity and concavity constraints are not satisfied at all 48 sets of average prices and quantities.

In this paper, because the chosen $\boldsymbol{\theta}^0$ was considered likely to be in a region of the parameter space with low probability, the burn-in period, s, was set at 35,000. A further $M = 50{,}000$ observations were then drawn for purposes of inference. Inferences based on MCMC samples can, of course, be made to any degree of accuracy by simply drawing as many observations as required.

DATA

The data comprises $IT = 1776$ annual observations on U.S. agricultural input and output prices and/or quantities, covering the $I = 48$ contiguous states over the $T = 37$ years from 1960 to 1996 inclusive. The data set is the same as that described in Ball et al. (1999) in all but two respects: it has been extended from 1990 to 1996; and the EKS technique due to Eltetö and Köves (1964) and Szulc (1964) has been used to convert standard Tornquist indices into multilateral indices that are transitive. Each observation in the data set comprises EKS price and quantity index numbers for two outputs (livestock and crops) and four inputs (capital, land, labor, and materials). For more details concerning the components of inputs and outputs and the construction of the standard Tornquist indices, see Ball et al. To aid in the interpretation of the empirical results reported in the next section, table 6.1 presents averages over time of price and quantity indices for each state, while table 6.2 presents corresponding averages of output-input quantity ratios.

RESULTS

This section mainly reports means and standard deviations of samples of observations on various functions of $\boldsymbol{\theta}$. A sample of observations on a function of $\boldsymbol{\theta}$ (e.g., an elasticity) is obtained by simply using each MCMC observation on $\boldsymbol{\theta}$ to evaluate the function (e.g., equation (11)). Each sample mean is a con-

Table 6.1. Average Prices and Quantities in Selected States

		Price of				Quantity of	
i	State	Materials w_1	Capital w_2	Land w_3	Labor w_4	Livestock q_1	Crops q_2
1	Alabama	0.775	0.596	0.542	0.660	0.848	0.959
2	Arizona	0.793	0.605	0.763	0.861	0.516	1.146
3	Arkansas	0.697	0.602	0.780	0.732	1.025	2.061
4	California	0.860	0.616	0.252	0.857	2.928	12.836
5	Colorado	0.737	0.614	0.722	0.772	1.418	1.439
6	Connecticut	1.018	0.605	1.893	0.856	0.110	0.177
7	Delaware	0.973	0.611	0.776	0.761	0.168	0.158
8	Florida	0.813	0.602	0.644	0.828	0.580	4.496
9	Georgia	0.811	0.596	0.614	0.713	1.067	2.340
10	Idaho	0.681	0.615	0.728	0.767	0.596	1.908
11	Illinois	0.766	0.609	1.338	0.925	1.774	6.769
12	Indiana	0.767	0.601	0.854	0.856	1.235	3.676
13	Iowa	0.711	0.609	1.565	0.981	4.065	7.136
14	Kansas	0.696	0.614	0.656	0.795	2.328	3.644
15	Kentucky	0.835	0.605	0.766	0.796	0.899	2.077
16	Louisiana	0.618	0.606	0.775	0.730	0.347	1.375
17	Maine	0.859	0.614	0.802	0.710	0.164	0.337
18	Maryland	0.919	0.611	0.991	0.796	0.331	0.496
19	Massachusetts	0.923	0.609	1.282	0.841	0.091	0.250
20	Michigan	0.741	0.613	0.555	0.918	0.840	2.423
21	Minnesota	0.778	0.611	0.967	0.901	2.356	4.441
22	Mississippi	0.615	0.599	0.693	0.657	0.674	1.677
23	Missouri	0.769	0.606	0.751	0.746	1.516	2.760
24	Montana	0.654	0.616	0.701	0.988	0.522	1.129
25	North Carolina	0.904	0.601	0.576	0.775	1.075	3.005
26	North Dakota	0.632	0.622	0.705	0.834	0.473	2.879
27	Nebraska	0.704	0.624	0.453	0.890	2.692	4.235
28	Nevada	0.872	0.602	0.472	0.819	0.105	0.148
29	New Hampshire	0.977	0.609	0.890	0.758	0.050	0.071
30	New Jersey	0.945	0.615	1.407	0.586	0.125	0.506
31	New Mexico	0.781	0.602	0.552	0.851	0.532	0.448
32	New York	0.831	0.618	0.648	0.721	1.176	1.509
33	Ohio	0.867	0.607	0.947	0.843	1.117	3.013
34	Oklahoma	0.725	0.604	0.440	0.792	1.229	1.664
35	Oregon	0.825	0.604	0.667	0.796	0.422	1.707
36	Pennsylvania	0.927	0.616	0.899	0.883	1.259	1.682
37	Rhode Island	1.008	0.605	1.207	0.882	0.010	0.033
38	South Carolina	0.778	0.600	0.529	0.608	0.272	1.073
39	South Dakota	0.628	0.615	0.805	0.884	1.256	2.063
40	Tennessee	0.707	0.601	0.567	0.718	0.657	1.356
41	Texas	0.748	0.604	0.542	0.687	3.576	5.292
42	Utah	0.788	0.603	0.934	0.840	0.298	0.328
43	Vermont	0.905	0.621	1.043	0.583	0.216	0.127
44	Virginia	0.883	0.599	0.657	0.686	0.621	1.010
45	Washington	0.692	0.621	0.942	0.800	0.617	2.859
46	West Virginia	0.725	0.596	0.288	0.592	0.136	0.178
47	Wisconsin	0.754	0.620	0.923	0.728	2.536	2.567
48	Wyoming	0.668	0.606	0.375	0.646	0.337	0.320

sistent estimate of the corresponding posterior mean (the posterior mean is an optimal Bayesian point estimate under weighted-squared-error loss). In this paper, interest centers on the parameters; elasticities of shadow input-demands with respect to shadow prices; and various measures of technical, allocative, and cost efficiency (e.g., levels and rankings). Graphs of some estimated marginal posterior distributions are provided to give a visual appreciation of likely and unlikely values of some of these quantities.

Parameters

Table 6.3 reports means and standard deviations of the MCMC observations on elements of $\gamma = (\delta_1, \delta_2, \beta_2, \beta_3, \beta_4, \gamma_2, \beta_{22}, \beta_{23}, \beta_{24}, \beta_{33}, \beta_{34}, \beta_{44}, \gamma_{22}, \phi_{22}, \phi_{32}, \phi_{42})'$. The sample means of observations on the elements of $\boldsymbol{\lambda}$ can be obtained from the constraints (8) and (9), or (15). The remaining parameters in the model are efficiency effects, and these will be discussed further below.

The numbers reported in table 6.3 are unremarkable in all but three respects. First, the standard deviations are surprisingly small, reflecting the use of a large amount of equality and inequality information in the estimation process. Second, the point estimates of β_2, β_3, and β_4 suggest that at unit shadow prices and quantities (values consistent with those reported in table 6.1) the cost-shares of materials, capital, land, and labor are 0.20, 0.27, 0.21, and 0.31 respectively. These estimated shares are not dissimilar to the average observed shares over all states and time periods (0.20, 0.20, 0.22, and 0.39). Finally, estimates of δ_1 and δ_2 can be used to obtain an estimate of the average annual proportionate change in costs due to technical change:

$$TC_t = \frac{\partial \ln(c_{it})}{\partial t} = \delta_1 + 2\delta_2 t. \tag{24}$$

The point estimate of δ_1 reported in table 6.3 suggests that at the start of the sample period technical change was reducing costs by approximately 1.27% per annum. The estimate of δ_2 (0.00017 when rounded to five decimal places) suggests that this effect accelerated over the sample period at a rate of approximately 0.034% per annum, so that by 1996 technical change was reducing costs at an annual average rate of 2.5%. Figure 6.1 presents the estimated marginal pdfs for the technical change effects in 1960 and 1996. Note that neither pdf appears to be truncated, implying that the technical regress constraint (7) is not binding. Moreover, there is negligible probability that technical change reduced costs by more than 3% or less than 1% per annum at any time during the sample period.

Table 6.2. Average Output:Input Quantity Ratios in Selected States

i	State	Ratio of Livestock to			
		Materials	Capital	Land	Labor
		q_1:m_1	q_1:m_2	q_1:m_3	q_1:m_4
1	Alabama	0.966	0.905	0.740	0.661
2	Arizona	0.880	1.026	0.259	0.727
3	Arkansas	0.736	0.784	0.802	0.605
4	California	0.819	0.995	0.195	0.412
5	Colorado	0.939	1.214	1.043	1.004
6	Connecticut	1.247	0.532	1.719	0.326
7	Delaware	1.072	1.440	1.458	1.088
8	Florida	0.634	0.569	0.226	0.277
9	Georgia	0.882	0.882	0.813	0.666
10	Idaho	0.693	0.575	0.563	0.438
11	Illinois	0.654	0.365	0.590	0.451
12	Indiana	0.727	0.430	0.578	0.435
13	Iowa	0.884	0.721	1.660	0.724
14	Kansas	0.849	0.826	0.822	0.800
15	Kentucky	1.132	0.528	0.676	0.386
16	Louisiana	0.533	0.408	0.353	0.302
17	Maine	0.936	0.580	1.474	0.356
18	Maryland	0.852	0.622	0.837	0.475
19	Massachusetts	0.912	0.392	0.829	0.198
20	Michigan	0.759	0.413	0.567	0.304
21	Minnesota	0.923	0.590	1.161	0.464
22	Mississippi	0.632	0.612	0.563	0.469
23	Missouri	0.885	0.558	0.650	0.400
24	Montana	0.759	0.470	0.352	0.414
25	North Carolina	0.922	0.616	0.705	0.462
26	North Dakota	0.476	0.248	0.281	0.275
27	Nebraska	0.855	0.911	0.664	0.843
28	Nevada	1.183	0.783	0.314	0.775
29	New Hampshire	1.174	0.518	0.752	0.308
30	New Jersey	0.776	0.341	0.451	0.157
31	New Mexico	1.162	1.288	0.356	0.831
32	New York	1.160	0.647	1.687	0.419
33	Ohio	0.832	0.390	0.641	0.307
34	Oklahoma	0.967	0.813	0.379	0.586
35	Oregon	0.751	0.515	0.352	0.261
36	Pennsylvania	1.163	0.637	1.108	0.397
37	Rhode Island	0.472	0.348	0.537	0.202
38	South Carolina	0.671	0.419	0.430	0.321
39	South Dakota	0.904	0.738	0.912	0.653
40	Tennessee	0.862	0.477	0.399	0.324
41	Texas	0.930	0.835	0.352	0.585
42	Utah	1.170	0.813	0.473	0.625
43	Vermont	1.472	0.881	2.010	0.540
44	Virginia	1.053	0.543	0.511	0.398
45	Washington	0.617	0.583	0.590	0.313
46	West Virginia	0.829	0.419	0.197	0.204
47	Wisconsin	1.268	0.804	2.257	0.473
48	Wyoming	0.978	0.855	0.277	0.627

State	Ratio of Crops to			
	Materials	Capital	Land	Labor
	q_2:m_1	q_2:m_2	q_2:m_3	q_2:m_4
Alabama	1.117	1.021	0.817	0.723
Arizona	1.978	2.304	0.578	1.600
Arkansas	1.556	1.609	1.596	1.175
California	3.526	4.303	0.863	1.825
Colorado	0.949	1.227	1.069	1.028
Connecticut	2.005	0.860	2.774	0.526
Delaware	1.048	1.367	1.357	1.000
Florida	4.832	4.372	1.782	2.162
Georgia	1.955	1.931	1.760	1.433
Idaho	2.194	1.815	1.799	1.408
Illinois	2.488	1.355	2.263	1.885
Indiana	2.130	1.255	1.730	1.374
Iowa	1.561	1.235	2.911	1.341
Kansas	1.317	1.286	1.286	1.254
Kentucky	2.608	1.211	1.563	0.894
Louisiana	2.050	1.593	1.431	1.271
Maine	1.930	1.206	2.912	0.714
Maryland	1.275	0.930	1.254	0.711
Massachusetts	2.649	1.110	2.379	0.620
Michigan	2.131	1.192	1.670	0.929
Minnesota	1.692	1.091	2.195	0.903
Mississippi	1.591	1.507	1.394	1.159
Missouri	1.610	1.003	1.194	0.768
Montana	1.628	1.013	0.766	0.908
North Carolina	2.776	1.722	1.807	1.101
North Dakota	2.793	1.501	1.730	1.729
Nebraska	1.319	1.423	1.043	1.330
Nevada	1.588	1.068	0.443	1.109
New Hampshire	1.769	0.756	1.148	0.483
New Jersey	3.355	1.399	1.892	0.673
New Mexico	1.006	1.090	0.299	0.688
New York	1.482	0.825	2.155	0.533
Ohio	2.216	1.033	1.748	0.876
Oklahoma	1.333	1.100	0.512	0.788
Oregon	2.939	2.049	1.448	1.068
Pennsylvania	1.556	0.850	1.480	0.531
Rhode Island	1.631	1.175	1.680	0.733
South Carolina	2.670	1.634	1.597	1.129
South Dakota	1.462	1.230	1.510	1.121
Tennessee	1.778	0.982	0.834	0.688
Texas	1.407	1.241	0.517	0.853
Utah	1.280	0.886	0.522	0.695
Vermont	0.861	0.518	1.140	0.305
Virginia	1.742	0.892	0.813	0.606
Washington	2.850	2.685	2.728	1.454
West Virginia	1.089	0.545	0.256	0.266
Wisconsin	1.261	0.806	2.302	0.486
Wyoming	0.920	0.808	0.263	0.596

Table 6.3. Selected Parameters

Parameter	Point Estimate	Standard Deviation
β_2	0.2745	0.0022
β_3	0.2123	0.0038
β_4	0.3139	0.0055
γ_2	0.4731	0.0142
β_{22}	0.1153	0.0035
β_{23}	0.0003	0.0010
β_{24}	-0.0582	0.0014
β_{33}	0.0620	0.0010
β_{34}	-0.0532	0.0014
β_{44}	0.0953	0.0031
γ_{22}	0.1337	0.0169
ϕ_{22}	0.0090	0.0023
ϕ_{32}	-0.0002	0.0013
ϕ_{42}	0.0051	0.0022
δ_1	-0.0127	0.0011
δ_2	-0.0002	0.0000[a]

[a] 0.00002 when rounded to 5 decimal places

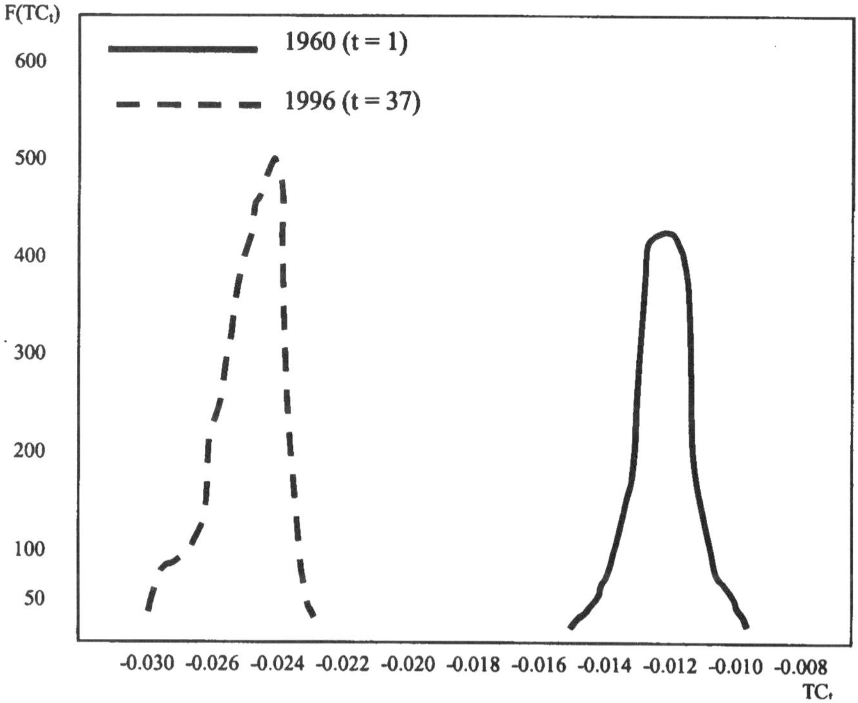

Figure 6.1. Estimated pdpf of the annual proportionate change in cost due to technical change, 1960 and 1996.

Elasticities

Meaningful information about the production technology can be conveyed by way of elasticities of shadow input-demand with respect to shadow prices. These elasticities are given by equation (11) and are identical to the usual elasticities of input-demand if the firm (state) is allocatively efficient. For illustrative purposes, table 6.4 reports means and standard deviations of samples of these shadow elasticities evaluated at average prices and quantities in six representative states: California, Florida, Iowa, Nevada, South Dakota, and Texas.

The estimated own-price elasticities are generally consistent with the findings of other U.S. studies: the estimated own-price elasticities for materials (which range from -0.55 to -0.35) are consistent with the elasticity estimate obtained by Vasavada and Chambers (1986) using a normalized quadratic functional form (-0.34), but somewhat higher than two normalized quadratic estimates reported by Shumway, Saez, and Gottret (1998) (-0.08 and -0.11); the estimated own-price elasticities for capital range from -0.32 to -0.29 and are consistent with the translog estimate reported by Antle (1984) (-0.25); the estimated own-price elasticities for land (-0.50 to -0.22) are slightly lower than the translog estimate reported by Ball (1988) (-0.58), but have a range that is bounded closely by two estimates reported by Antle (-0.18 for the pre-war period and -0.58 for the post-war period); and, finally, the estimated own-price elasticities for labor lie within the narrow range -0.38 to -0.36, slightly lower than the estimate reported by Vasavada and Chambers (-0.51), slightly higher than the estimate of O'Donnell, Shumway, and Ball (-0.24), and between two estimates reported by Shumway, Saez, and Gottret (-0.10 and -0.40).

There is somewhat less consistency between the estimated cross-price elasticities in table 6.4 and the results of earlier U.S. and Canadian studies. The estimates in table 6.4 suggest that three input pairs are substitutes in production: materials and labor, capital and land, and capital and labor. These conclusions are consistent only with the findings of Vasavada and Chambers (in the case of capital and land), O'Donnell, Shumway, and Ball (in the case of materials and labor), and (except in the case of materials and labor) Lopez and Tung (1982). There is mixed evidence concerning the complementarity and substitutability of remaining input pairs, since point estimates of the relevant cross-price elasticities appear sensitive to the prices and quantities at which they are evaluated. Moreover, the estimated marginal posterior pdfs of many cross-price elasticities span zero. To illustrate, figure 6.2 presents the estimated marginal posterior pdfs of the land-labor cross-price elasticity (η_{34}) evaluated at average Florida and Iowa prices and quantities. Observe from figure 6.2 and table 6.4 that differences in average prices and quantities between these states (notably the price of land and the output mix – see table 6.1) cause a sufficiently large change in the position and shape of the estimated marginal pdf that the point

Table 6.4. Elasticities at Average Prices in Selected States[a]

Elasticity	California ($n = 4$)	Florida ($n = 8$)	Iowa ($n = 10$)	Nevada ($n = 31$)	S.Dakota ($n = 39$)	Texas ($n = 41$)
η_{11}	-0.381 (0.036)	-0.347 (0.067)	-0.518 (0.017)	-0.550 (0.008)	-0.535 (0.010)	-0.544 (0.009)
η_{12}	-0.133 (0.051)	-0.196 (0.080)	-0.051 (0.009)	0.016 (0.009)	-0.030 (0.007)	0.007 (0.008)
η_{13}	-0.003 (0.011)	0.081 (0.012)	0.218 (0.011)	0.070 (0.005)	0.157 (0.005)	0.048 (0.007)
η_{14}	0.516 (0.026)	0.462 (0.031)	0.351 (0.018)	0.464 (0.012)	0.408 (0.009)	0.489 (0.013)
η_{21}	-0.027 (0.008)	-0.037 (0.010)	-0.024 (0.003)	0.011 (0.006)	-0.017 (0.004)	0.004 (0.005)
η_{22}	-0.287 (0.017)	-0.294 (0.020)	-0.319 (0.013)	-0.315 (0.010)	-0.317 (0.011)	-0.319 (0.009)
η_{23}	0.091 (0.009)	0.181 (0.005)	0.277 (0.015)	0.115 (0.005)	0.211 (0.008)	0.097 (0.006)
η_{24}	0.223 (0.011)	0.150 (0.015)	0.066 (0.006)	0.189 (0.009)	0.123 (0.007)	0.217 (0.013)
η_{31}	-0.004 (0.011)	0.041 (0.010)	0.122 (0.013)	0.125 (0.010)	0.129 (0.010)	0.094 (0.011)
η_{32}	0.468 (0.018)	0.449 (0.028)	0.323 (0.016)	0.299 (0.008)	0.305 (0.009)	0.316 (0.011)
η_{33}	-0.217 (0.044)	-0.476 (0.007)	-0.499 (0.006)	-0.342 (0.015)	-0.495 (0.003)	-0.258 (0.037)
η_{34}	-0.247 (0.045)	-0.014 (0.016)	0.054 (0.012)	-0.082 (0.016)	0.061 (0.011)	-0.151 (0.027)
η_{41}	0.143 (0.008)	0.149 (0.010)	0.219 (0.020)	0.246 (0.012)	0.223 (0.015)	0.227 (0.010)
η_{42}	0.298 (0.015)	0.239 (0.016)	0.086 (0.006)	0.145 (0.005)	0.119 (0.005)	0.168 (0.007)
η_{43}	-0.063 (0.008)	-0.010 (0.011)	0.060 (0.010)	-0.024 (0.004)	0.041 (0.007)	-0.036 (0.004)
η_{44}	-0.378 (0.011)	-0.378 (0.009)	-0.365 (0.018)	-0.367 (0.011)	-0.382 (0.010)	-0.360 (0.015)

[a] Numbers in parentheses are standard deviations of MCMC samples. Subscripts refer to inputs (1 = materials; 2 = capital; 3 = land; 4 = labor).

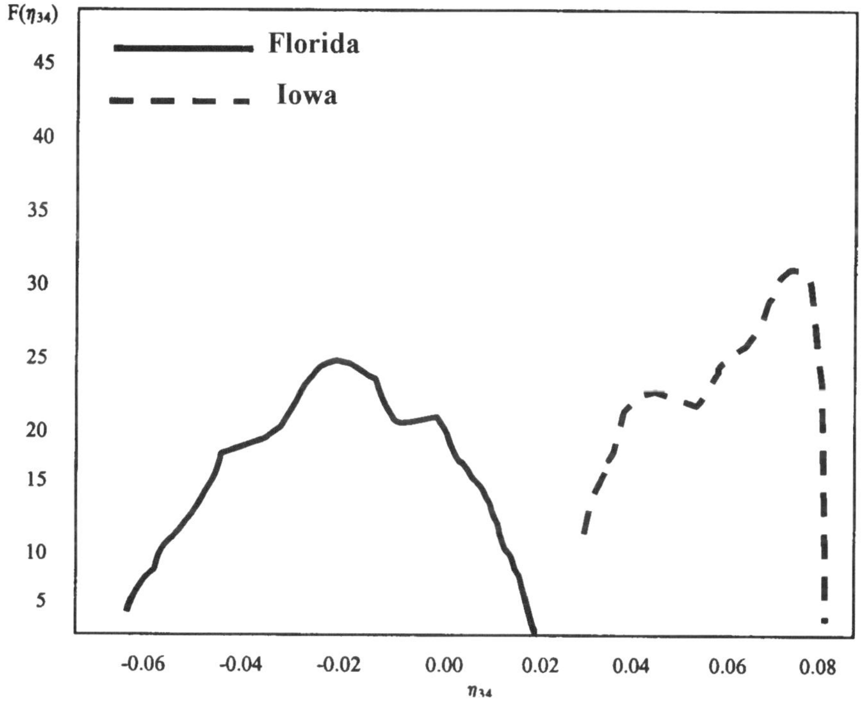

Figure 6.2. Estimated pdfs of the land-labor cross-price elasticity evaluated at average Florida and Iowa prices and quantities.

estimate of the elasticity changes sign. Also note from the estimated Florida pdf that the probability is approximately 0.25 that land and labor are substitutes.

Technical Efficiency

Technical efficiency is measured by first normalizing b_i to one for the most efficient state (Atkinson and Cornwell). Then an upper bound on the relative technical efficiency of state i can be calculated as

$$TE_i = b_i = \frac{1}{\exp(\mu_i - \min\{\mu_1, \ldots, \mu_I\})} \tag{25}$$

Table 6.5 reports means and standard deviations of samples of observations on these relative efficiency scores. The point estimates reported in this table range from a low of 0.26 (West Virginia) to a high of only marginally less than one (Florida), with an average value of 0.64. The estimates are somewhat lower than expected, and significantly lower than the regional estimates re-

Table 6.5. Measures of Relative Technical Efficiency [a]

i	State	Score (TE_i) Estimate	Score (TE_i) St. Deviation	Ranking (TR_i)
1	Alabama	0.589	0.018	28
2	Arkansas	0.709	0.025	14
3	Arizona	0.798	0.020	7
4	California	0.963	0.021	2
5	Colorado	0.775	0.023	8
6	Connecticut	0.646	0.019	22
7	Delaware	0.843	0.038	5
8	Florida	1.000 [a]	0.001	1
9	Georgia	0.827	0.021	6
10	Iowa	0.865	0.028	3
11	Idaho	0.726	0.027	12
12	Illinois	0.741	0.026	9
13	Indiana	0.666	0.021	19
14	Kansas	0.741	0.028	10
15	Kentucky	0.632	0.015	23
16	Louisiana	0.591	0.019	27
17	Massachusetts	0.558	0.013	38
18	Maryland	0.559	0.020	36
19	Maine	0.703	0.031	17
20	Michigan	0.560	0.020	35
21	Minnesota	0.684	0.025	18
22	Missouri	0.562	0.017	33
23	Mississippi	0.609	0.023	25
24	Montana	0.534	0.020	42
25	North Carolina	0.705	0.017	16
26	North Dakota	0.630	0.017	24
27	Nebraska	0.735	0.025	11
28	New Hampshire	0.478	0.019	43
29	New Jersey	0.550	0.014	40
30	New Mexico	0.576	0.017	30
31	Nevada	0.560	0.018	34
32	New York	0.652	0.022	20
33	Ohio	0.556	0.019	39
34	Oklahoma	0.558	0.015	37
35	Oregon	0.647	0.019	21
36	Pennsylvania	0.574	0.022	31
37	Rhode Island	0.462	0.014	46
38	South Carolina	0.598	0.012	26
39	South Dakota	0.723	0.022	13
40	Tennessee	0.462	0.016	45
41	Texas	0.578	0.026	29
42	Utah	0.539	0.016	41
43	Virginia	0.472	0.014	44
44	Vermont	0.569	0.021	32
45	Washington	0.847	0.025	4
46	Wisconsin	0.707	0.025	15
47	West Virginia	0.261	0.007	48
48	Wyoming	0.440	0.013	47

[a] 0.9998 when rounded to 4 decimal places

ported by O'Donnell, Shumway, and Ball, almost certainly because a single cost frontier has been used in conjunction with disaggregate data (i.e., state-level rather than regional-level) to measure the characteristics of what might be several different technologies. Two other factors may be contributing to these low estimates. First, the Bayesian point estimate of the efficiency score for the most efficient state will be one if and only if the marginal pdf of the score is degenerate (i.e., collapses to a point), and this is uncommon in empirical work. Second, in models with few inputs and outputs, simple diagrams can be used to illustrate how the estimated efficiency scores of outlier firms or states (i.e., states with output and input mixes and levels that are a long way from the norm) will be low even though they may be quite close to the efficient frontier. One state that arguably falls into this "outlier" class is Rhode Island, which has the lowest level of livestock output per unit of labor input of all states – see tables 6.1 and 6.2.

Corresponding to these relative technical efficiency scores is a set of technical efficiency rankings, denoted TR_i. The efficiency rankings reported in table 6.5 have been assigned on the basis of the magnitudes of estimated relative efficiency scores. There is clearly some uncertainty surrounding both the estimated scores and the estimated rankings, not least because they are based on estimates of unknown parameters. One of the advantages of the Bayesian approach is that it allows us to express this uncertainty in a convenient way, using marginal pdfs. For illustrative purposes, figures 6.3 to 6.5 represent the uncertainty surrounding efficiency estimates for Iowa (ranked 3rd on the basis of a relative technical efficiency estimate of 0.87) and Washington (ranked 4th on the basis of an estimate of 0.85). Figure 6.3 presents the estimated marginal pdfs of the relative efficiency scores (the TE_i), and reveals that there is zero probability that the relative technical efficiency of Iowa is greater than 0.95, and zero probability that the relative technical efficiency of Washington is greater than 0.92. Figures 6.4 and 6.5 present estimated probability distributions for the corresponding rankings (the TR_i). It can be seen from figures 6.4 and 6.5 that Iowa and Washington are ranked 4th with probabilities 0.20 and 0.39 respectively, there is a 2% probability that Iowa is ranked as low as 6th, and there is a 17% probability that Washington is ranked as high as 3rd. There is zero probability that either state is ranked as high as 1st.

Allocative Efficiency

Following Atkinson and Cornwell, an upper bound on the relative allocative efficiency of state i can be calculated as:

$$AE_i = \exp[\mathrm{f}(\mathbf{x}, \mu_i, \boldsymbol{\beta}, \mathbf{j}_N) - \mathrm{f}(\mathbf{x}, \mu_i, \boldsymbol{\beta}, \mathbf{a}_i)] \qquad (26)$$

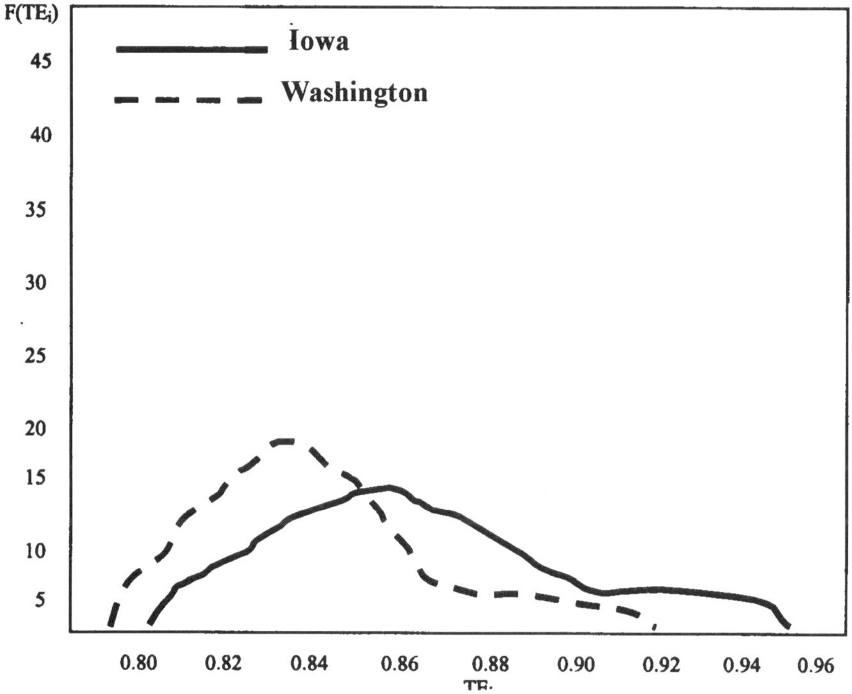

Figure 6.3. Estimated pdfs of the relative technical efficiencies of Iowa and Washington.

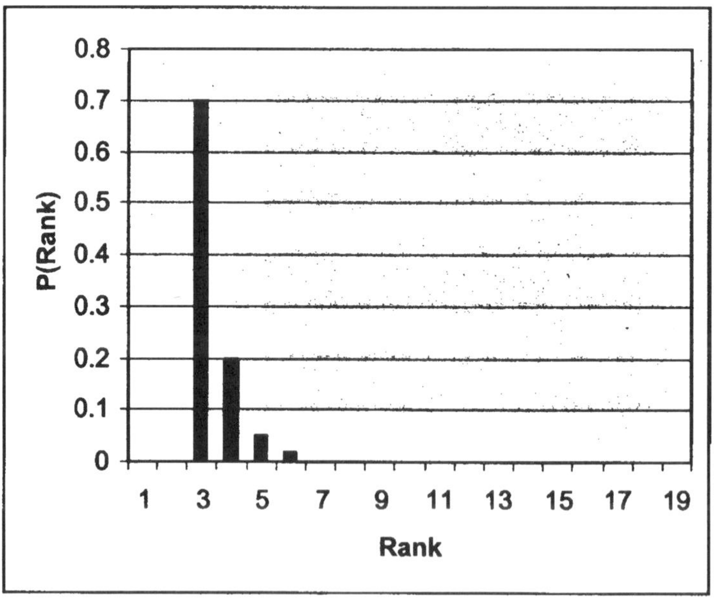

Figure 6.4. Estimated probability distribution of the technical efficiency ranking of Iowa.

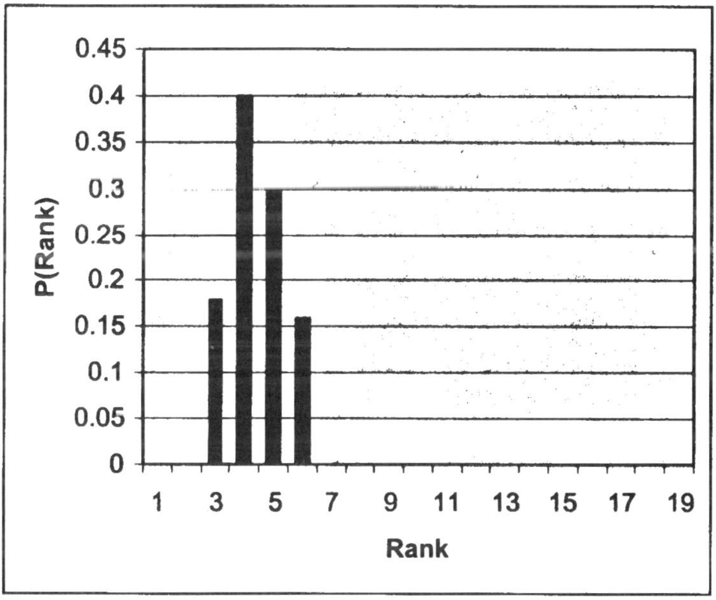

Figure 6.5. Estimated probability distribution of the technical efficiency ranking of Washington.

where $\mathbf{j}_N$ is an $N \times 1$ unit vector and $\bar{\mathbf{x}}$ is the average of $\mathbf{x}_{it}$ over i and t. That is, the relative allocative efficiency of state i can be obtained as the ratio of predicted cost (at mean prices and quantities) when $\mathbf{a}_i = \mathbf{j}_N$ to predicted cost when $\mathbf{a}_i$ is unconstrained. Table 6.6 reports estimates of these AE_i scores, together with a corresponding set of allocative efficiency rankings, denoted AR_i. The AE_i estimates reported in table 6.6 range from 0.65 (Arizona) to 0.94 (South Dakota), with an average value of 0.88.

Underpinning these allocative efficiency scores and rankings are the $\mathbf{a}_i$ parameters. These parameters are of particular interest because they point to over- or under-utilization of inputs. Means and standard deviations of MCMC samples of observations on the unnormalized elements of $\mathbf{a}_i$ (i.e., a_{1i}, a_{2i} and a_{3i}) are reported in table 6.7. Parameter values less than (greater than) one imply over-utilization (under-utilization) of that input relative to labor (the input on which allocative effects were normalized). It appears from table 6.5 that the most allocatively efficient state, South Dakota, over-utilizes materials and land, and under-utilizes capital, relative to labor. More generally, 19 out of 48 states appear to over-utilize materials, only 4 states appear to over-utilize capital (Illinois, Indiana, North Dakota, and Montana), and 18 states appear to over-utilize land. Once again, the uncertainty associated with these estimates can be represented using pdfs. To illustrate, the estimated marginal posterior pdfs for a_{1i}, a_{2i}, and a_{3i} for South Dakota ($i = 39$) are depicted in figure 6. 6. Note that the

Table 6.6. Measures of relative allocative efficiency

i	State	Score (TE_i)		Ranking (TR_i)
		Estimate	St. Deviation	
1	Alabama	0.921	0.047	11
2	Arkansas	0.897	0.038	25
3	Arizona	0.648	0.027	48
4	California	0.741	0.026	46
5	Colorado	0.895	0.045	26
6	Connecticut	0.855	0.042	37
7	Delaware	0.883	0.046	32
8	Florida	0.821	0.032	44
9	Georgia	0.900	0.046	23
10	Iowa	0.931	0.051	4
11	Idaho	0.922	0.042	10
12	Illinois	0.928	0.047	5
13	Indiana	0.935	0.046	3
14	Kansas	0.924	0.047	8
15	Kentucky	0.914	0.048	18
16	Louisiana	0.915	0.045	16
17	Massachusetts	0.848	0.041	41
18	Maryland	0.926	0.045	6
19	Maine	0.870	0.048	35
20	Michigan	0.893	0.048	27
21	Minnesota	0.926	0.051	7
22	Missouri	0.937	0.045	2
23	Mississippi	0.904	0.043	22
24	Montana	0.851	0.039	39
25	North Carolina	0.889	0.046	29
26	North Dakota	0.918	0.044	15
27	Nebraska	0.919	0.045	14
28	New Hampshire	0.913	0.045	20
29	New Jersey	0.913	0.043	19
30	New Mexico	0.739	0.034	47
31	Nevada	0.850	0.041	40
32	New York	0.851	0.044	38
33	Ohio	0.912	0.041	21
34	Oklahoma	0.862	0.037	36
35	Oregon	0.882	0.042	33
36	Pennsylvania	0.872	0.043	34
37	Rhode Island	0.915	0.042	17
38	South Carolina	0.920	0.045	12
39	South Dakota	0.938	0.047	1
40	Tennessee	0.920	0.042	13
41	Texas	0.822	0.043	43
42	Utah	0.830	0.039	42
43	Virginia	0.923	0.046	9
44	Vermont	0.887	0.045	30
45	Washington	0.884	0.044	31
46	Wisconsin	0.891	0.043	28
47	West Virginia	0.898	0.041	24
48	Wyoming	0.816	0.039	45

Table 6.7. Allocative efficiency parameters

		Materials (a_{1i})		Capital (a_{2i})		Land (a_{3i})	
i	State	Estimate	St. Dev.	Estimate	St. Dev.	Estimate	St. Dev.
1	Alabama	0.653	0.042	1.900	0.136	1.051	0.065
2	Arkansas	0.601	0.050	3.326	1.065	1.558	0.436
3	Arizona	0.483	0.037	5.284	0.548	0.230	0.018
4	California	0.573	0.046	4.945	0.455	0.437	0.041
5	Colorado	0.462	0.024	1.940	0.187	0.786	0.089
6	Connecticut	3.689	0.235	4.611	0.295	5.348	0.324
7	Delaware	0.400	0.032	2.834	0.256	1.435	0.115
8	Florida	0.716	0.079	4.440	0.706	0.736	0.073
9	Georgia	0.682	0.049	3.442	0.344	1.854	0.179
10	Iowa	0.989	0.073	2.005	0.309	1.704	0.274
11	Idaho	0.982	0.082	2.322	0.381	1.202	0.156
12	Illinois	1.052	0.066	0.575	0.063	0.494	0.034
13	Indiana	1.051	0.079	0.669	0.047	0.601	0.040
14	Kansas	0.614	0.048	1.312	0.126	0.724	0.077
15	Kentucky	2.088	0.129	2.037	0.172	2.009	0.224
16	Louisiana	0.854	0.072	1.411	0.168	0.570	0.063
17	Massachusetts	4.019	0.292	5.826	0.464	5.781	0.458
18	Maryland	1.004	0.084	2.352	0.227	1.578	0.127
19	Maine	1.690	0.119	4.507	0.210	7.660	0.447
20	Michigan	2.193	0.157	2.939	0.230	3.872	0.463
21	Minnesota	1.288	0.079	1.984	0.227	2.293	0.347
22	Missouri	1.243	0.107	1.474	0.188	1.112	0.110
23	Mississippi	0.839	0.069	3.455	0.456	1.722	0.191
24	Montana	1.480	0.120	0.547	0.051	0.197	0.013
25	North Carolina	1.605	0.121	5.397	0.530	4.964	0.415
26	North Dakota	1.188	0.094	0.473	0.047	0.412	0.031
27	Nebraska	0.666	0.040	1.488	0.181	0.705	0.148
28	New Hampshire	2.212	0.157	1.737	0.190	1.929	0.224
29	New Jersey	1.880	0.143	2.008	0.189	1.205	0.081
30	New Mexico	0.778	0.049	3.345	0.244	0.258	0.020
31	Nevada	1.197	0.081	1.821	0.157	0.369	0.026
32	New York	2.294	0.203	4.070	0.362	9.010	0.624
33	Ohio	2.132	0.179	2.036	0.207	2.161	0.190
34	Oklahoma	0.924	0.059	1.439	0.110	0.315	0.044
35	Oregon	1.503	0.101	3.393	0.469	1.144	0.170
36	Pennsylvania	2.661	0.258	5.283	0.651	6.034	0.625
37	Rhode Island	0.956	0.083	2.878	0.467	1.537	0.186
38	South Carolina	1.252	0.100	2.715	0.270	2.298	0.218
39	South Dakota	0.922	0.064	1.450	0.100	0.995	0.067
40	Tennessee	1.797	0.164	1.805	0.169	1.209	0.089
41	Texas	0.703	0.068	1.359	0.154	0.190	0.024
42	Utah	1.229	0.088	1.315	0.089	0.232	0.012
43	Virginia	1.670	0.116	1.989	0.213	1.394	0.179
44	Vermont	1.962	0.143	3.694	0.151	4.884	0.190
45	Washington	1.029	0.073	4.804	0.358	1.918	0.161
46	Wisconsin	1.715	0.162	3.231	0.345	4.657	0.398
47	West Virginia	2.328	0.156	2.103	0.169	1.274	0.196
48	Wyoming	1.017	0.073	1.817	0.154	0.256	0.019

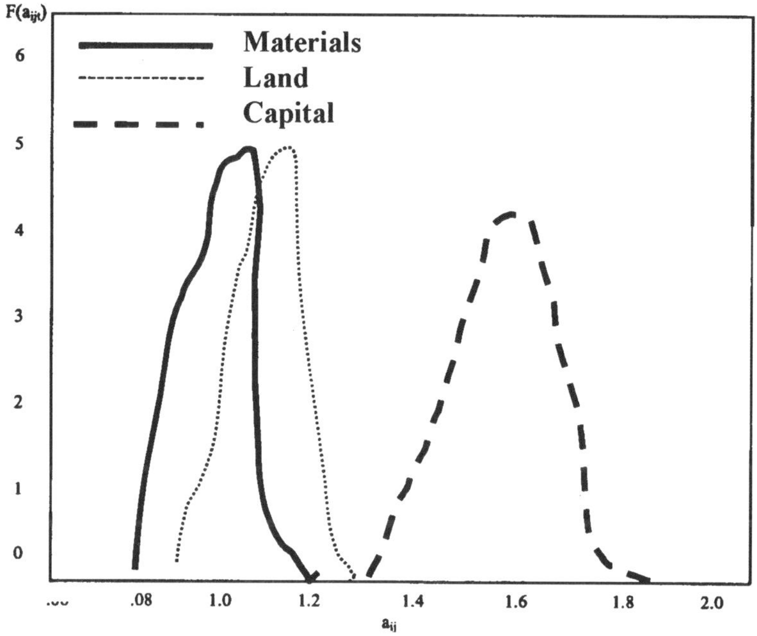

Figure 6.6. Estimated probability distribution of the technical efficiency ranking of Washington.

the pdfs for materials and land both span one — South Dakota is 7.98 times more likely to over-utilize than under-utilize materials relative to labor, and 1.25 times more likely to over-utilize than under-utilize land.

Cost Efficiency

The relative cost efficiency of state i in period t can be obtained as the ratio of predicted cost when $\mathbf{a}_i = b_i = 1$ to predicted cost when $\mathbf{a}_i$ and b_i are both unconstrained. Equivalently:

$$CE_i = TE_i \times AE_i \tag{27}$$

where TE_i and AE_i are given by (25) and (26). Estimates of the CE_i and associated rankings, denoted CR_i, are reported in table 6.8. The CE_i estimates range from 0.24 (West Virginia) to 0.82 (Florida) with an average value of 0.56. Values of these efficiency scores can obviously be traced back to the values reported in tables 6.5 and 6.6.

Table 6.8. Measures of relative cost efficiency

i	State	Score (CE_i)		Ranking (CR_i)
		Estimate	St. Deviation	
1	Alabama	0.542	0.023	26
2	Arkansas	0.636	0.027	13
3	Arizona	0.517	0.023	29
4	California	0.714	0.029	6
5	Colorado	0.693	0.035	7
6	Connecticut	0.552	0.023	23
7	Delaware	0.744	0.031	5
8	Florida	0.821	0.032	1
9	Georgia	0.744	0.036	4
10	Iowa	0.805	0.044	2
11	Idaho	0.669	0.024	12
12	Illinois	0.688	0.030	8
13	Indiana	0.622	0.023	17
14	Kansas	0.684	0.026	9
15	Kentucky	0.578	0.028	20
16	Louisiana	0.541	0.018	27
17	Massachusetts	0.472	0.023	39
18	Maryland	0.517	0.016	30
19	Maine	0.611	0.026	18
20	Michigan	0.499	0.024	35
21	Minnesota	0.632	0.026	14
22	Missouri	0.526	0.022	28
23	Mississippi	0.550	0.020	24
24	Montana	0.454	0.017	40
25	North Carolina	0.627	0.028	16
26	North Dakota	0.578	0.020	19
27	Nebraska	0.675	0.028	11
28	New Hampshire	0.436	0.019	42
29	New Jersey	0.502	0.023	33
30	New Mexico	0.425	0.018	44
31	Nevada	0.476	0.021	37
32	New York	0.555	0.025	22
33	Ohio	0.507	0.020	31
34	Oklahoma	0.481	0.018	36
35	Oregon	0.570	0.022	21
36	Pennsylvania	0.500	0.019	34
37	Rhode Island	0.422	0.017	46
38	South Carolina	0.550	0.028	25
39	South Dakota	0.678	0.030	10
40	Tennessee	0.425	0.018	45
41	Texas	0.474	0.017	38
42	Utah	0.447	0.021	41
43	Virginia	0.435	0.017	43
44	Vermont	0.505	0.022	32
45	Washington	0.748	0.026	3
46	Wisconsin	0.629	0.025	15
47	West Virginia	0.235	0.012	48
48	Wyoming	0.359	0.016	47

Conclusion

Parametric and non-parametric methodologies can both be used to estimate levels of technical and allocative efficiency, and both are quite popular in empirical work. This paper uses the parametric approach because it has formal and well-developed mechanisms for accounting for statistical noise. The particular parametric methodology used, Bayesian methodology, also has well-developed mechanisms for incorporating inequality information (e.g., concavity information) into the estimation process. Moreover, Bayesians have intuitively appealing ways of summarizing uncertainty about unknown parameters (or functions of parameters, including measures of relative efficiency), as demonstrated in figures 6.1 to 6.6.

The paper uses U.S. state-level agricultural data to estimate a system of nonlinear equations derived from a translog shadow cost frontier. Estimates of the parameters are used to obtain estimates of elasticities of shadow input demands with respect to shadow prices, and state-level estimates of relative technical, allocative, and cost efficiency. The estimated own-price elasticities range from -0.55 to -0.22 and are consistent with the findings of other U.S. and Canadian studies. Average levels of technical, allocative, and cost efficiency are estimated to be 0.64, 0.88 and 0.56, slightly lower than expected.

There is strong evidence that Florida is the most technically efficient state, with an efficiency level of marginally less than one. I find that Florida is also the most cost-efficient state (efficiency level 0.82), while South Dakota is the most allocatively efficient (0.94). Two of the least efficient states are West Virginia (only 26% technically efficient and 24% cost efficient) and Arizona (only 65% allocatively efficient). While there is mixed evidence concerning the under-utilization of materials and land relative to labor (in an allocative sense), the results suggest that almost every state under-utilizes capital. There is uncertainty surrounding these estimates, and a distinguishing feature of the paper is the way in which this uncertainty has been summarized (e.g., using graphs of probability density functions).

The empirical results have been generated by making numerous tractable assumptions concerning, for example, the production technology (e.g., constant returns to scale), technical change (e.g., Hicks neutral), inefficiency effects (i.e., fixed rather than random), the distributions of random error terms, and functional form. Varying these assumptions provides several interesting opportunities for further empirical research. Arguably the most important of these involves the use of disaggregated data (e.g., county- or firm-level data) to estimate cost frontiers for meaningful subsets of states (e.g., USDA farm production regions). The motivation for using disaggregated data is the lack of prior empirical evidence that firms can be aggregated to the state level in a manner that preserves curvature and other properties implied by economic theory (i.e., "consistent" aggregation). The motivation for estimation of regional frontiers is that there appear to be variations in non-transferable input endowments (e.g., soil and climate), and hence production technologies, across some states. Accounting for these variations can be expected to lead to in an increase in estimated technical, allocative, and cost efficiencies.

Notes

1. The author wishes to thank Eldon Ball for providing the data, and Tom Cox and Richard Shumway for helpful comments.

2. Moreover, relaxation of the CRS assumption led to implausibly low empirical estimates of technical, allocative, and cost efficiency in all but two states (two of the smallest states).

3. Bayesians are not the only statisticians who use Monte Carlo simulation techniques to evaluate mathematically intractable integrals. Sampling theorists use these techniques to simulate likelihood functions, which are multi-dimensional integrals of (usually) normal density functions. Maximum simulated likelihood (MSL) estimates are then obtained by maximizing the (log) simulated likelihood instead of the intractable actual likelihood. For more details see Hyslop (1999).

References

Antle, J. M. 1984. "The Structure of U.S. Agricultural Technology, 1910-1978." *American Journal of Agricultural Economics,* 66(4):414-21.

Atkinson, S. E., and C. Cornwell. 1994. "Parametric Estimation of Technical and Allocative Inefficiency with Panel Data." *International Economic Review,* 35(February):231-43.

Atkinson, S. E., and R. Halvorsen. 1984. "Parametric Efficiency Tests, Economies of Scale, and Input Demand in U.S. Electric Power Generation." *International Economic Review,* 25:647-62.

Atkinson, S. E., and R. Halvorsen. 1990. "Tests of Allocative Efficiency in Regulated Multi-Product Firms." *Resources and Energy,* 12:65-77.

Ball, V. E. 1988. "Modelling Supply Response in a Multiproduct Framework." *American Journal of Agricultural Economics,* 70(November):813-25.

Ball, V. E., F. Gollop, A. Kelly-Hawke, and G. Swinand. 1999. "Patterns of State Productivity Growth in the U.S. Farm Sector: Linking State and Aggregate Models." *American Journal of Agricultural Economics,* 81(1):164-79.

Bauer, P. W. 1990. "Recent Developments in the Econometric Estimation of Frontiers." *Journal of Econometrics,* 46:39-56.

Casella, G., and E. I. George. 1992. "Explaining the Gibbs Sampler." *The American Statistician,* 46(3):167-73.

Chambers, R. G. 1988. *Applied Production Analysis.* Cambridge: University Press.

Chib, S., and E. Greenberg. 1996. "Markov Chain Monte Carlo Simulation Methods in Econometrics." *Econometric Theory,* 12:409-31.

Eltetö, O., and P. Köves. 1964. "On a Problem of Index Number Computation Relating to International Comparison." (in Hungarian) *Statisztikai Szemle,* 42:507-18.

Ferrier, G., and K. Lovell. 1990. "Measuring Cost Efficiency in Banking: Econometric and Linear Programming Evidence." *Journal of Econometrics,* 46:229-45.

Greene, W. 1980. "On the Estimation of a Flexible Frontier Production Model." *Journal of Econometrics,* 13:101-15.

Griffiths, W. E. 1988. "Bayesian Econometrics and How to Get Rid of Those Wrong Signs." *Review of Marketing and Agricultural Economics*, 56(1):36-56.

Griffiths, W. E., and D. Chotikapanich. 1997. "Bayesian Methodology for Imposing Inequality Constraints on a Linear Expenditure System with Demographic Factors." *Australian Economic Papers*, December: 321-41.

Griffiths, W. E., R. C. Hill, and G. G. Judge. 1993. *Learning and Practicing Econometrics.* New York: John Wiley and Sons.

Griffiths, W. E., C. J. O'Donnell, and A. Tan Cruz. 2000. "Imposing regularity conditions on a system of cost and cost-share equations: a Bayesian approach." *Australian Journal of Agricultural Economics*, (March): 107-27.

Hyslop, D. R. 1999. "State Dependence, Serial Correlation and Heterogeneity in Intertemporal Labour Force Participation of Married Women." *Econometrica*, 67(6):1255-94.

Judge, G. G., W. E. Griffiths, R. C. Hill, H. Lutkepohl, and T. C. Lee. 1985. *The Theory and Practice of Econometrics*, 2nd ed. New York: John Wiley and Sons.

Kumbhakar, S. 1988. "Estimation of Input-Specific Technical and Allocative Inefficiency in Stochastic Frontier Models." *Oxford Economic Papers*, 40:535-49.

Kumbhakar, S. 1991. "Estimation of Technical Inefficiency in Panel Data Models with Firm- and Time-Specific Effects." *Economics Letters*, 36(2):43-48.

Kumbhakar, S. 1992. "Allocative Distortions, Technical Progress, and Input Demand in U.S. Airlines: 1970-1984." *International Economic Review*, 3(3):723-37.

Lopez, R. E., and F.-L. Tung. 1982. "Energy and non-energy input substitution possibilities and scale effects in Canadian agriculture." *Canadian Journal of Agricultural Economics*, 30(July):115-32.

Lovell, C. A. K., and R. C. Sickles. 1983. "Testing Efficiency Hypotheses in Joint Production: A Parametric Approach." *Review of Economics and Statistics*, 65:51-58.

O'Donnell, C. J. 1996. "Inefficiency, Uncertainty and the Structure of Cost, Cost-Share and Input-Demand Functions." *University of New England Dept. of Econometrics Working Papers in Econometrics and Applied Statistics.* No. 89, August.

O'Donnell, C. J., and A. D. Woodland. 1995. "Estimation of Australian Wool and Lamb Production Technologies Under Uncertainty: An Error-Components Approach." *American Journal of Agricultural Economics*, 77(August):552-65.

O'Donnell, C. J., C. R. Shumway, and V. E. Ball. 1999. "Input demands and inefficiency in U.S. agriculture." *American Journal of Agricultural Economics*, 81(November):865-80.

Poirier, D. J. 1995. *Intermediate Statistics and Econometrics.* Cambridge, Mass.: The MIT Press.

Pope, R. D., and J.-P. Chavas. 1994. "Cost Functions Under Production Uncertainty." *American Journal of Agricultural Economics*, 76(May):196-204.

Shumway, C. R., and H. Lim. 1993. "Functional Form and U.S. Agricultural Production Elasticities." *Journal of Agricultural and Resource Economics*, 18(December):266-76.

Shumway, C. R., R. R. Saez, and P. E. Gottret. 1988. "Multiproduct Supply and Input Demand in U.S. Agriculture." *American Journal of Agricultural Economics*, 70(May):330-37.

Szulc, B. J. 1964. "Indices for Multiregional Comparisons."(in Hungarian) *Prezeglad Statystyczny*, 3:239-54.

Toda, Y. 1976. "Estimation of a Cost Function When Cost is Not Minimum: The Case of Soviet Manufacturing Industries, 1958-1971." *Review of Economics and Statistics*, 58:259-68.

Vasavada, U., and R. G. Chambers. 1986. "Investment in U.S. Agriculture." *American Journal of Agricultural Economics*, 68(4):950-60.

7

Public R&D and Infrastructure Policies: Effects on Cost of Midwestern Agriculture

Wallace E. Huffman, V. Eldon Ball, Munisamy Gopinath, and Agapi Somwaru[1]

Abstract

U.S. competitiveness in agricultural products and the competitiveness of individual states can be shifted by changes in public policies on R&D, extension, infrastructure, and other programs. States are heterogeneous in agro-climatic conditions and resource-base, and they have different public R&D, extension, and infrastructure policies. The objective of this paper is to assess the impacts of public investment in research and infrastructure on cost of production and input demands in five Midwestern states (Minnesota, Iowa, Illinois, Missouri, and Indiana). We represent technology via the variable cost function. Our model identifies two outputs (crops and livestock) and four variable inputs (capital, land, hired labor, and materials). Fixed factors include self-employed and unpaid family labor, own R&D, potential spillovers from other states' investment in research, and infrastructure (i.e., roads and highways).

We estimate the variable cost function using pooled time series and cross section observations. The parameter estimates are used to construct measures of price responsiveness. The own-price elasticities of demand, evaluated at sample means, for capital, land, hired labor, and materials are -0.86, -0.70, -0.59, and -0.04, respectively. Pair-wise substitution dominates. Capital and land and capital and hired labor exhibit strong substitutability. The impacts on variable costs of increasing endowments of the fixed factors are large and negative. Finally, the implied internal rate of return to investment in agricultural research exceeds 40 percent per annum.

Introduction

The nominal value of U.S. agricultural exports in 1996, when the FAIR Act was enacted, was $59.9 billion, 33 percent larger than during the early 1990s. However, the composition of U.S. agricultural exports has two interesting trends. First, crop products accounted for 81 percent of total exports. Corn, wheat and soybean exports accounted for more than 50 percent of total crop product exports. Second, a majority of exports are sourced from a few of the 50 states. For instance, the top 7 states (California, Iowa, Illinois, Nebraska, Texas, Kansas, and Minnesota) accounted for about 50% of total U.S. agricultural exports in 1996 (USDA, 1998). Moreover, export share of farm receipts, and exports per acre of land (accounting for scale effects) reveal large differences among states in export performance or competitiveness (Appendix table 1).

Given the homogeneity of the states with respect to domestic and trade policies, the differences in competitiveness are likely to arise from heterogeneity in their agro-climatic conditions, their resource-base including infrastructure capital, and differences in relative productivity arising from research and development (R&D). Previous investigations of the impact of public R&D on U.S. agriculture have, with few exceptions, focused on the aggregate farm sector. And public investment in infrastructure (i.e., roads and highways) has not been the target of any of these investigations. Moreover, previous studies of productivity growth have very limited implications for U.S. agricultural trade because TFP is measured as the ratio of an index of aggregate output to an index of all inputs under the control of farmers. Knowing that TFP is increasing tells one nothing about changes in the composition of output (input). Studies by Alston, Craig, and Pardey (1999) and Ball, Gollop, Kelly-Hawke, and Swinand (1999) show large interstate variation in TFP, but lack implications for particular commodities or commodity groups. Huffman and Evenson (1993) disaggregated the agricultural sector of each state into crop and livestock sub-sectors in order to gain insights about the seemingly slow growth rate of TFP for Midwestern states. They found that TFP growth rates across 42 states were generally faster for the crop than for the livestock sub-sectors. For some states/ regions, large differences existed in crop and livestock sub-sector TFP growth rates. In Iowa, for example, the average annual rate of TFP growth in the crop sub-sector was a robust 2.5 percent, while the livestock sub-sector exhibited little or nor growth in productivity. This difference is one likely source of change in regional comparative advantages.

Binswanger (1974) and Deininger (1995) have used single output cost functions to examine input substitution and technical change bias. However, the single aggregate output measure precluded drawing inferencesfor particular output groups. Huffman and Evenson (1989) and Shumway and Alexander (1988) use multi-output profit functions and associated supply and demand functions to fit their models to state or regional data. Huffman and Evenson (1989) limited their analysis to cash grain farms in the Census of Agriculture, 1949-74, and estimated supply functions for wheat, feed grains, and soybeans. Shumway and Alexander (1988) estimated supply functions from annual data, 1951-82, for 5 commodity groups (feed grains, food grains, oil crops, other crops, and livestock). Huffman and Evenson (1989) included a public research stock variable to proxy the effects of own-state and spillover public research effects on the supply of outputs and demand for inputs, while Shumway and Alexander (1988) used only a trend to represent technology. Thus, only one prior study has used a consistent profit/cost function framework to examine the impacts of public research on state or regional level supply functions for disaggregated output, and none has included infrastructure.

The objective of this paper is to assess the impact of public R&D and infrastructure on the cost of production and input demands in Midwestern agriculture. The results will provide some evidence on competitiveness and ultimately on the rate of return on public investments in research and infrastructure.

METHODOLOGY

We represent technology via the dual cost function. The cost function provides a useful summary of behavioral responses to changes in relative input prices.[2] Moreover, this approach allows us to estimate the effects on cost of R&D expenditures, as well as R&D capital accumulation by other states.

The Model

We assume that each state produces outputs $y = (y_1, \ldots, y_M)$ using variable inputs $x = (x_1, \ldots, x_N)$ and fixed inputs $K = (K_1, \ldots, K_L)$. The vector of variable input prices is denoted by $w = (w_1, \ldots, w_N)$ with $w \cdot x = \sum_{n=1}^{N} w_n x_n > 0$.

Let $S(y, x, K)$ denote the set of feasible input and output combinations. We assume that S is a closed, non-empty, and convex set. The variable cost function is defined as:

$$C(y, w, K) = \min_x \{w \cdot x : (y, x, K) \in S\} \qquad (1)$$

$C(y, w, K)$ is non-decreasing in w and y, non-increasing in K, concave and homogeneous of degree one in w, and convex in K. If, in addition, $C(y, w, K)$ is differentiable, it satisfies Shephard's lemma:

$$x = \nabla_w C(y, w, K) \qquad (2)$$

where x is the vector of cost-minimizing variable input demands. Hence, $C(y, w, K)$ is assumed twice continuously differentiable in all its arguments.

Translog Cost Function

The variable cost function in (1) is approximated by the following translog function[3]

$$\ln C = \alpha_0 + \sum_{i=1}^{N} \alpha_i \ln w_i + \sum_{j=1}^{M} \beta_j \ln y_j + \sum_{l=1}^{L} \gamma_l \ln K_l \tag{3}$$

$$+ 1/2 \sum_{i=1}^{N}\sum_{j=1}^{N} \alpha_{ij} \ln w_i \ln w_j + 1/2 \sum_{i=1}^{M}\sum_{j=1}^{M} \beta_{ij} \ln y_i \ln y_j + 1/2 \sum_{i=1}^{L}\sum_{j=1}^{L} \gamma_{ij} \ln K_i \ln K_j$$

$$+ \sum_{i=1}^{N}\sum_{j=1}^{M} \delta_{ij} \ln w_i \ln y_j + \sum_{i=1}^{N}\sum_{j=1}^{L} \theta_{ij} \ln w_i \ln K_j + \sum_{i=1}^{M}\sum_{j=1}^{L} \phi_{ij} \ln y_i \ln K_j$$

$$+ \zeta_s \ln S(\ln K_3 + \sum_{i=1}^{N} \rho_{is} \ln w_i)$$

where K_3 is the level of own-state R&D capital, and S is the R&D spillover capital (going to other states). The translog function is viewed as a second-order Taylor's expansion of a more general cost function. At the point of expansion, the parameters of the variable cost function are equal to the first- and second-order derivatives of the approximation. For this interpretation, the symmetry constraints are

$$\alpha_{ij} = \alpha_{ji}, \beta_{ij} = \beta_{ji}, \gamma_{ij} = \gamma_{ji}. \tag{4}$$

Homogeneity of degree one in variable input prices requires

$$\sum_{i=1}^{N} \alpha_i = 1, \sum_{i=1}^{N} \alpha_{ij} = \sum_{i=1}^{N} \delta_{ij} = \sum_{i=1}^{N} \theta_{ij} = \sum_{i=1}^{N} \rho_{ij} = 0. \tag{5}$$

The approximating function to an underlying concave function will exhibit this property at the point of approximation. In order for $C(y, w, K)$ to be a concave function in w, it is necessary and sufficient that the matrix

$$\mathrm{A} = \nabla^2_{ww} C(w, y, K)$$

be negative semi-definite. Lau (1978) has shown that every negative semi-definite matrix has a Cholesky factorization. The matrix A can thus be written in terms of the Cholesky decomposition as $\mathrm{A} = LDL'$ where L is a unit triangular matrix ($L_{ii} = 1; L_{ij} = 0; j > i$) and D is a diagonal matrix with typical element D_{ii} referred to as a Cholesky value. Lau demonstrates that if the matrix A is to be negative semi-definite, then $D_{ii} \le 0$. Thus imposing $D_{ii} \le 0$ is sufficient to impose concavity.

Rather than reparameterizing the function to be estimated in terms of the Cholesky parameters and then restricting the D_{ii}, we express the Cholesky decomposition in terms of the parameters of the Hessian matrix and restrict a function of these parameters equivalent to the D_{ii}. That is:

$$D_{ii} = g(w, \Theta) \le 0, i = 1,\dots,N, \tag{6}$$

where Θ is the system parameter vector.

Using Shephard's lemma, we obtain:

$$\frac{\partial \ln C}{\partial \ln w_i} = \frac{w_i x_i}{C} = s_i \tag{7}$$

which when applied to (3),[4] yields:

$$s_i = \alpha_i + \sum_{j=1}^{N} \alpha_{ij} \ln w_j + \sum_{j=1}^{M} \delta_{ij} \ln y_j + \sum_{j=1}^{L} \theta_{ij} \ln K_j, i = 1,\dots,N \tag{8}$$

Two commonly used summary measures of price responsiveness are the Allen partial elasticities of substitution (σ_{ij}) and the price elasticity (η_{ij}) of demand. For the translog cost function, these measures can be calculated as:

$$\sigma_{ii} = \frac{\alpha_{ii} + s_i^2 - s_i}{s_i}, i = 1,\dots,N, \tag{9}$$

$$\sigma_{ij} = \frac{\alpha_{ij} + s_i s_j}{s_i s_j}, i, j = 1,\dots,N, i \ne j, \tag{10}$$

$$\eta_{ii} = s_i \sigma_{ii} \tag{11}$$

$$\eta_{ij} = s_j \sigma_{ij} \tag{12}$$

The effects on variable cost of a change in the fixed inputs, expressed as elasticities, are:

$$\frac{\partial \ln C}{\partial \ln K_l} = \gamma_l + \sum_{i=1}^{L} \gamma_{il} \ln K_l + \sum_{i=1}^{N} \theta_{il} \ln w_i + \sum_{i=1}^{M} \phi_{il} \ln y_i, l = 1,\dots,K \tag{13}$$

Estimation Procedures

The parameters to be estimated are contained in the system of equations (3) and (8). The data used in estimation are obtained by pooling time series (1960-96) and cross section (Minnesota, Iowa, Illinois, Missouri, and Indiana) information. It is unlikely that the data are drawn from a homogeneous sample due to differences in output composition and production technologies. We assume that differences in technologies (resulting from state variation) imply that the parameters of the cost function are state specific. However, to conserve degrees of freedom, we restrict state parameter variation to the zero- and first-order terms in the second-order expansion. This implies state-specific intercepts in the cost share equations.[5]

Our estimation procedure is described in Ball (1988) and is discussed only briefly here. The system of equations to be estimated can be written:

$$z_t - f(y_t, w_t, K_t \,|\, \Theta) = u_t \,, \tag{14}$$

where z_t is the vector of jointly dependent variables; y_t, w_t, and K_t are vectors of exogenous variables previously defined; and Θ is the system parameter vector. Assuming that the errors $(u_{1t}, \ldots, \mu_{nt})'$ are temporally independent, each with mean zero, the same distribution, and positive-definite error covariance matrix Σ,[6] the Aitken-type estimator Θ is obtained by minimizing

$$S(\Theta) = 1/T \sum_{t=1}^{T} (z_t - f_t)' (\hat{\Sigma} \otimes I)^{-1} (z_t - f_t). \tag{15}$$

Equation (15) is minimized with respect to Θ, given a prior consistent estimate of Σ. Using $\hat{\Theta}$, a new estimate of Σ is obtained based on the inner product of the estimated errors. The estimates are iterated until the coefficient vector $\hat{\Theta}$ and the covariance matrix $\hat{\Sigma}$ stabilize. It is well known that such iteration does not improve the asymptotic variance of the estimator (Madansky, 1966). However, when estimating a system of equations with constraints across one of the endogenous variables, such iteration results in estimates that are invariant to the equation deleted under the assumptions of the error structure (Berndt and Savin, 1976).

The estimation procedure must permit imposition of the theoretical curvature restrictions. This procedure is characterized by the nonlinear constrained optimization problem:

$$\min_{\Theta} S(\Theta) = 1/T \sum_{t=1}^{T} (z_t - f_t)' (\hat{\Sigma} \otimes I)^{-1} (z_t - f_t) \tag{16}$$

subject to:

$$D_{ii} = g_i(w, \Theta) \leq 0, i = 1, \ldots, N,$$

where $g_i(w, \Theta)$ are curvature restrictions. The objective function (16) is minimized subject to the constraint set using the Generalized Algebraic Modeling System (GAMS 2.25).

We modify the Aitken procedure by imposing the inequality constraints in the first stage, replacing $(\hat{\Sigma} \otimes I)^{-1}$ in equation (16) with the identity matrix, and then solving for $\hat{\Theta}$. Using Θ, a new estimate of Σ is based on the inner product of residuals. The problem in (16) is then solved with $(\hat{\Sigma} \otimes I)^{-1}$ employed as the weighting matrix. Finally, we iterate over steps one and two until the estimated parameter vector $\hat{\Theta}$ and the error variance-covariance matrix $\hat{\Sigma}$ stabilize.

Also see Greene (2000) for a discussion of SURE and Rothenberg (1974) for a discussion of efficiency gained through restrictions on the parameters to be estimated.

DATA

The USDA's Economic Research Service recently constructed state accounts for the farm sector consistent with a gross output model of production. Output is defined as gross production leaving the farm, as opposed to real value added. Inputs are not limited to labor and capital but include intermediate inputs as well. An overview of the sources and methods used to construct the annual production accounts for the period 1960 to 1996 is provided in Ball et al. (1999). These data are used to construct two output indices — livestock and crops — for each state. Variable inputs include service flows from land , hired labor, and materials. Self-employed and unpaid family labor (family labor),[7] infrastructure (i.e., roads and highways, and agricultural research capital)[8] are fixed factors.

Bell and McGuire (1997) have constructed the capital stock for highways. Their annual highway investment series consists of annual outlays for new highway construction and maintenance. The Federal Highway Administration's composite price index for highways was used to deflate investment expenditures. The highway capital stock for each state is the accumulation of past real annual investments adjusted for discards of worn-out assets and loss of efficiency of highway assets over their service lives. They assumed that highway asset discards were calculated based on an assumed mean service life for a group of assets and a distribution of actual discards around this mean service life. The frequency of occurrence of a particular service life was obtained as the ordinate under the standard normal curve. They assumed that efficiency loss is a function of asset age and can be approximated by a rectangular hyperbola concave to the origin.

The current stock of technical knowledge is determined largely by past agricultural research expenditures. Therefore, we could proceed by including, as separate regressors, a finite number of lagged research expenditures in equation (3). However, lagged research expenditures tend to be highly correlated. It is unlikely that one could estimate the separate contribution of each of these lagged expenditures with any precision (Griliches, 1979, 1998).

An alternative is to assume that a relationship exists between the level of technical knowledge, K_2, and past real agricultural research expenditures. This relationship can be represented by:

$$K_3 = W(B)R \tag{17}$$

where $W(B)$ is a lag polynomial describing the relative contributions of past research expenditures to the stock of K_2 and B is the lag operator. To approximate $W(B)$, Griliches (1979, 1998) provides useful insights that Huffman and Evenson (1993, 1994) adopted. They assumed a trapezoidal lag structure over 33 years. After a gestation period (2 years) during which the impacts of research are negligible, blossoming is assumed to occur over the next five years with increasing weights. This period of increase is followed by a period of maturity (5 years) during which the weights are constant. Finally, we observe a period of obsolescence and declining weights, spanning 19 years. After 33 years, the weights are zero. Over the first 33 years the weights sum to one.

We adopt the same set of weights for each of the five states. First, we construct the own-state stock component (K_3) for each of the five states using a state's own past public agricultural research expenditures. Second, we construct the agricultural-research spillin (or spillover) stock (S) for each state by adding together all the own-state research stocks for the five states and then subtracting a state's own agricultural research stock. The research capital variable for a state is the summation of its own contribution and the spillin contribution (Khanna, Huffman, and Sandler, 1994; McCunn and Huffman, 2000).

The annual agricultural research expenditure data for this study are taken from Huffman, McCunn, and Xu (2001). Nominal annual agricultural research expenditures are then deflated by an agricultural research price index reported in Huffman and Evenson's (1993) for 1927 to 1990, and their methods are applied to extend this price index to 1996. This index assumes that roughly 70% of research expenditures are labor costs, an assumption that is broadly consistent with available data on the composition of research expenditures.[9]

The Results

Table 7.1 contains the parameter estimates obtained by applying the non-linear constrained optimization procedure described above. The system of equations (3) and (8) is estimated subject to the constraints (4), (5), and (6). We also incorporate constraints consistent with convexity in the fixed factors.[10] Note that all parameter estimates are statistically significant at the 1-percent level.

Table 7.2 presents summary measures of the elasticites of demand. From the results in that table, a number of facts emerge. First, all own-price elasticites are negative. Moreover, all factor demands are price inelastic. Pair-wise substitution dominates. Capital and land, and capital and hired labor, exhibit strong substitutability. There also appears to be scope for substitution between materials and hired labor.

The estimates of the elasticity of variable cost with respect to the fixed factors evaluated at the sample mean are reported in Panel B of table 7.2. All elasticites are negative implying that an increase in the factor endowment results in lower variable cost. The magnitudes of the elasticities range from –0.42, the elasticity of variable cost with respect to self-employed labor, to –0.87, the elasticity of variable cost with respect to own R&D capital.

The estimated cost function can be used to measure the rate of return to investment in public agricultural research. Our cost function specification includes two research variables — the state's own R&D capital stock and the potential spillover from other states' R&D investments. This specification permits the assessment of the separate contributions of a state's own research investment and that of the remaining states.

The discounted benefits from a one-unit increase in public agricultural research expenditures in a given state in time t is R_t:

$$\Delta R_t = -\sum_{i=0}^{m} \frac{\Delta C_{t+i}}{(1+r)^i} \tag{18}$$

Hence, the discounted benefits are obtained by discounting the future variable cost saving over years t to m. The internal rate of return for a given state can then be computed as r, a percentage rate of return, by solving the following equation:

$$1 = \sum_{i=0}^{m} \frac{\Delta C_{t+i}}{\Delta R_t} \cdot \frac{1}{(1+r)^i} = \sum_{i=0}^{m} \frac{\Delta C_{t+i}}{\Delta K_{3,t+i}} \cdot \frac{\Delta K_{3,t+i}}{\Delta R_t} \cdot \frac{1}{(1+r)^i} \tag{19}$$

where $K_{3,t-i}$ is its own public research stock and

$$\frac{\Delta K_{3,t+i}}{\Delta R_t} = \omega_i \tag{20}$$

Table 7.1. Iterative SURE Parameter Estimates of the Translog Cost Function with Symmetry, Homogeneity, and Concavity Restrictions Imposed, 1960-1996: Five Midwestern States (n=925)

Parmeter and variables		Coefficients	Asymptotic	
			Standard errors	t-value
α_0	Intercept	-0.096	0.025	-3.902
α_i	Price of capital services[1]	0.214	0.026	8.262
P1.St1-IL	Illinois	-0.016	0.005	-3.404
P1.St2-IN	Indiana	-0.032	0.005	-6.153
P1.St3-MN	Minnesota	0.056	0.006	9.180
P1.St4-MO	Missouri	-0.255	0.042	-5.944
α_2	Price of land services	0.108	0.025	4.268
P2.St1-IL	Illinois	-0.017	0.004	-4.358
P2.St2-IN	Indiana	-0.036	0.004	-8.372
P2.St3-MN	Minnesota	-0.018	0.005	-3.461
P2.St4-MO	Missouri	0.024	0.005	4.897
α_3	Wage rate of hired labor	0.193	0.024	7.845
P3.St1-IL	Illinois	-0.078	0.008	-9.873
P3.St2-IN	Indiana	0.050	0.005	9.433
P3.St3-MN	Minnesota	-0.004	0.001	-6.666
P3.St4-MO	Missouri	0.624	0.065	9.468
β_1	Livestock output	1.336	0.284	4.689
β_2	Crops output	-0.260	0.039	-6.532
γ_1	Self-employed labor	0.087	0.025	3.466
γ_2	Highway stock	0.140	0.018	7.407
γ_3	Public research stock (own)	0.255	0.077	3.277
α_{ij}	Variable input price Interaction terms			
P1.P1		-0.019	0.002	-9.047
P1.P2		0.095	0.025	3.754
P1.P3		0.104	0.007	13.684
P2.P2		0.022	0.002	8.800
P2.P3		-0.113	0.024	-4.612
P3.P3		0.042	0.012	3.281
β_{ij}	Output interaction terms			
Y1.Y1		2.273	0.289	7.862
Y2.Y1		0.424	0.030	14.133
Y2.Y2		0.113	0.019	5.947

[1]The first α_i row of coefficients is for Iowa. $P_i = \ln w_i$, $Y_i = \ln Y_i$, $K_i = \ln K_i$, and $S = \ln S$.

Table 7.1, *continued*

	Parmeter and variables	Coefficients	Asymptotic Standard errors	Asymptotic t-value
γ_{ij}	Quasi-fixed factor interaction terms			
K1.K1		0.079	0.020	3.930
K1.K2		-0.012	0.001	-7.500
K1.K3		-0.022	0.002	-7.586
K2.K2		0.120	0.008	13.483
K2.K3		-0.036	0.008	-4.500
K3.K3		0.190	0.019	9.595
δ_{ij}	Output quantity and variable input price interaction terms			
Y1.P1		-0.225	0.019	-11.658
Y1.P2		-0.018	0.002	-7.200
Y1.P3		0.249	0.029	8.412
Y2.P1		-0.006	0.001	-5.000
Y2.P2		-0.012	0.002	-4.800
Y2.P3		-0.048	0.017	-2.807
θ_{ij}	Variable input price and fixed factor interaction terms			
P1.K1		0.150	0.018	7.936
P1.K2		0.052	0.012	4.262
P1.K3		-0.107	0.025	-4.147
P2.K1		-0.004	0.001	-3.636
P2.K2		0.054	0.013	4.090
P2.K3		0.037	0.004	8.604
P3.K1		0.002	0.000	3.333
P3.K2		-0.045	0.010	-4.245
P3.K3		0.230	0.025	9.387
ϕ_{ij}	Output quantity and fixed factor interaction terms			
Y1.K1		-0.200	0.044	-4.535
Y1.K2		-0.701	0.077	-9.103
Y1.K3		1.446	0.390	3.706
Y2.K1		0.133	0.015	8.750
Y2.K2		-0.200	0.048	-4.115
Y2.K3		0.061	0.007	7.922
ρ_{is}	Prices and spillover			
P1.S		0.474	0.089	5.326
P2.S		2.750	0.271	10.147
P3.S		9.401	1.642	5.725
ζ_s	Spillover and own research			
S.K3		0.011	0.003	3.666

Table 7.2. Elasticities

Panel A. Variable input demand elasticities associated with the cost function (evaluated at the overall sample mean of the data)

Inputs Materials	Prices Capital services	Land services	Hired Labor	Material
Capital services	-0.868	0.622	0.707	-0.461
Land services	0.731	-0.697	-0.387	0.353
Hired labor	0.681	-0.317	-0.589	0.225
Material	-0.267	0.174	0.135	-0.042

Panel B. Elasticity of variable cost due to an increase in fixed factors (evaluated at the overall sample means of the data)

Family Labor	- 0.420
Highway Stock	-0.477
Public Research Stock:	
Own	-0.866
Spillover	-0.665

where the research expenditure weights used to construct the stock of public agricultural research are the T_i's. The impacts on variable cost from an increase in a state's public agricultural research stock (ignoring spillovers) is:

$$\frac{\Delta C_{t+i}}{\Delta K_{3,t+i}} = \left[\gamma_3 + \sum_{i=1}^{L} \gamma_{i3} \ln K_l + \sum_{i=1}^{N} \theta_{i3} \ln w_i + \sum_{j=1}^{M} \phi_{j3} \ln y_j \right] \cdot \frac{C_{t+i}}{K_{3,t+i}} \tag{21}$$

Hence, the average own-state rate of return-ignoring spillovers (r_l) can be obtained by substituting equations (20) and (21) into (19), giving equation (22), and then solving for r_l:

$$1 = \sum_{i=0}^{m} \left[\gamma_3 + \sum_{i=1}^{L} \gamma_{i3} \ln K_l + \sum_{i=1}^{N} \theta_{i3} \ln w_i + \sum_{j=1}^{M} \phi_{j3} \ln y_j \right] \cdot \frac{C_{t+i}}{K_{3,t+i}} \omega_i \frac{1}{(1+r_1)^i} \tag{22}$$

Turning to spillover (spill) effects, we modify (22), by replacing ln C, ln w_i, ln y_j, C_{t+i}, and K_3 with their respective sample mean values:

(23)

$$1 = \left[\gamma_3 + \sum_{i=1}^{L} \gamma_{i3} \ln \overline{K}_l + \sum_{i=1}^{N} \theta_{i3} \ln \overline{w}_i + \sum_{j=1}^{M} \phi_{j3} \ln \overline{y}_j \right] \cdot \frac{\overline{C}}{\overline{K}_3} . \sum_{i=0}^{m} \left[\omega_i \frac{1}{(1+r_1)^i} \right]$$

where r_1 is the rate of return within a state of its research. The marginal internal rate of return, looking only at public agricultural research spillover (spillin) benefits (r_2), is obtained by solving:

(24)

$$1 = \left[\left(\gamma_3 + \sum_{i=1}^{L} \gamma_{i3} \ln \overline{K}_l + \sum_{i=1}^{N} \theta_{i3} \ln \overline{w}_i + \sum_{j=1}^{M} \phi_{j3} \ln \overline{y}_j \right) \frac{\overline{C}}{\overline{K}_3} + (n-1)\zeta_s \frac{\overline{C}}{\overline{S}} \right] . \sum_{i=0}^{m} \left[\omega_i \frac{1}{(1+r_2)^i} \right]$$

where n is the number of states in our sample and (n-1) is the number of states where spillover/spillin benefits, ζ_s, is the coefficient associated with research spillover stocks in equation (3), and $\overline{S}$ is the average research spillin/spillover stock for the 5 sample states.

Using parameter estimates from table 7.1 and sample mean values for all the variables, a one-unit increase in agricultural research expenditure yields an (average) marginal internal rate of return from within-state benefits (r_1) of 11 percent. The (average) marginal rate of return due to spillover benefits (r_2) is 43 percent. Because the one-unit cost is charged twice rather than only once (probably the best approach), the marginal rate of return, including average within-state and average-spillover benefits combined, will significantly exceed 40 percent.

CONCLUSIONS

In the study we examined the cost and input demand structure of agricultural production in five midwestern states – Illinois, Indiana, Iowa, Minnesota, and Missouri. Specifically, we estimated the impact of a state's public agricultural research stock on its cost of production and its effects on neighboring states with similar crop and livestock patterns.

For this purpose, we estimated a translog cost function with two outputs (livestock and crops), four variables inputs (capital services, land services, hired labor, and materials) and four fixed factors (own R&R capital stock, research capital of other states, infrastructure, and self-employed and unpaid family labor). The estimation was carried out using inequality constrained optimization procedures where the constraints were those implied by theory.

The own-price elasticities of demand, evaluated at sample means, were negative. Moreover, all factor demands were price inelastic. Pair-wise substitution dominates.

The impacts on variable costs of increasing the factor endowments are large and negative. The internal rate of return to investment in agricultural research was 11 percent per annum. Including research spillovers boosts the estimated rate of return to 43 percent. This result has important implications for research planning. A state's research stock confers distinct competitive advantages to its producers, but it also reduces cost of production of states with similar production patterns. Hence, independent state planning of agricultural research may be socially inefficient.

Notes

1. The authors are Professor, Department of Economics, Iowa State University; Economist, ERS-USDA; Associate Professor, Department of Agricultural and Resource Economics, Oregon State University; and Economist, ERS-USDA. They received helpful comments from two reviewers of an earlier draft of the paper, and participants in the joint NC208, ERS, and Farm Foundation sponsored conference, "Agricultural Productivity: Data, Methods, and Measures," Washington, D.C., March 1999. Financial assistance has been obtained from ERS, the Iowa Agricultural and Home Economics Experiment Station, and the Oregon Agricultural Experiment Station.

2. A profit function methodology is an alternative route to an estimable model. It does get free of potential endogeneity of output quantities. A tradition remains of using cost functions for aggregate data, e.g., Morrison and Schwartz (1996); Pope and Just (1996). However, in a cost function, it is easy to assess returns to scale in outputs.

3. Alternative specifications of the cost function include generalized Leontief and generalized quadratic (see Diewert, 1973). However, the generalized Leontief cost function with its fixed-proportion foundation is unduly restrictive on input substitution. Because the translog is less restrictive, it seems to be more realistic when aggregated data are used for estimation. However, permitting greater substitution may require that curvature restrictions be imposed on the cost function.

4. The term involving the parameter z_s from the cost function, equation (3), was ignored in these share equations. The consequence will be that all estimated parameters are asymptotically unbiased, but they are not asymptotic most efficient.

5. However, we do permit the intercept terms in equation (8), which means that the intercept terms in the cost-share equation differ across states. None of the other coefficients in (8) are permitted to differ across states. Freeing the intercept term in the share equation also implies that state dummy variables are added to the cost function, equation (3), in the parts that involves a_i, $i = 1, 2, 3$.

6. Note that å is a symmetric matrix. Diagonal elements are variances of the set variances for disturbances in the equations being estimated, and the off-diagonal elements are the covariances' cross-equation disturbances of these equations. No attempt is made to incorporate inter-spatial correlation across states.

7. The idea of fixed fixity comes from sluggish response to opportunity cost change or from the input being outside the control of farmers. Perspectives differ on the issue of whether family labor is a fixed or variable input. In this paper, we assume that family labor responds sluggishly to changes in opportunity costs. Hence, family labor (self-employed and unpaid family labor) is treated as a fixed input.

8. Although private agricultural research undoubtedly affects the stock of agricultural technology, further work is required to create this variable. Excluding private agricultural research might bias the coefficients associated with public agricultural research and other variables.

9. Of course alternative methods exist for constructing agricultural research stock measures. Other methods used to create an agricultural research price index, e.g., Alston and Pardey, 1996, and Fuglie et al., 1996, yields an agricultural research price index that is highly correlated with one constructed using the Huffman and Evenson method. Weights or lag structure differ greatly and these differences will make the agricultural research stock variables behave very differently over time, even when the annual agriculture-research expenditure data are the same.

10. See Lau (1974) for a discussion of the economics of convexity.

REFERENCES

Alston, J. M., and P. G. Pardey. 1996. *Making Science Pay*. Washington, D.C.: American Enterprise Institute.

Alston, J., B. Craig, and P. Pardey. 1999. "Research Lags and Research Returns." In *Paying for Agricultural Productivity*, ed. J. M. Alston, P. G. Pardey, and V. H. Smith.. Baltimore, Md.: The Johns Hopkins University Press, 333-65.

Anselin, L., A. Varga, and Z. Acs. 1997. "Local Geographic Spillovers between University Research and High Technology Innovations." *Journal of Urban Economics*, 42:422-48.

Ball, V. E. 1988. "Modeling Supply Response in a Multiproduct Framework." *American Journal of Agricultural Economics,* 70(November):813-25.

Ball, V. Eldon, F. Gollop, A. Kelly-Hawke, and G. Swinand. 1999. "Patterns of Productivity Growth in the U.S. Farm Sector: Linking State and Aggregate Models." *American Journal of Agricultural Economics,* 81:164-79.

Barten, A. P. 1969. "Maximum Likelihood Estimation of a Complete System of Demand." *European Economic Review*, 1(Fall):7-73.

Bell, M. E., and T. J. McGuire. 1997. "Macroeconomic Analysis of the Linkages Between Transportation Investments and Economic Performance," NCHRP Report 389, Transportation Research Board, National Research Council, Washington, D.C.: National Academy Press.

Berndt, E. R., and N. E. Savin. 1976. "Estimation and Hypothesis Testing in Singular Equation Systems with Autoregressive Disturbances." *Econometrica,* 43:937-57.

Bernstein, J. I., and X. Yan. 1997. "International R&D Spillovers between Canadian and Japanese Industries." *Canadian Journal of Economics*, 30:276-94.

Binswanger, H. 1974. "The Measurement of Technical Change Biases with Many Factors of Production." *American Journal of Agricultural Economics,* 64(December):964-76.

Committee on the Future of the Colleges of Agriculture in the Land Grant College System, Board on Agriculture. 1995. *Colleges of Agriculture at Land Grant Universities: A Profile*. Washington, D.C.: National Academy Press.

Deininger, K. W. 1995. *Technical Change, Human Capital, and Spillovers in United States Agriculture, 1949-1985*. New York: Garland Pub. Inc.

Diewert, W. E. 1973. "Functional Forms for Profit and Transformation Functions." *Journal of Economic Theory*, 6:284-316.

Evenson, Robert E. 2001, forthcoming. "Economic Impact Studies of Agricultural Research and Extension." Chapter for *Handbook of Agricultural Economics*, volume 1, ed. B. L. Gardner and G. Rausser. Amsterdam, Netherlands: Elsevier Science/North Holland.

Fuglie, K., N. Ballenger, K. Day, C. Klotz, M. Olinger, J. Reilly, U. Vasavada, and J. Yee. 1996. *Agricultural Research and Development. Public and Private Investments Under Alternative Markets and Institutions.* Agricultural Economic Research #735, Economic Research Service, USDA.

Greene, W. H. 2000. *Econometic Analysis.* 4th edition. Upper Saddle River, N. J.: Prentice Hall, Inc.

Griliches, Z. 1998. *R&D and Productivity: The Econometric Evidence*. Chicago: The University of Chicago Press.

Griliches, Z. 1979. "Issues in Assessing the Contribution of Research and Development to Productivity Growth." *Bell Journal of Economics,* 10(Spring):92-116.

Griliches, Z. 1995. "R&D and Productivity." In *Handbook of the Economics of Innovation and Technological Change*, ed. P. Stoneman. Cambridge, Mass.: Blackwell Publishers.

Huffman, W. E., and R. E. Evenson. 1989. "Supply and Demand Functions for Multiproduct U.S. Cash Grain Farms: Biases Caused by Research and Other Policies." *American Journal of Agricultural Economics,* 71(August):761-73.

Huffman, W. E., and R. E. Evenson. 2000. *The Development of U.S. Agricultural Research and Education: An Economic Perspective*. Ames: Iowa State University, Department of Economics (available at the USDA, National Agricultural Library).

Huffman, W. E., and R. E. Evenson. 1994. *The Development of U.S. Agricultural Research and Education: An Economic Perspective*. Ames: Iowa State University, Department of Economics (available at the USDA, National Agricultural Library).

Huffman, W. E., and R. Evenson. 1993. *Science for Agriculture*. Ames: Iowa State University Press.

Huffman, W. E., A. McCunn, and J. Xu. 2001, forthcoming. "Public Agricultural Research Expenditures with an Agricultural Productivity Emphasis: Data for 48 States, 1927-1995." Iowa State University, Department of Economics, Staff Paper.

Khanna, J., W. E. Huffman, and T. Sandler. 1994. "Agricultural Research Expenditure in the United States: A Public Goods Perspective." *Review of Economics and Statistics,* 66(May):267-77.

Lau, L. J. 1974. "Economics of Monotonicity, Convexity, and Quasi-Convexity." Technical Report No. 123, The Economic Series, Institute for Mathematical Studies in the Social Sciences, Stanford University, California.

Lau, L. J. 1978. "Testing and Imposing Monotonicity, Convexity, and Quasi-Convexity Constraints." In *Production Economics: A Dual Approach to Theory and Applications,* ed. D. McFadden and M. Fuss. Amsterdam: North Holland Publishing Co.

Lim, H., and C. R. Shumway. 1997. "Technical Change and Model Specification: U.S. Agricultural Production." *American Journal of Agricultural Economics,* 79(May):543-54.

Madansky, A. 1966. "On the Efficiency of Three-Stage Least Squares Estimation." *Econometrica,* 34:851-61.

McCunn, A., and W. E. Huffman. 2000. "Convergence in U.S. TFP Growth for Agriculture: Implications of Interstate Research Spillovers for Funding Agricultural Research." *American Journal of Agricultural Economics,* 82(May):370-88.

Morrison, C. J., and A. E. Schwartz. 1996. "State Infrastructure and Productivity Performance." *American Economics Review,* 85(December):1095-111.

Pope, R., and R. E. Just. 1996. "Empirical Implementation of ex ante Cost Functions." *Journal of Econometrics,* 72:231-49.

Rothenberg, T. J. 1974. *Efficient Estimation with* a priori *Information*. New Haven, Conn.: Yale University Press.

Shumway, C. R., and W. P. Alexander. 1988. "Agricultural Product Supplies and Input Demands: Regional Comparisons." *American Journal of Agricultural Economics,* 80 (February):153-61.

U. S. Department of Agriculture, Economic Research Service. 1998. *Agricultural Exports: Estimated Value by Commodity Group and State* (available at http://www.ers.usda.gov/Data/FATUS/).

Appendix Table 7.1. Agricultural Output, Exports, Export Share of Farm Receipts, Productivity, and Public R&D Expenditures for Selected States

State	Output index[a]	Exports[b] mil.$	Export Share (%)	Exports/ acre ($)	TFP level[c]	R&D Exp.[d]
Florida	2.474	900	19	87	1.819	93.8
Iowa	5.731	4100	32	124	1.571	70.0
Washington	2.180	1900	35	121	1.556	36.5
Arkansas	2.343	1900	33	128	1.432	33.7
Kentucky	1.786	1100	31	79	1.428	29.7
California	8.674	7700	30	257	1.386	148.7
Minnesota	3.903	2600	32	87	1.369	55.4
Illinois	3.804	3700	40	132	1.322	36.1
Louisiana	0.970	704	34	83	1.299	39.4
Indiana	2.355	1900	34	120	1.258	56.7
Mississippi	1.475	821	27	66	1.250	35.8
Kansas	3.784	2700	29	56	1.248	44.8
Average [e]	**1.1791**	**1200**	**23**	**74**	**1.242**	**40.6**
Alabama	1.173	535	17	55	1.209	35.7
Michigan	1.655	890	25	85	1.187	49.3
New Mexico	0.785	70	4	2	1.171	14.9
Arizona	0.859	370	17	11	1.119	43.8
Virginia	1.027	520	22	61	1.108	39.0
Utah	0.437	125	13	11	1.104	20.4
Nevada	0.161	6	2	1	1.034	9.4
Tennessee	1.041	540	24	46	0.937	26.9
Montana	2.388	327	13	6	0.855	16.8
Oklahoma	1.605	385	15	11	0.845	29.5
Wyoming	0.385	40	5	1	0.761	5.8
W. Virginia	0.194	31	8	8	0.586	6.7

[a] Output index for 1996 and relative to 1987 level of Alabama.
[b] Export data are for 1997.
[c] Table is sorted by level of TFP, which are relative to 1987 level of Alabama.
[d] R&D expenditures for 1995.
[e] Average for 43 states for which data are available.
Data sources: FAS/USDA, Ball et al., and Huffman and Evenson.

8

Sources of Agricultural Productivity Growth at the State Level, 1960-1993

Jet Yee, Wallace E. Huffman, Mary Ahearn, and Doris Newton

Abstract

This paper examines the contributions of public agricultural research, extension, and highway infrastructure to agricultural productivity of regions composed of groups of states for the period 1960 to 1993. It estimates separate contributions of a state's own public agricultural research, spillovers of agricultural research from other states, and the interaction effects between a state's own public agricultural research and agricultural extension.

Introduction

Aggregate U.S. agriculture has experienced relatively high levels of total factor productivity (TFP) compared to other sectors of the economy (Jorgenson and Gollop 1992). For the period 1960-1993, the average annual rate of growth for the U.S. agricultural sector was 1.99 percent (Ball et al., 1997). Aggregate U.S. agricultural productivity growth rates, however, mask variations across states and regions (Huffman and Evenson, 1993, p.199; Ball et al., 1999). The source of generally high TFP growth in agriculture is often identified with the relatively large investment in the public-supported USDA-Land Grant University System of research (R&D) and extension. Having a better understanding of the sources of TFP growth and the rate of return to public resources invested in research and extension is important for sound public policy decision-making.

Prior efforts to explain the sources of state TFP growth have been undertaken by Huffman and Evenson (1993) and Alston, Craig, and Pardey (1998). The Huffman and Evenson data set covered the period 1950-1982 for 42 U.S. states. They used public and private research stocks and agricultural extension stocks to explain TFP. Their public agricultural research stock variable was constructed from a subset of all USDA, state agricultural experiment station, and veterinary college research expenditures. Their subset contained research expenditures on identifiable farm-level outputs that might reasonably be expected to impact farm-level agricultural productivity (Huffman and Evenson, 1994). Furthermore, they allocated the public expenditures on agricultural research to state agricultural sub-regions (Huffman and Evenson, 1993, p. 195). These expenditures were then used to create one public agricultural research stock variable for each state that contained both a state's own contribution and spill-in contributions from other sub-regions in adjacent states. Although the impacts of public agricultural research were generally positive on agricultural productivity, applied livestock research had a negative impact on livestock sector productivity. Also, impacts of agricultural extension were positive in the crop sector but negative in the livestock sector, and interaction effects between public agricultural research and extension were positive in the livestock sector and negative in the crop sector.

Later, Alston, Craig, and Pardey (1998) constructed a state level productivity data set for 48 states, 1949-1991, and examined the impacts of a single combined public agricultural research and extension variable on TFP. They used essentially all of the public agricultural research expenditures of state agricultural experiment stations to construct the research and extension stock variable, irrespective of whether the research was focused on production agriculture. Interstate spillovers were generally excluded. They found positive effects of the combined public agricultural research and extension variable on agricultural productivity.

The objective of this chapter is to examine the contributions of public agricultural research, extension, and highway infrastructure to agricultural productivity of regions composed of groups of states for the 1960 to 1993 period. In addition, we examine separate contributions of a state's own public agricultural research, spillovers from agricultural research in other states, and the interaction effects between a state's own public agricultural research and agricultural extension. Our results add new information about the relative importance of research spillovers and interaction effects between public agricultural research and extension.

The study uses newly constructed state TFP, public agricultural research expenditures, and agricultural extension expenditures data (See chapter 2 in this volume). One notable feature of the new output and input indices used to calculate the TFPs is the careful attention paid to quality changes in the labor, capital, and materials inputs. However, conventional inputs such as land, labor, capital, and materials will likely not fully account for changes in productivity over time even if the conventional inputs have been adjusted for changes in quality. Huffman, McCunn, and Xu (2000) recently made a major revision of the earlier Huffman and Evenson state public agricultural research data set covering 1927-1982 and extended it to 1995.

Background on Sources of Agricultural Productivity

In previous econometric studies of agricultural productivity, public and private research, extension, and farmers' schooling have been identified as the primary sources of productivity growth. We begin by focusing on each of these three potential sources of productivity gains.

Research and Development

Agricultural research is performed both by the private and the public sectors. One social benefit of this research is an increase in agricultural productivity. The output of agricultural research includes higher-yielding crop varieties, better livestock-breeding practices, more effective fertilizers and pesticides, and better farm-management practices. Agricultural research is required not only to increase agricultural productivity but also to offset biological effects of pests that would depreciate earlier gains. For example, yield gains for a particular wheat variety tend to be lost over time because pests and diseases evolve, making the variety susceptible to attack even though the variety was immune when it was originally introduced. Thus, a significant share of agricultural research expenditures is devoted to maintenance research (Huffman and Evenson, 1993, p. 114).

Public agricultural research is performed in state agricultural experiment stations, land grant and other universities, and the USDA's Agricultural Research Service and Economic Research Service. Agricultural research is also performed by the private sector, mainly in the areas of farm machinery, agrichemicals and pharmaceuticals, and food processing. Private research expenditures have increased dramatically during the past three decades and now surpass those of the public sector (Fuglie et al., 1996; Fuglie, 2000). By contrast, the rate of growth in public research expenditures has slowed significantly since the mid-1970s. While public research expenditures have stagnated, the demands on the agricultural research system have expanded beyond reducing the cost of agricultural production to also include protection of the environment and food safety.

When the benefits of public agricultural research are measured as impacts on agricultural productivity, the implied social rate of return is generally large. Early summaries include Ruttan (1980, 1982); Echeverria (1990); Huffman and Evenson (1993); and Fuglie et al., and a more recent summary of more than 300 studies internationally is in Evenson (2000). For U.S. agriculture, the real rate of return is high, between 20 percent and 60 percent. The high rates of return to public agricultural research emerge regardless of the level of aggregation (individual commodities or more aggregate measures) or geographical area considered.

Extension

Agricultural extension is engaged in decoding scientific and market information and in training. With heterogeneous farming conditions, extension activity helps farmers make good production and marketing decisions. Extension agents disseminate information on crops, livestock, and management practices to farmers and demonstrate new techniques as well as consult directly with farmers on specific production and management problems. In particular, giving farmers good information on new technologies can speed the adoption process (Wozniak, 1993; Huffman, 2000), and thus generally increases the rate of return on research expenditures. Unlike research, agricultural extension input can be expected to have an almost immediate impact on agricultural productivity.

Little growth in real public extension expenditures has occurred since 1980. The federal share of public extension expenditures has steadily fallen during the past few decades. The bulk of public extension funding now comes from state and county governments rather than the federal government. There is increasing involvement by the private sector in extension activities, as well. For example, private crop consultants offer advice on pest and nutrient management practices (Board on Agriculture, 1997).

The empirical evidence on the social rate of return to public agricultural extension is quite mixed. Estimates range from negative values to over 100 percent (Fuglie et al., 1996; Evenson, 2000). Huffman and Evenson (1993) found a positive return to extension in the crop sector but a negative rate of return in the livestock sector. In recent years, a major problem in estimating the rate of return to extension has been the frequent change in definitions of extension activities by the USDA. Over time, a larger proportion of extension expenditures has gone to non-production related activities such as non-farm natural resources and human nutrition.[1]

Infrastructure

The transportation of agricultural inputs and outputs in modern agriculture requires good infrastructure, especially roads and communications. Aschauer (1989) argued that declining public capital stocks were a drag on productivity in the non-farm sector during the 1970s. Since that time, several studies have investigated the impact of public infrastructure (highways and streets, water and sewer systems, schools, hospitals, conservation structures, mass transit, etc.) on productivity outside of agriculture.

For the non-farm sector, the empirical evidence is that public infrastructure has a positive and statistically significant impact on output and productivity. This finding is even more impressive given that much public infrastructure spending goes for improving the environment and other objectives that are not captured in output or productivity measures (as conventionally measured). This finding also implies that the rate of return to public infrastructure investment may be under estimated because of the neglect of environmental and other benefits.

Little research, however, has examined the effects of public infrastructure on agricultural productivity in the United States. Antle (1983) found a positive contribution of transportation and communication infrastructure on agricultural productivity for a cross-section of 66 countries. More recently, Gopinath and Roe (1997) found a significant positive relationship between infrastructure and U.S. agricultural productivity at the national level. No previous studies have considered infrastructure as a determinant of agricultural productivity at the state level.

The Model

Assume an aggregate production function with disembodied technical change: where Q is all types of outputs aggregated into one output index, A is a constant, F( ) is a well-behaved function, K is physical capital input, L is quality-adjusted labor input, M is quality-adjusted materials input, ownrd is a state's own public research stock, spillin is the public research stock from other states, ext is public extension stock, and hiway is highway stock. (Since the measure of labor input we use accounts for the changing educational attainment of the farm workforce over time, we do not explicitly consider education as a determinant of output growth.)

$$Q = AF(K, L, M)(ownrd)^{\delta_1}(spillin)^{\delta_2}(\text{ext})^{\delta_3}(\text{hiway})^{\delta_4} \tag{1}$$

Now define total factor productivity (P) as:

$$P \equiv \frac{Q}{F(K,L,M)} = A\ (ownrd)^{\delta_1}(spillin)^{\delta_2}(\text{ext})^{\delta_3}(\text{hiway})^{\delta_4} \tag{2}$$

Taking natural logarithms of both sides of equation (2) and adding an interaction between ownrd and ext, dummy variables for extreme weather conditions, and a random disturbance term, we obtain the following model that is linear in the unknown parameters (δ_j's):

$$\begin{aligned} \ln P = \ln A + \delta_1 \ln(ownrd) + \delta_2 \ln(spillin) + \delta_3 \ln(ext) \\ + \delta_4 \ln(hiway) + \delta_5 \ln(ownrd) \cdot \ln(ext) + \delta_6 drought + \delta_7 flood + u. \end{aligned} \tag{3}$$

With the interaction term included, the output elasticity of ownrd is $\delta_1 + \delta_5$ ln(ext) and of ext is $\delta_3 + \delta_5$ ln(ownrd). Since we estimate equation (3) for each of seven regions, we also include state dummy variables in equation (3) to obtain our final econometric model.

Given estimates of the parameters in equation (3), own-state (private) and social (including inter-state spillovers) impacts of public agricultural research on agricultural productivity/output can be evaluated. The numbers can then be used to compute rate of return estimates to an increment in public agricultural research.

For simplicity, first assume no interstate spillovers of public agricultural research in one state on agricultural productivity in other states. The internal rate of return (r) for an additional research expenditure of ΔR_t in period t is the discount rate that results in the following equality:

$$\Delta R_t = \sum_{i=0}^{m} \frac{\Delta Q_{t+i}}{(1+r)^i} \tag{4}$$

or

$$1 = \sum_{i=0}^{m} \frac{\Delta Q_{t+i}}{\Delta R_t} \cdot \frac{1}{(1+r)^i} = \sum_{i=0}^{m} \frac{\Delta Q_{t+i}}{\Delta T_{t+i}} \cdot \frac{\Delta T_{t+i}}{\Delta R_t} \cdot \frac{1}{(1+r)^i} \tag{5}$$

where Q is a state's output and T is its own public research stock (ownrd). Since the output elasticity of ownrd is $\delta_1 + \delta_5$ ln(ext),

$$\frac{\Delta Q_{t+i}}{\Delta T_{t+i}} = [\delta_1 + \delta_5 \ln(ext_{t+i})] \cdot \frac{Q_{t+i}}{T_{t+i}} \tag{6}$$

and

$$\frac{\Delta T_{t+i}}{\Delta R_t} = w_i \tag{7}$$

where the w_i's are the weights used to construct the research stock in equation (3). Hence, we can rewrite equation (5) as

$$1 = \sum_{i=0}^{m} \left[\delta_1 + \delta_5 \overline{\ln(ext)}\right] \cdot \frac{\overline{Q}}{\overline{T}} \cdot w_i \cdot \frac{1}{(1+r)^i} = \left[\delta_1 + \delta_5 \overline{\ln(ext)}\right] \cdot \frac{\overline{Q}}{\overline{T}} \cdot \left[\sum_{i=0}^{m} w_i \cdot \frac{1}{(1+r)^i}\right] \tag{8}$$

where we have replaced ln(ext_{t+i}), Q_{t+i}, and T_{t+i} with their respective sample mean values.

With spillovers, equation (8) must be modified. An additional research expenditure of $\Delta R_{i,t}$ by state i in period t has two impacts. First, it affects state i's own research stock with a direct impact on state i's output. Second, it affects the research spillover stock available to all other states in the same region as state i with indirect impacts on those other states' outputs. By a similar derivation, we obtain

$$1=\left[\left[\delta_1+\delta_5\overline{\ln(ext)}\right]\cdot\frac{\overline{Q}}{\overline{T}}+(n-1)\cdot\delta_2\cdot\frac{\overline{Q}}{\overline{S}}\right]\cdot\left[\sum_{i=0}^{m}w_i\cdot\frac{1}{(1+r)^i}\right] \tag{9}$$

where $\overline{S}$ is the average research stock from other states in the same region (spillin) and n is the number of states in the region. With positive interstate spillovers effects of public agricultural research on productivity of other states and other things equal, we see that the internal rate of return, r, must increase from equation (9) relative to equation (8). That is, the social (accounting for impacts in all states) rate of return from the investment in public agricultural research by one state must be greater than the private rate of return, or return as any one state sees it. Furthermore in equation (9), the internal rate of return must increase as n, the number of states in the region, increases, other things equal. This is just the regional public good attribute playing itself out. When Q, T, and S are in real terms, the r's computed are marginal real internal rates of return.

The Data and Variables

Considerable effort has gone into the construction of the new data used in this study, and the states are grouped together into seven regions using regional boundaries defined by McCunn and Huffman (2000) and Khanna, Huffman, and Sandler (1994). The choice of regional boundaries is always somewhat subjective, but the McCunn and Huffman study found these boundaries to be adequate for a study of convergence in state agricultural TFP growth rates, and Khanna, Huffman, and Sandler found them adequate for a study of state government decisions on funding state agricultural experiment stations.[2] See table 8.1 for an abbreviated description of the variables.

Data on total factor productivity (TFP) by state are the ERS estimates of Ball and Nehring (1998), reported on in chapter 2 of this volume.[3] The TFP numbers for each state are spatially adjusted so that they are comparable across states.

Public agricultural research expenditure data were compiled by Huffman, McCunn, and Xu (2000). The Huffman and Evenson (1993, 1994) research variables stopped in 1982, while the new series is extended to 1995, uses re-

Table 8.1. Variable Definitions for State Agricultural Sector Data

Variable	Definition
P	Total factor productivity
ownrd	Own research stock
spillin	Spill-in research stock
ext	Extension stock per farm
hiway	Highway stock
drought	Drought dummy
flood	Flood dummy

fined methods, incorporates new historical data on non-USDA federal expenditure on SAES research 1955-65, and builds on actual annual SAES expenditure/receipts over the 1927-1948 period.

The public agricultural research sector in each state consists of the state institutions — state agricultural experiment stations and schools/colleges of veterinary medicine — and the USDA agencies of the Agricultural Research Service and Economic Research Service.[4] Using the USDA's Current Research System (USDA-CRIS, 1993) classification of research, agricultural research expenditures to enhance and maintain agricultural productivity are taken to be a subset of all CRIS research commodity categories. They are research expenditures on all farm commodities (e.g., corn, wheat, fruits, vegetables, beef, swine, dairy) in CRIS research problem areas (RPAs) that have a focus on enhancing or maintaining productivity (see Huffman and Evenson, 1994, pp. 120-121), plus all expenditures on the research commodity categories of structures and inputs, farm management, insects, and general science. Although some reporting errors by public scientists in their CRIS reports undoubtedly occur, we do not consider these errors to be serious in major aggregates.[5] The annual nominal agricultural research expenditures by state are converted to real (1984 = 1.00) expenditures using Huffman and Evenson's agricultural research price index (Huffman and Evenson, 1993, table 4.1 and pp. 236-37).[6]

Research expenditures in a given year are expected to have an impact on productivity for many years. One might proceed by including as separate variables a finite number of lagged research expenditures in either equations (1) or (3). However, a large number of lagged research expenditures uses up a large number of degrees of freedom. Also, the lagged values of the research expenditures tend to be highly correlated. Griliches (1998) emphasizes that near-multicollinearity problems make it impossible to estimate research timing weights without some type of prior restrictions or imposition of a plausible structure on the weights. Another alternative is to construct a research stock variable as weighted past research expenditures:

$$T_t = \sum_{i=0}^{m} w_i R_{t-i} \tag{10}$$

The science of constructing research stock variables remains in its infancy (Griliches 1998, Ch. 12). The main reason is that the "true research stock" is unknown, so we need a good proxy or instrumental variable for it (Greene 1997, pp. 441-443). In studies of the impact of private research in manufacturing, the stock of research capital is frequently created from research expenditures using the perpetual inventory method. Griliches (1998), however, concludes that the usual declining balance or geometric depreciation doesn't fit very well the likely gestation, blossoming, and eventual obsolescence of knowledge. In agriculture, where none of the public agricultural research is conducted on farms or under the direct control of farmers, the perpetual inventory approach is implausible (see Evenson, 2000). Alternatively, a few researchers have included many lags without much structure on the lag pattern or weights, but one does not get very far in causal inference with non-experimental data unless a significant amount of structure is imposed on the analysis (Klette, Moen, and Griliches, 2000).

To approximate the likely impacts on state TFP of discoveries from public agricultural research expenditures, Huffman and Evenson (1993, 1994) impose the following pattern: after a gestation period of 2 years during which impacts are negligible, impacts are assumed to occur over the next 7 years and are represented by increasing weights, followed by 6 years of maturity during which weights are constant, and then 20 years of obsolescence with declining weights, i.e., a trapezoid-shaped lagged impact of research. We adopt the Huffman and Evenson trapezoidal-weight pattern to derive the proxy variable for the public agricultural research stock.

Two public research stock variables are used, an own-state and a spillin/spillover variable. For example, some of the public agricultural research discoveries in Iowa may spillover to one or more of the surrounding states or Iowa may benefit from public agricultural research conducted in surrounding states. Consistent with our regional grouping of states, we assume that discoveries from public agricultural research in a given state are an impure public good (Cornes and Sandler, 1996; Khanna, Huffman, and Sandler, 1994). In particular, we impose the simplifying assumption that benefits are regionally confined and apply a simple but reasonable technology of public supply aggregation (see Sandler, 1998). For a given state in a region, the spillover (or spillin) stock is defined as the total public agricultural research stock of all states in the region less the state's own public agricultural research stock.

Although private agricultural research may be a source of agricultural productivity growth, we have not included private research. This omission is due primarily to the difficulty in obtaining meaningful state level private agricultural research expenditures data. For example, a firm may perform its research at one or two research centers and sell new technology nationally. The impact of private research should be measured by where the technology is utilized, not where the technology is created. Also, to the extent that private research is appropriable, the prices of enhanced quality inputs will reflect it. Thus, quality adjusted inputs remove some of the problems associated with exclusion of private research stocks.[7]

Our agricultural extension variable is constructed from production-oriented agricultural extension expenditures in each state. Over 1955-1993, extension was organized into four program areas: agriculture and natural resources (ANR), community resource development (CRD), 4-H youth (4-H), and home economics (HE). The ANR program area includes crop production and management, livestock production and management, farm business management, agricultural marketing and supply, and natural resources. We collected data on professional extension full-time equivalents (FTE's) by state and major program areas from annual State Accomplishment Reports. To obtain state-level estimates of agricultural production-related extension expenditures, we multiplied each state's total expenditures on extension by the share of ANR full-time equivalent staff to total professional staff in all areas. An extension capital stock for each state is obtained by applying the perpetual inventory method with a 50 percent depreciation rate and dividing by the number of farms.

Infrastructure refers to highways in this study. Data are available for 1931-1992 on capital stock from capital outlay and capital stock from maintenance (both in 1987 dollars) from the U.S. Department of Transportation, State Transportation Economic Division.[8] The Federal Highway Administration's composite price index was used to deflate the capital expenditure and maintenance outlay series to 1987 dollars. In this data set, the standard perpetual inventory technique was used to generate the highway capital stock from expenditures data. Following Eberts, Park, and Dalenberg (1986), discards are assumed to follow a truncated normal distribution, with the truncation occurring at one half the average life and one and one half times the average life. We extended the highway capital series to 1993. Our highway capital stock measure is total highway capital stock, which equals capital stock from capital outlays plus capital stock from maintenance.[9]

Extreme weather conditions (droughts and floods) affect agricultural productivity, and if they are controlled for, we may get better estimates of the impacts of other sources of productivity.[10] We employed the USDA's precipitation data weighted by harvested crop acreage to create a variable (pre-plant) equal to cumulative February to July rainfall. We then created a drought dummy variable (drought) equal to 1 if pre-plant is less than 1 standard deviation below normal (and 0 otherwise) and a flood dummy variable (flood) equal to 1 if pre-plant is more than 1 standard deviation above normal (and 0 otherwise).

The Empirical Results

We report econometric estimates of the TFP equation for each of the seven regions, and estimates of the marginal internal rate of return to investments in public agricultural research.

The TFP Equations

For each region, we report an estimate of equation (3) with and without an interaction variable between the stock of own public research and extension. The equations were first estimated by ordinary least squares with state dummy variables included.[11] Where first-order autocorrelation was a significant problem (as judged by the Durbin-Watson statistic), the productivity equation was re-fitted as a time series model with an AR(1) process, or quasi-differenced, with ρ allowed to differ across states.[12] Wooldridge (1999, pp. 387-388) suggests that this procedure is satisfactory when no lagged dependent variable is included as a regressor, as in equation (3). The Southern Plains was the only region that did not require a quasi-differencing. The estimation procedure also took into account within-state heteroskedasticity as well as contemporaneous covariance between states in the same region. These results are reported in tables 8.2-8.8.

For the Northeast region, all the estimated coefficients of the public policy variables are significantly different from zero at the 5 percent level when the research-extension interaction variable is included, and the estimated coefficients of spillin research and highways are positive (see table 8.2). When the research-extension interaction variable is excluded, extension performs poorly. Thus, the TFP model with the research-extension interaction variable is judged to be the best for this region. The estimated coefficient of the research-extension interaction variable is, however, negative, suggesting own-research and extension are substitutes for impacting agricultural productivity in the Northeast region. Also, as expected, drought and flood have statistically significant negative impacts on agricultural productivity.

For the Southeast region, the TFP equation including the research-extension interaction variable performs poorly in the sense that the coefficients of the public policy variables are not individually significantly different from zero at the 5 percent level, except for spillin research (see table 8.3). When the research-extension interaction term is excluded, the estimated coefficients of

Table 8.2. TFP Equation: Northeast – CT, DE, MA, MD, ME, NH, NJ, NY, PA, RI, VT
(with AR(1) process)

Variable	Non-interaction Model	Interaction Model
Constant	-12.444	-31.150
	(-12.10)	(-9.13)
DE	0.313	0.026
	(1.69)	(0.13)
ME	0.373	0.286
	(5.00)	(3.63)
MD	-0.697	-0.816
	(-5.46)	(-5.92)
MA	-0.242	-0.195
	(-2.48)	(-1.84)
NH	0.309	0.161
	(3.02)	(1.49)
NJ	-0.752	-0.824
	(-5.66)	(-6.08)
NY	-1.120	-1.094
	(-9.41)	(-9.90)
PA	-1.005	-1.074
	(-8.47)	(-8.61)
RI	0.429	0.184
	(3.32)	(1.37)
VT	0.608	0.519
	(5.71)	(4.69)
ln(ownrd)	0.161	1.386
	(3.15)	(6.93)
ln(spillin)	0.239	0.331
	(3.07)	(3.86)
ln(ext)	0.033	1.430
	(1.67)	(6.40)
ln(hiway)	0.350	0.235
	(8.11)	(5.17)
ln(ownrd)ln(ext)		-0.090
		(-6.36)
drought	-0.018	-0.016
	(-3.43)	(-2.95)
flood	-0.015	-0.017
	(-3.30)	(-3.50)

Note: T-ratio in parentheses.

Table 8.3. TFP Equation: Southeast – AL, FL, GA, KY, NC, SC, TN, VA, WV
(with AR(1) process)

Variable	Non-interaction Model	Interaction Model
Constant	-10.029	-11.476
	(-26.15)	(-3.39)
FL	0.309	0.310
	(8.20)	(8.27)
GA	0.159	0.161
	(8.05)	(7.98)
KY	0.142	0.145
	(6.19)	(6.14)
NC	0.120	0.121
	(3.98)	(3.98)
SC	0.002	0.003
	(0.06)	(0.10)
TN	-0.169	0.166
	(-8.36)	(-7.86)
VA	-0.164	-0.162
	(-8.41)	(-8.30)
WV	-0.612	-0.605
	(-18.46)	(-16.87)
ln(ownrd)	0.084	0.178
	(4.11)	(0.82)
ln(spillin)	0.384	0.380
	(11.28)	(10.97)
ln(ext)	0.070	0.187
	(3.42)	(0.69)
ln(hiway)	0.033	0.033
	(0.94)	(0.94)
ln(ownrd)ln(ext)		-0.007
		(-0.43)
drought	-0.023	-0.023
	(-3.18)	(-3.18)
flood	-0.003	-0.003
	(-0.53)	(-0.54)

Note: T-ratio in parentheses.

own research, spillin research, and extension are all positive and individually significantly different from zero at the 5 percent level. We conclude that this is the best specification for the Southeast region. However, highways have a positive but not statistically significant effect on agricultural productivity in the Southeast region.

In the Central region, when the own-research and extension interaction variable is excluded from the TFP equation, the estimated coefficients of own research and highways are not significantly different from zero at the 5 percent level (see table 8.4). However, when the research-extension interaction variable is included, all of the public policy variables have coefficients that are significantly different from zero and the estimated coefficients for spillin research and highways are positive. Furthermore, the estimated coefficient of the interaction variable is positive, suggesting that own research and extension are complements for impacting agricultural productivity in the Central region. In this region, which has very little irrigation, both drought and flood variables have statistically significant negative impacts on agricultural productivity.

In the Northern Plains region, the own-research and extension variables perform poorly when no own research-extension interaction variable is included (see table 8.5). The estimated coefficients of own-research and extension are not significantly different from zero. However, when the research-extension interaction variable is included, all the estimated coefficients of the public policy variables are significantly different from zero at the 5 percent level. The interaction model is the preferred model for the Northern Plains region. Furthermore, own-research and extension are complements in the Northern Plains region. Drought has a statistically significant negative impact on productivity here.

In the Southern Plains region, all the estimated coefficients of the public policy variables are positive and significantly different from zero when no research-extension interaction variable is included in the TFP equation (table 8.6). When the interaction variable is included, its estimated coefficient is negative but not significantly different from zero. Hence, we conclude that in the Southern Plains region, the TFP regression without the research-extension variable is best. The impacts of droughts and floods on TFP are negative, but drought is statistically stronger and more frequently occurring in this region.

In the Mountain region, the estimated coefficient of own-research is not significantly different from zero and that of extension is negative (and significant) when no research-extension interaction variable is included (see table 8.7). However, when the interaction variable is included, the estimated coefficients of all research and extension variables are significantly different from zero, and the coefficient for own research is positive. For the Mountain region, we conclude that the TFP equation including the research-extension interaction variables is best. However, the estimated coefficient for the research-ex-

Table 8.4. TFP Equation: Central – IN, IL, IA, MI, MO, MN, OH, WI
(with AR(1) process)

Variable	Non-interaction Model	Interaction Model
Constant	-11.738	15.960
	(-13.52)	(2.26)
IL	0.077	-0.034
	(1.38)	(-0.51)
IA	0.240	0.210
	(12.55)	(9.52)
MI	-0.203	-0.210
	(-3.24)	(-5.22)
MO	-0.102	-0.083
	(-4.79)	(-4.55)
MN	0.082	0.035
	(3.29)	(1.22)
OH	-0.069	-0.153
	(-1.33)	(-2.56)
WI	0.181	0.164
	(7.81)	(6.36)
ln(ownrd)	0.011	-1.589
	(0.33)	(-3.72)
ln(spillin)	0.530	0.397
	(8.45)	(5.78)
ln(ext)	0.074	-2.304
	(2.19)	(-3.68)
ln(hiway)	0.041	0.142
	(0.71)	(2.12)
ln(ownrd)ln(ext)		0.142
		(3.84)
drought	-0.039	-0.038
	(-4.17)	(-4.17)
flood	-0.027	-0.027
	(-3.16)	(-3.18)

Note: T-ratio in parentheses.

tension interaction variable is negative, suggesting that own research and extension are substitutes. Here, highways do not have a statistically significant effect on agricultural TFP, and drought has a statistically significant negative impact on agricultural productivity.

For the Pacific region, neither TFP model performs well. In particular, extension does not have a statistically significant impact in either equation (see

Table 8.5. TFP Equation: Northern Plains – KS, NE, ND, SD (with AR(1) process)

Variable	Non-interaction Model	Interaction Model
Constant	-10.515	23.205
	(-6.41)	(1.39)
NE	0.059	0.189
	(0.81)	(1.97)
ND	0.145	0.249
	(1.06)	(1.61)
SD	0.396	0.286
	(2.69)	(2.00)
ln(ownrd)	0.259	-2.204
	(1.95)	(-1.83)
ln(spillin)	0.134	0.352
	(1.40)	(3.10)
ln(ext)	-0.132	-3.087
	(-0.86)	(-2.08)
ln(hiway)	0.347	0.516
	(2.11)	(2.67)
ln(ownrd)ln(ext)		0.181
		(2.01)
drought	-0.070	-0.070
	(-4.95)	(-5.03)
flood	0.010	0.007
	(0.67)	(0.52)

Note: T-ratio in parentheses.

table 8.8). Extension appears to be a different activity in this region than in the other six regions. Hence, between the two equations reported, the one without the research-extension interaction variable is best. In this equation, the estimated coefficients of own-research and spillin-research are positive, but the coefficient for own-research is only significant at the 7 percent level. Highways have a positive, but not statistically significant, impact on agricultural TFP in the Pacific region. Because the region is one of normally low precipitation and heavy reliance on irrigation, the negative but no significant effect of the drought and flood variables is not too surprising.

Next, we summarize the TFP elasticities associated with the public policy variables (own research, spillin research, extension, and highways) and report them in table 8.9. In the models without the research-extension interaction variable, the TFP (or output) elasticities are the estimated coefficients. In the

Table 8.6. TFP Equation: Southern Plains – AR, LA, MS, OK, TX (ordinary least squares)

Variable	Non-interaction Model	Interaction Model
Constant	-10.527	-20.139
	(-18.67)	(-3.89)
LA	-0.402	-0.373
	(-6.83)	(-6.18)
MS	-0.211	-0.197
	(-6.25)	(-5.77)
OK	-0.198	-0.178
	(-6.29)	(-5.41)
TX	-0.517	-0.470
	(-5.11)	(-4.55)
ln(ownrd)	0.096	0.680
	(2.14)	(2.15)
ln(spillin)	0.245	0.241
	(4.37)	(4.33)
ln(ext)	0.141	0.962
	(3.88)	(2.17)
ln(hiway)	0.172	0.151
	(3.04)	(2.63)
ln(ownrd)ln(ext)		-0.048
		(-1.87)
drought	-0.039	-0.040
	(-2.88)	(-2.99)
flood	-0.017	-0.018
	(-1.37)	(-1.47)

Note: T-ratio in parentheses.

models containing interaction variables, the TFP (output) elasticity of ownrd is $\delta_1 + \delta_5$ ln(ext) and of ext is $\delta_3 + \delta_5$ ln(ownrd), and these elasticities are evaluated by region at the respective regional sample mean values.

For the non-interaction model, TFP elasticities for ownrd are positive for all regions, except for the Mountain region (where the estimated coefficient was not significant). The TFP elasticity of public research spillin is positive for all regions. Perhaps surprising, the TFP elasticity with respect to public research spillin is greater than for ownrd, except for the Northern Plains. The elasticity of TFP with respect to hiway is positive for all regions. Ext, however, has a negative TFP elasticity for three regions: the Northern Plains, Mountain, and Pacific regions.

Table 8.7. TFP Equation: Mountain – AZ, CO, ID, MT, NV, NM, UT, WY
(with AR(1) process)

Variable	Non-interaction Model	Interaction Model
Constant	-6.731	-28.525
	(-15.68)	(-7.03)
CO	0.156	0.111
	(4.74)	(3.93)
ID	0.128	0.140
	(2.36)	(3.72)
MT	-0.222	-0.245
	(-5.46)	(-6.22)
NV	0.015	0.011
	(0.23)	(0.23)
NM	-0.194	-0.071
	(-5.01)	(-2.20)
UT	-0.072	-0.010
	(-1.82)	(-0.28)
WY	-0.261	-0.164
	(-5.59)	(-4.90)
ln(ownrd)	-0.058	1.529
	(-1.45)	(5.68)
ln(spillin)	0.360	0.241
	(7.57)	(6.33)
ln(ext)	-0.054	1.696
	(-2.24)	(5.46)
ln(hiway)	0.119	0.013
	(2.64)	(0.37)
ln(ownrd)ln(ext)		-0.109
		(-5.56)
drought	-0.015	-0.014
	(-3.30)	(-2.70)
flood	-0.004	-0.001
	(-0.85)	(-0.12)

Note: T-ratio in parentheses.

In the TFP models including the research-extension interaction variable, the TFP elasticity of ownrd is positive for all regions, except for the Northern Plains. The TFP elasticity of spillin is positive for all regions. In general, the interaction model, when evaluated at the regional sample means, increases the TFP elasticities for ownrd and decreases the TFP elasticities for spillin, com-

Table 8.8. TFP Equation: Pacific – CA, OR, WA (with AR(1) process)

Variable	Non-interaction Model	Interaction Model
Constant	-12.613	-22.838
	(-20.93)	(-2.05)
OR	-0.110	-0.161
	(-0.36)	(-0.50)
WA	-0.006	-0.062
	(-0.03)	(-0.25)
ln(ownrd)	0.159	0.806
	(1.84)	(1.17)
ln(spillin)	0.485	0.484
	(4.01)	(3.93)
ln(ext)	-0.069	0.747
	(-0.90)	(0.84)
ln(hiway)	0.128	0.076
	(1.00)	(0.55)
ln(ownrd)ln(ext)		-0.047
		(-0.92)
drought	-0.015	-0.016
	(-1.36)	(-1.37)
flood	-0.010	-0.009
	(-0.93)	(-0.88)

Note: T-ratio in parentheses.

pared to the non-interaction model. Spillin, however, still has a higher TFP elasticity than ownrd. Hiway has a positive TFP elasticity for all regions. Ext has a negative TFP elasticity for two regions (Northern Plains and Pacific). The coefficient of the interaction term is positive (and significant) for the Northern Plains region, implying that a greater level of investment in a state's own research and extension efforts is needed to make the TFP elasticities positive. As noted, the TFP model with the own research and extension interaction variable is preferred to the non-interaction model for most of the regions.

The Social Rate of Return to Public Agricultural Research

The private and social real internal rates of return are computed for each of the seven regions. For each region, we compute the sample mean of Q, T, S, and ln(ext) to obtain values for a "representative state in the region" and for a "representative spillover effect" from other states within the region. Table 8.10

Table 8.9. TFP or Output Elasticities with Respect to Public Policy Variables (interaction terms evaluated at the respective regional means)

	Non-interaction Term Model			
Region	**ownrd**	**spillin**	**ext**	**hiway**
Northeast	0.16	0.24	0.03	0.35
Southeast*	0.08	0.38	0.07	0.03
Central	0.01	0.53	0.07	0.04
Northern Plains	0.26	0.13	-0.13	0.35
Southern Plains*	0.10	0.25	0.14	0.17
Mountain	-0.06	0.36	-0.05	0.12
Pacific*	0.16	0.49	-0.07	0.13

	Interaction Term Model			
Region	**ownrd**	**spillin**	**ext**	**hiway**
Northeast*	0.19	0.33	0.05	0.26
Southeast	0.09	0.38	0.07	0.03
Central*	0.10	0.40	0.03	0.14
Northern Plains*	-0.01	0.35	-0.19	0.52
Southern Plains	0.09	0.24	0.16	0.15
Mountain*	0.13	0.24	0.004	0.01
Pacific	0.20	0.48	-0.06	0.08

Note: Computation from the preferred model is denoted by (*).

In the models without the research-extension interaction variable, the TFP/output elasticities are the estimated coefficients. In the models containing interaction variables, the TFP/output elasticity of ownrd is $\delta_1 + \delta_5$ ln(ext) and of ext is $\delta_3 + \delta_5$ ln(ownrd). These elasticities are evaluated at the regional sample means.

presents the calculated rates of return for each of the 7 regions for both the non-interaction and interaction models.

The marginal private real rate of return is obtained using equation (8) [or equation (9) with d_2 set to zero]. The marginal social real rate of return takes into account own-state and spillover effects and is obtained using equation (9). As expected, for all regions the marginal social real rate of return is always greater than the own-state/private real rate of return. The own-state/private rates of return are generally sizeable but less than 100 percent and within the range found by other researchers for the U.S. (see Evenson, 2000). The social rates of return are several times the private rates of return. Some of the social rates of return are surprisingly high, e.g., for the Northeast and Central regions the real social rate of return exceeds 600 percent. Regions having a large number of states and large farm output per state have the largest social rate of return, which is as expected with a regional public good (i.e., summation technology for publicness within the region).

Table 8.10. Own-State/Private and Social Real Rate of Return (r) to Public Agricultural Research in a State (percent)

Region	Interaction Private	Interaction Social	Non-interaction Private	Non-interaction Social	Number of States in the region
Northeast	106	>600*	92	>600	11
Southeast	46	290	43	290*	9
Central	89	>600*	12	>600	8
Northern Plains	0.0	390*	510	>600	4
Southern Plains	52	210	55	220*	5
Mountain	73	260*	0.0	250	8
Pacific	127	>600	>600	>600*	3

*Denotes the computation from the preferred model.

Note: In the Northern Plains and Mountain regions, when the estimated output elasticity was negative, the output elasticity was set to zero in the rate of return computation.

Concluding Remarks

This paper uses new data to provide evidence on the contribution of public own-state research, spillin research, extension, and highway infrastructure to agricultural productivity change from 1960 to 1993. The results are an addition to the literature because few studies have focused on agricultural productivity at the state level, and none has considered the effects of infrastructure or the direct effect of public agricultural research spillovers.

Where our results overlap with prior studies, the results are largely as expected. Public agricultural research and highways have positive impacts on agricultural productivity, and the marginal real social rate of return to public agricultural research is large. The results for public agricultural extension are mixed, consistent with Huffman and Evenson (1993), and extension does not have a positive impact on agricultural productivity in two regions (in the interaction model). The model containing an own-research-and-extension interaction variable was preferred to the non-interaction model for most regions.

Spillin research stocks also impact agricultural productivity positively in almost all regions. Furthermore, the computed real rates of return to investments in public agricultural research for any one state are less than the social rate of return for all states in its region. Thus, state level planning for public agricultural research would be socially sub-optimal. This outcome is consistent with the regional public-good nature of public agricultural research, and implies a need for regional coordination of public agriculture research funding and provision. Furthermore, our estimates of the marginal social rate of return are large compared to results from other studies reported in Evenson (2000), but this result seems to be due largely to the direct incorporation of research spillin in our model of state agricultural productivity.

Notes

1. Schooling enhances the cognitive and analytical skills of students. It also increases the ability of individuals to acquire and process information efficiently, an ability that is important to good decision-making (Huffman, 2000). Thus, schooling has the potential to increase the productivity of farmers (as decision makers) and of other farm workers. Huffman (2000) summarizes evidence showing strong positive effects of farmers' schooling on the adoption of appropriate farm-level technologies. The evidence on effects of farm worker schooling, excluding farm operators, is, however, mixed. For example, there is no significant impact of field worker schooling in fresh fruit and vegetable harvesting.

While the number of workers employed in U.S. agriculture and total hours worked have declined over time, the quality per hour, measured by average schooling, has increased, especially for farmers. For example, in 1964, only about one-third of all farmers had completed high school, compared with more than three-quarters of farmers by 1990. The evidence shows that investing in schooling of farmers is a good social investment when agricultural technologies are frequently changing.

2. Several regional groupings of states were considered in our work. The oldest grouping consists of the 10 Economic Research Service (ERS) farm production regions (Northeast, Lake States, Corn Belt, Northern Plains, Appalachian, Southeast, Delta, Southern Plains, Mountain, and Pacific). A National Research Council (NRC) study on colleges of agriculture at the land grant universities (Committee on the Future of the Colleges of Agriculture in the Land Grant University System 1995) included a cluster analysis to classify state agricultural experiment stations (SAES's) expenditures into 9 commodity research clusters. See figure 7.4 in the NRC report for the grouping of states by research cluster. The NRC regions differ from the ERS and Huffman and McCunn's (HM) regions in not having non-contiguous states in clusters. For example, California, Oregon, and Washington are in the same ERS region (Pacific), while California and Florida are in the same NRC region. The ERS and HM grouping are more similar to each other than to the NRC grouping. In fact, the Pacific, Mountain, Northern Plains, and Northeast regions are the same for both the ERS and HM groupings. In addition, HM Central = ERS Corn Belt + ERS Lake States, HM Southern Plains = ERS Southern Plains + ERS Delta, and HM Southeast = ERS Southeast + ERS Appalachian.

3. The data in the publication are also available electronically from the ERS homepage at URL: http://usda.mannlib.cornell.edu/data-sets/inputs/98003

4. The project information reported to CRIS contains a location code for where the research is conducted. When an ARS or ERS project is conducted in Iowa for example, the funds would go into the public agricultural research expenditures in Iowa. If the location is given as Washington, D.C., for ERS, the funds are not counted in our public research expenditures data. Part but not all of the ARS research at Beltsville was included in the public research expenditures in Maryland.

5. The annual research expenditure series, 1970-95, were derived from USDA-CRIS electronic data files. For 1927-1969, the total public agricultural research expenditures series were derived from (1) reported expenditures by the state agricultural experiment stations to the USDA and (2) a projected ratio of total public agricultural research expenditures to total SAES expenditures by state. The projected ratios were obtained by fitting simple time series models to the actual data by state, 1970-95, and then projecting backward. A major adjustment was made 1948-1950 to reflect the dramatic shift in the share of U.S. total public agricultural

research between state institutions and the USDA (see Huffman and Evenson, 1993, table 4.1 and figure 4.1).

6. The Huffman and Evenson research price index gives about 70 percent weight to faculty salaries and behaves similar to Alston and Pardey's and Fuglie's agricultural research price index over 1970-1990.

7. See Huffman and Evenson (2001) for one approach to the construction of private R&D stock variables at the state level.

8. The state highway database is located at the web page: http://www.bts.gov

9. Alicia Munnell at Boston College also has a database on infrastructure including the components highways, water systems, and others by state. Munnell's database has been employed in several studies of the manufacturing sector (e.g., Morrison and Schwartz, 1996). We, however, do not employ Munnell's database in our study primarily because her database is only from 1970 to 1986.

10. Weather data are available from the ERS homepage as an ERS data product: Weather in U.S. Agriculture. Monthly temperatures and precipitation data for farm production regions and States for 1950-94. (URL: http://usda.mannlib.cornell.edu/data-sets/general/92008). Weather for certain months (e.g., February to July) may be more important than that of other months in explaining agricultural productivity.

11. A concern might arise about *lnP* or *u* in equation (3) because our data is stationary. Nonstationarity is a problem of the statistical process generating the data and *not in any particular realization* of the process(es). Furthermore, a time series that is nonstationary need not be explosive (Wooldridge, 1999, pp. 347-349).

12. An informal method for providing guidance about whether a time series is weakly dependent [i.e., the process $\{y(t)$: t =1, 2, ..., n$\}$ is such that $y(t)$ and $y(t+h)$ are "almost independent" as h increases without bound] can be obtained from the estimate of ρ in $y(t) = \rho y(t-1) + \varepsilon(t)$ or $u(t) = \rho^* u(t-1) + \varepsilon(t)^*$, t =1, 2, ..., n. When the estimate of $\rho << 1$, then the series need not be differenced (Wooldridge, 1999, pp. 376-395). For the estimated equations reported in tables 8.2-8.8, the estimate of ρ was less than 0.8 in all cases.

References

Alston, J., B. Craig, and P. Pardey. 1998. "Dynamics in the Creation and Depreciation of Knowledge, and the Returns to Research." International Food Policy Research Institute, EPTD Discussion Paper No. 35, August.

Antle, J. 1983. "Infrastructure and Aggregate Agricultural Productivity: International Evidence." *Economic Development and Cultural Change,* 31(3):609-19.

Aschauer, D. 1989. "Is Public Expenditure Productive?" *Journal of Monetary Economics,* 23:177-200.

Ball, E., and R. Nehring. 1998. *Patterns of State Productivity Growth in the U.S. Farm Sector*. Economic Research Service Staff Paper Number 9804.

Ball, V. E., J. Bureau, R. Nehring, and A. Somwaru. 1997. "Agricultural Productivity Revisited." *American Journal of Agricultural Economics*, 79:1045-63.

Ball, V. E., F. Gollop, A. Kelly, and G. Swinand. 1999. "Patterns of Productivity Growth in the U.S. Farm Sector: Linking State and Aggregate Models." *American Journal of Agricultural Economics*, 81:164-79.

Board on Agriculture, National Research Council. 1997. *Precision Agriculture in the 21st Century: Geospatial and Information Technologies in Crop Management*. Washington, D.C.: National Academy Press.

Bredahl, M., and W. Peterson. 1976. "The Productivity and Allocation of Research: U.S. Agricultural Experiment Stations." *American Journal of Agricultural Economics*, 58:684-92.

Committee on the Future of the Colleges of Agriculture in the Land Grant University System, National Research Council. 1995. *Colleges of Agriculture at the Land Grant Universities: A Profile*. Washington, D.C.: National Academy Press.

Cornes, R., and T. Sandler. 1996. *The Theory of Externalities, Public Goods and Club Goods*. New York: Cambridge University Press.

Eberts, R., C. S. Park, and D. Dalenberg. 1986. *Public Infrastructure Data Development*. National Science Foundation, Washington, D.C.

Echeverria, R. G. 1990. "Assessing the Impact of Agricultural Research." in *Methods for Diagnosing Research System Constraints and Assessing the Impact of Agricultural Research*, ed. Ruben G. Echeverria. The Hague: International Service for National Agricultural Research.

Evenson, R. E. 2000. "Economic Impact Studies of Agricultural Research and Extension." Chapter in *Handbook of Agricultural Economics*, vol. I, ed. B.L. Gardner and G. Rausser. Amsterdam: Elsevier Science/North-Holland.

Evenson, R. E. 1989. "Spillover Benefits of Agricultural Research: Evidence from U.S. Experience." *American Journal of Agricultural Economics*, 71(2):447-52.

Fuglie, K. 2000. "Trends in Agricultural Research Expenditures in the United States." Chapter in K.O. Fuglie and D.E. Schimmelpfennig, *Public-Private Collaboration in Agricultural Research*. Ames: Iowa State University Press, 9-24.

Fuglie, K., et al., 1996. *Agricultural Research and Development: Public and Private Investments Under Alternative Markets and Institutions*. Economic Research Service, Agricultural Economic Report No. 735.

Gopinath, M., and T. Roe. 1997. "Sources of Sectoral Growth in an Economy Wide Context: The Case of U.S. Agriculture." *Journal of Productivity Analysis*, 8(3):293-310.

Greene, W. H. 1997. *Econometric Analysis*. 3rd ed. Upper Saddle River, N.J.: Prentice Hall, Inc.

Griliches, Z. 1998. *R&D and Productivity: The Econometric Evidence*. Chicago, Ill.: The University of Chicago Press.

Huffman, W. E. 2000. "Human Capital: Education and Agriculture." Chapter in *Handbook of Agricultural Economics*, vol. I, ed. B.L. Gardner and G. Rausser. Amsterdam: Elsevier Science/North-Holland.

Huffman, W. E., and R. E. Evenson. 2001. "Structural Adjustment and Productivity Change in U.S. Agriculture, 1950-1982." *Agricultural Economics*, 23: 127-47.

Huffman, W. E., and R. E. Evenson. 1994. *The Development of U.S. Agricultural Research and Education: An Economic Perspective*. Iowa State University, Department of Economics.

Huffman, W. E., and R. E. Evenson. 1993. *Science for Agriculture*, Ames: Iowa State University Press.

Huffman, W. E., A. McCunn, and J. Xu. 2000. "Public Agricultural Research with an Agricultural Productivity Emphasis: Data for 48 States, 1927-1995." Iowa State University, Staff Paper.

Jorgenson, D. W., and F. Gollop. 1992. "Productivity Growth in U.S. Agriculture: A Postwar Perspective." *American Journal of Agricultural Economics*, 74(August):745-50.

Khanna, J., W. Huffman, and T. Sandler. 1994. "Agricultural Research Expenditures in the United States: A Public Goods Perspective." *Review of Economics and Statistics*, 76(2):267-77.

Klette, T. K., J. Moen, and Z. Griliches. 2000. "Do Subsidies to Commercial R&D Reduce Market Failures? Microeconometric Evaluation Studies." *Research Policy*, 29:471-95.

McCunn, A., and W.E. Huffman. 2000. "Convergence in U.S. Productivity Growth for Agriculture: Implications of Interstate Research Spillovers for Funding Agricultural Research." *American Journal of Agricultural Economics*, 82:370-88.

Morrison, C. J., and A. E. Schwartz. 1996. "State Infrastructure and Productive Performance." *American Economic Review*, 86(5):1095-111.

Norton, G. 1981. "The Productivity and Allocation of Research: U.S. Agricultural Experiment Stations, Revisited." *North Central Journal of Agricultural Economics*, 3:1-12.

Ruttan, V. W. 1980. "Bureaucratic Productivity: The Case of Agricultural Research." *Public Choice*, 35:529-47.

Ruttan, V. W. 1982. *Agricultural Research Policy*. Minneapolis: University of Minnesota Press.

Sandler, T. 1998. "Global and Regional Public Goods: A Prognosis for Collective Action." *Fiscal Studies*, 19:221-47.

U.S. Department of Agriculture (USDA), CSRS, CRIS. 1993. *Manual of Classification of Agricultural and Forestry Research*. Revision V. Beltsville, Md: USDA-CSRS.

Wooldridge, J. M. 1999. *Introductory Econometrics: A Modern Approach*. Mason, Ohio: South-Western College Publishing.

Wozniak, G. W. 1993. "Joint Information Acquisition and New Technology Adoption: Later versus Early Adoption." *Review of Economics and Statistics*, 75:438-45.

Part III.
Productivity Growth and the Environment

9

Environmental Indicators of Pesticide Leaching and Runoff from Farm Fields

Robert L. Kellogg, Richard F. Nehring, Arthur Grube,
Donald W. Goss, and Steven Plotkin

Abstract

The purpose of this chapter is to present an indicator of how changes in pesticide use in agriculture have changed the potential for risk to human health and the environment from pesticide contamination of water leaving farm fields. Environmental indicators are designed to be *relative* estimates of potential risk that are based on pesticide use and on the *factors* that are known to be important determinants of pesticide loss from farm fields, such as the intrinsic potential of soils to leach or runoff pesticides, the chemical properties of the pesticides, annual rainfall and its relationship to leaching and runoff, and changes in cropping patterns. The analytical framework consists of about 4,700 resource polygons representing the intersection of 48 states, 280 watersheds at the 6-digit Hydrologic Unit level, and 1,400 combinations of climate and soil groups. Twelve crops are included in the analysis: corn, soybeans, wheat, cotton, sorghum, barley, rice, potatoes, oats, sugarbeets, tobacco, and peanuts. Model estimates of pounds of pesticides applied, mass loss, and annual concentrations leaving the farm field (edge of field and bottom of root zone) were obtained for pesticides used on each of the 12 crops in each of the resource polygons for each year from 1960 through 1997. Indicators of potential risk are constructed from estimates of annual concentrations that exceed "safe" thresholds for chronic exposure to four target groups — humans, fish, crustaceans, and algae. It is expected that temporal and spatial trends of these indicators will closely track the change in potential risk to human health and the environment from agricultural use of pesticides.

Introduction

Pesticides are a vital input in today's agriculture, protecting food and fiber from damage by insects, weeds, diseases, nematodes, and rodents. U.S. agriculture spends about 8 billion dollars annually on pesticides — about 70 percent of all pesticides sold within the country (Aspelin, 1997). It is estimated that each dollar invested in pest control returns approximately 4 dollars in crops saved (Pimentel, 1992). Nevertheless, pests still destroy nearly 13 percent of all potential food and fiber crops in the United States. Farmer expenditures on pesticides are about 4-5 percent of total farm production costs.

The dependence of agriculture on chemical pesticides developed over the last 60 years as the agricultural sector shifted from labor-intensive production methods to more capital- and chemical- intensive production methods. Sixty years ago, most crops were produced largely without the use of chemicals. Insect pests and weeds were controlled by crop rotations, destruction of crop refuse, timing of planting dates to avoid high pest population periods, mechanical weed control, and other farming practices. While these practices are still in use, changes in technology, changes in prices, and government policies resulted in the development of today's chemically intensive agriculture.

Usage of conventional pesticides on farms increased from about 400 million pounds (active ingredient) in the 1960s to over 800 million pounds in the late 1970s and early 1980s, primarily due to the widespread adoption of herbicides in corn production (Aspelin, 1997). Since that time, usage has been somewhat lower, ranging from about 700 to 780 million pounds per year. Pesticide usage in agriculture can vary considerably from year to year depending on weather, pest outbreaks, crop acreage, and economic factors such as pesticide prices and crop prices. Whereas the quantity of pesticides used by agriculture has fallen off slightly in recent years, total expenditures on pesticides by farmers are still increasing.

During the 1960s, agricultural pesticide use was dominated by insecticides, which accounted for about half of all pesticides used. The quantity of insecticides applied fell as the organochlorines (DDT, aldrin, and toxaphene) were replaced by pyrethroids and other chemicals that required lower application rates. Today, 70 percent of the quantity of pesticides used in agriculture is herbicides. Corn leads all other crops — by a substantial margin — in total pesticide use. Rice, potatoes, vegetables, and fruits, however, actually use pesticides more intensively than do corn and other crops. Minimum tillage is practiced by many farmers, reducing the need for machinery, labor, and energy, but increasing agriculture's pesticide dependency. Pesticide use trends vary markedly from one part of the country to another as farmers respond to local pest problems and as crop production patterns vary.

Even as today's chemically intensive agriculture is partly responsible for providing abundant low-cost supplies of food and fiber, it has also created water quality problems. When the chemical revolution first started, there was little concern about environmental consequences. Scientific testing indicated that DDT and other agricultural chemicals were generally not harmful to humans if used as directed. By the mid-1960s, however, there was a growing awareness that some agricultural chemicals were damaging the environment, and may have been affecting humans as well. Awareness that agricultural chemicals were not staying on the fields, but were being washed into streams and rivers and seeping into ground water, came about with the development of sensitive chemical-testing procedures. These procedures did not become available for organochlorine pesticides (DDT, DDE, aldrin, dieldrin, heptachlor, and chlordane) until the late 1960s. The DDT problem was known before that time (Rachel Carson's book *Silent Spring* was released in 1962) largely because of bioaccumulation, resulting in detectable levels in animals high in the food chain.

Today, pesticide levels in water are routinely monitored. Pesticide residues have been found in ground water, surface water, and rainfall. EPA began to emphasize ground-water monitoring for pesticides in 1979 following discovery of DBCP and aldicarb in ground water in several states. In 1985, 38 states reported that agricultural activity was a known or suspected source of ground water contamination within their borders (ASIWPCA, 1985). Since then, several Federal and State agencies have developed programs to sample water resources and test for the presence of agricultural chemicals. Results published to date have shown that chemicals used in agricultural production have been found in ground water, sometimes at levels exceeding EPA's drinking water criteria (USEPA, 1990; Hamilton and Shedlock, 1992; Monsanto, 1990; Kolpin et al., 1994; Barbash et al., 1999). Monitoring for pesticides in surface water was frequent in the 1960s and 1970s; such studies were conducted that led to the banning of chlorinated hydrocarbon insecticides. Sampling in the 1980s and 1990s found that the four leading herbicides in use during that time — atrazine, metolachlor, alachlor, and cyanazine — were frequently detected in surface waters in agricultural regions (USDA, 1992; Goolsby and Battaglin, 1993; Goolsby et al., 1993; Larsen et al., 1999; Baker and Richards, 1995). The highest levels occurred after planting and during the early part of the growing season. Most of the pesticides commonly used presently and in the past have also been found in the atmosphere, including DDT, toxaphene, dieldrin, heptachlor, organophosphorous insecticides, triazine herbicides, alachlor, and metolachlor (USGS, 1995). These airborne pesticides return to the earth with rainfall to further contribute to water contamination. A recent report by USGS of a survey of pesticides in the nation's waters concluded that pesticides were common in surface and shallow ground water in both urban and agricultural areas, but investigators were not able to determine if contamination is lessening or worsening (USGS, 1999).

Chemical testing can detect the presence of a pesticide, and often can measure how much of the pesticide is in the water, but cannot identify the source of the pesticide. It is not known what portion of observed residues originate from quasi-point sources within agriculture — such as applicator loading and mixing sites — or from nonagricultural sources. Since agriculture is the largest user of pesticides, it is likely that much of the pesticide residue found in the environment originated from agriculture. However, a significant — but unknown — portion of the pesticide residue originates from non-agricultural sources. Non-agricultural uses include: home, lawn, and garden use; industrial use; pest control in forestry; weed control along roadsides; ditches, railways, and rights-of-way; pest control by municipalities and local governments; golf courses; and the military.

Concerns about potential risks to human health and the environment resulted in the 1972 amendments to the Federal Insecticide, Fungicide, and Rodenticide Act (FIFRA), which increased the stringency of health and safety data required to support a pesticide registration. EPA first banned use of some organochlorine pesticides for agricultural purposes in the 1970s, and has since imposed use limitations on many other pesticides. The amendments also required that all existing pesticides be re-registered using current health and environmental standards. Chemical companies have responded to these regulatory pressures by marketing new chemicals that are thought to be less harmful to humans and the environment, or less likely to migrate from farm fields to contaminate ground water and surface water.

The purpose of the research reported here is to develop an indicator of how changes in pesticide use in agriculture have changed the potential for risk to human health and the environment from pesticide contamination of water. The ideal situation would be to have consistent data for all regions in the country on contamination of surface water and ground water *that is exclusively due to use of pesticides in agricultural production*. Such a database does not exist, nor could it ever be realistically produced. An alternative — and more tractable — approach is to model the *potential* for risk associated with pesticide loss from farm fields. The environmental indicators created here have been designed to be *relative* estimates of *potential* risk that are based on pesticide use and the factors that are known to be important determinants of pesticide loss from farm fields, such as the intrinsic potential of soils to leach or runoff pesticides, the chemical properties of the pesticides, annual rainfall and its relationship to leaching and runoff, and changes in cropping patterns. Indicators are constructed from model estimates of annual concentrations leaving the farm field (edge of field and bottom of root zone) that exceed "safe" thresholds for chronic exposure to four target groups — humans, fish, crustaceans, and algae.

Modeling Pesticide Loss from Farm Fields

Even when the label instructions are carefully adhered to, a small portion of pesticides applied reaches surface and ground water, as evidenced by the detection of pesticides in water-quality monitoring studies. The greatest losses occur in the first few rainfall events after application. Losses in surface runoff are the greatest — up to 5 percent of the amount applied. Losses to subsurface drainage and percolation, however, are much less — typically less than 0.5 percent (Baker et al., 1992). Wind or water is required to transport pesticides from the field to surface or ground water. Whether and how much a pesticide migrates from the field where it was applied depends on a complex interaction of pesticide and soil properties with weather conditions and site characteristics.

Physical properties of pesticides (soil sorption propensity, vapor pressure, and solubility) and the persistence of the pesticide in the environment are important factors in the tendency of pesticides to move from the application site. *Persistence* is a pesticide's resistance to decompose through chemical, photochemical (sunlight), and microbial action. The half-life of a chemical is not a constant value, but can vary considerably depending on soil temperature and moisture, microbial populations present, amount and kind of organic material, pH, and soil type. Pesticides that persist longer in the environment are more likely to move off-site than are less persistent pesticides because they are exposed to more leaching and runoff events. *Sorption* is the binding of the chemical to the soil. Pesticides that are strongly adsorbed tend not to leach, but rather are lost with the soil through soil erosion processes. Pesticides that are weakly adsorbed are lost mainly in surface runoff water and percolation. *Vapor pressure* is the measure of a pesticide's tendency to evaporate. Losses to the atmosphere of sprayed pesticides during application can be significant. Water solubility, which is related to soil sorption propensity, also influences the amount of pesticide that is likely to be removed by runoff or by leaching.

Soil characteristics and their interaction with physical properties of pesticides also affect the potential for pesticides to migrate from the field. Three major soil characteristics that affect chemical movement are soil texture, soil permeability, and soil organic matter. *Soil texture* is an indication of the relative proportions of sand, silt, and clay in the soil. Pesticides tend to be adsorbed mostly on clay and organic matter. The higher the clay content, the greater the number of binding sites for pesticide retention. Coarse, sandy soils generally allow water to move rapidly downward and offer few opportunities for adsorption. *Soil permeability* is a general measure of how fast water can move downward in a particular soil. Permeability is controlled by soil structure, which is the way individual soil particles clump together, creating pore space between the soil aggregates. This pore space is key to water retention and movement through the soil. Organic matter content is the most important variable affecting sorption of pesticides onto soil particles. Organic matter provides binding sites, is very reactive chemically, and promotes microbial populations that contribute to pesticide degradation. Soil organic matter also influences how much water the soil can hold before movement occurs.

Pesticide loss from farm fields was estimated using a process model called GLEAMS, which incorporates these complex interactions of weather, pesticide properties, and soil characteristics (Knisel, 1993). GLEAMS was initially developed to simulate edge-of-field and bottom-of-root zone loading of sediment and chemicals in order to evaluate alternative management practices. It operates on a daily time step with daily climatic data. GLEAMS consists of three major components: hydrology, erosion/sediment yield, and pesticides. Soil-water accounting procedures represent the principal *hydrologic* processes of infiltration, runoff, water application by irrigation, soil evaporation, plant transpiration, and soil water movement within and through the root zone. A modification of the Universal Soil Loss Equation (USLE) is used to simulate storm-by-storm rill and interrill *soil erosion* in overland flow areas. The *pesticide* component of GLEAMS is designed to allow simulation of interactions among pesticide properties, climate, soils, and management. Adsorption characteristics are coupled with the hydrologic component to route pesticides within and through the root zone. Upward movement of pesticides and plant uptake are simulated with soil evaporation and plant transpiration, respectively. Pesticide degradation into metabolites is tracked for compounds that have potentially toxic daughter products; in the output, metabolites are included in the concentration calculation for the parent compound.

A National Pesticide Loss Database was constructed using GLEAMS estimates for 243 pesticides applied to 120 generic soils for 20 years of daily weather from each of 55 climate stations. This procedure resulted in 1,603,800 runs of 20 years each. Separate GLEAMS estimates were made for irrigated and nonirrigated conditions. The crop in these simulations was a generic row crop behaving similar to corn, soybeans, cotton, or sorghum, planted in straight rows. The model estimates irrigation timing and amounts depending on soil moisture.

Soil Parameters

The 120 generic soils were selected on the basis of soil texture and organic matter content. Twelve textures were combined with four organic matter contents for surface horizons. Four textures were used for subsurface horizons, which were not coarser in texture than the surface horizon. Organic matter content of a subsurface horizon was 20 percent of the surface horizon. The thirty horizon combinations times four organic matter contents made up 120 soils, summarized in table 9.1. The hydrologic group (HG) was assigned according to subsurface texture. Using tables found in the GLEAMS manual and relations in the USDA Soil Survey Handbook, the other soil parameters required by GLEAMS were estimated. The Curve Number (CN) is also required, and was assigned as shown in table 9.1.

Table 9.1. 120 Basic Soils Represented by Surface and Subsurface Texture

				CN by OM percent			
	Surface Texture	Subsurface Texture	HG	0.5	1.0	2.0	4.0
1	Sand	Fine Sandy Loam	A	66	65	63	58
2	Loamy Fine Sand	Fine Sandy Loam	A	69	68	67	64
3	Fine Sandy Loam	Fine Sandy Loam	A	72	71	70	67
4	Sand	Silt Loam > 15% clay	B	74	74	72	69
5	Loamy Fine Sand	Silt Loam > 15% clay	B	78	77	75	72
6	Fine Sandy Loam	Silt Loam > 15% clay	B	80	79	78	75
7	Sand	Silty Clay Loam	C	86	81	80	79
8	Loamy Fine Sand	Silty Clay Loam	C	88	85	83	82
9	Fine Sandy Loam	Silty Clay Loam	C	91	87	86	85
10	Sand	Clay > 70% clay	D	91	90	89	88
11	Loamy Fine Sand	Clay > 70% clay	D	92	91	90	89
12	Fine Sandy Loam	Clay > 70% clay	D	93	92	91	90
13	Loam	Silt Loam > 15% clay	B	74	74	72	69
14	Silt Loam < 15% clay	Silt Loam > 15% clay	B	78	77	75	72
15	Silt Loam > 15% clay	Silt Loam > 15% clay	B	80	79	78	75
16	Loam	Silty Clay Loam	C	86	81	80	79
17	Silt Loam < 15% clay	Silty Clay Loam	C	88	85	83	82
18	Silt Loam > 15% clay	Silty Clay Loam	C	91	87	86	85
19	Loam	Clay > 70% clay	D	91	90	89	88
20	Silt Loam < 15% clay	Clay > 70% clay	D	92	91	90	89
21	Silt Loam > 15% clay	Clay > 70% clay	D	93	92	91	90
22	Sandy Clay Loam	Silty Clay Loam	C	86	81	80	79
23	Clay Loam	Silty Clay Loam	C	88	85	83	82
24	Silty Clay Loam	Silty Clay Loam	C	91	87	86	85
25	Sandy Clay Loam	Clay > 70% clay	D	91	90	89	88
26	Clay Loam	Clay > 70% clay	D	92	91	90	89
27	Silty Clay Loam	Clay > 70% clay	D	93	92	91	90
28	Silty Clay	Clay > 70% clay	D	91	90	89	88
29	Clay < 70% clay	Clay > 70% clay	D	92	91	90	89
30	Clay > 70% clay	Clay > 70% clay	D	93	92	91	90

Climate Parameters

GLEAMS includes a climate generator that simulates daily rainfall and temperature. Fifty-five climatic stations across the United States were chosen. The program generates daily weather using the mean, standard deviation, and skew for monthly precipitation, maximum temperature, minimum temperature, and the mean and standard deviation of monthly solar radiation. Two other monthly values required are: 1) the probability of having a wet day after a wet day, and 2) the probability of having a dry day after a wet day. A twenty-year record was simulated with the climate generator from a 40-year frequency distribution to produce a distribution of pesticide loss estimates that would reasonably represent most weather conditions. Planting and harvest dates were estimated for each of the 55 climate stations based on mean and standard deviation of monthly low temperatures from the climatic record.

Pesticide Application Parameters

GLEAMS results were simulated for the 243 pesticides in the SCS/ARS/CES pesticide properties database (Wauchope et al., 1992). The pesticide database is included in the GLEAMS model pesticide parameter editor, including foliar characteristics constructed using the procedure by Willis and McDowell (1987). The *Insect Control Guide* (1995) and the *Weed Control Guide* (1996) were used to define the action of each compound, when applied, how frequently applied, and recommended rates and methods of application.

Pesticide application timing was based on the planting date, harvest date, and purpose. Pesticide application method was based on planting date and purpose. Some herbicides were designated only for pre-plant application and some only for post-plant application. Those herbicides with applications designated as "all methods" were included with the pre-emergent herbicides. Preplant pesticides were simulated with application seven days before planting. Pre-emergent pesticides were simulated with application on the planting date. Post-plant pesticides were simulated with application fourteen days after the planting date. Over-the-top insecticides, fungicides, and miticides were simulated in 3 repeat applications commencing after one third of the growing season was completed. For example, an insecticide with a recommended application every five days was first applied one-third of the way through the growing season, and then repeated every five days for a total of three applications. Growth regulators were applied after one-fourth of the growing season was completed. Defoliants were applied 5 days before harvest date.

Some soil insecticides and nematicides were incorporated in the soil, some surface applied, and some applied over-the-top of foliage. The SCS/ARS/CES pesticide properties database also includes growth regulators and defoliants, both of which are applied on foliage. The insecticides used as foliar applications were applied at label-recommended frequency. The *Insect Control Guide* included recommendations on the frequency of application, i.e., 3-5 days, 5 days, or 7 days. In GLEAMS, insecticides were applied every 3 days for the 3-5 day recommendation, every 5 days for the 5-day recommendation, and every 7 days for the 7-day recommendation.

Model Output

The daily mass of pesticide that was removed by leaching, runoff in solution, and runoff with sediment were recorded for each model run. Annual totals for each were obtained by summing the daily estimates. The daily volume of water that leached below the root zone or ran off as surface water was recorded and summed over the year, and the daily mass of sediment loss was recorded and summed over the year. Output from the GLEAMS simulations included annual estimates of:

1. Mass loss in leachate as a percent of the amount applied.
2. Mass loss dissolved in runoff as a percent of the amount applied.
3. Mass loss adsorbed to eroded soil as a percent of the amount applied.
4. Volume of water percolation in centimeters.
5. Volume of water runoff in centimeters.
6. Sediment loss in kilograms.

Annual concentrations were calculated as mass loss per volume of water. Mass loss was expressed as a percentage of the amount applied for the model runs so that loss estimates could be derived for any application rate. This procedure is necessary because, in practice, application rates often vary from the recommended rates used to make the model runs. The procedure provides the flexibility to apply different application rates across the country to reflect differences in use, or to simulate the effects of proposed policies on pesticide loss. The annual concentration was also expressed as a ratio of concentration to the mass of pesticide applied so that, when multiplied by the application rate, the concentration adjusted to reflect actual pesticide use rates would be obtained.

The 20-year distribution of mass loss and the concentration estimates were used to derive prediction equations corresponding to any percentile of the distribution. The 95th percentile, which is the mass loss or concentration that can be expected to be exceeded only five percent of the time, was used to develop the environmental risk indicators.

Development of Environmental Risk Indicators

Overview. A model was constructed to estimate environmental risk indicators for 1960 through 1997 using the following data and information:

- The National Pesticide Loss Database.
- Annual estimates of pesticide use by crop and state from farmer surveys, primarily the Doane farm panel surveys and the USDA pesticide use surveys.
- Annual county estimates of acres planted.
- Soil distribution from the National Resources Inventory.
- Irrigated acreage from the National Resources Inventory.
- Water quality thresholds corresponding to drinking water standards (or equivalent derived from mammalian chronic toxicity data) and the maximum safe levels for chronic exposure of fish, algae, and crustaceans to pesticides.

The analytical framework consists of about 4,700 resource polygons representing the intersection of 48 states (sufficient data do not exist to include Hawaii or Alaska), 280 watersheds at the 6-digit Hydrologic Unit level, and 1,400 combinations of climate and soil groups identified in the National Pesticide Loss Database. Twelve crops are currently included in the analysis: corn, soybeans, wheat, cotton, sorghum, barley, rice, potatoes, oats, sugarbeets, tobacco, and peanuts. Ninety-four pesticides were included in estimates for 1960-86, and up to 192 pesticides were included in estimates for 1987-97. Estimates of pounds of pesticides applied, mass loss, and annual concentrations were obtained for pesticides used on each of the 12 crops in each of the resource polygons for each of the 38 years.

Environmental risk was assessed using Threshold Exceedence Units (TEUs). TEUs account for both the extent that the annual concentration exceeds a water-quality threshold, and the number of acres on which pesticides are applied that produces the threshold exceedence. A concentration-threshold ratio was calculated for each pesticide used in each resource polygon, representing a per-acre estimate of potential risk. Thresholds differ considerably among pesticides, but, with this approach, all pesticides have the same per-acre estimate of potential risk (equal to 1) at the point where the threshold and the concentration coincide. As application rates increase, the concentration increases, and the per-acre estimate of risk increases; the per-acre risk score for pesticides with relatively high toxicity increases faster than that of pesticides with lower toxicity. Where the threshold is exceeded, the per-acre estimate of potential risk is multiplied by the acres treated to obtain TEU estimates that, when summed over a state or region, represent a regional estimate of potential risk. Where the concentration is below the threshold, the potential risk score is zero because the procedures used to derive the thresholds emulate those used by EPA to define "safe" thresholds for chronic exposure.

TEUs were derived in this manner specifically for the purpose of comparing risk from multiple pesticides over space and time. They are a relative measure of potential risk (the higher the TEU score, the higher the potential risk), but they do not measure actual risk. It is expected that temporal and spatial trends of these indicators will closely track the change in potential risk to human health and the environment from agricultural use of pesticides.

Management factors (such as tillage practices or grass waterways and stream buffers) that may mitigate pesticide loss from farm fields are not included in the indicators.

The general algorithm for the potential risk index aggregated to the national level is:

$$TEU_t = \sum_r \sum_k \sum_p \text{Exceedence per acre treated}_{t,r,k,p} * \text{Acres treated}_{t,r,k,p}$$

Exceedence per acre treated = [(RELCONC*APPRATE)/THRESH)-1], negative values discarded.

Acres treated = ACRES*PCTTREAT*PCTSOIL

where:

TEU_t	=	Threshold Exceedence Units for a given year.
RELCONC	=	Relative annual concentration associated with the 95th percentile pesticide loss per acre for a specific pesticide on a specific soil group and a specific climate group.
APPRATE	=	Application rate for a specific pesticide.
THRESH	=	Threshold concentration below which the pesticide loss concentration is defined to be "safe" for chronic exposure.
ACRES	=	Acres of crop planted in the state-watershed combination.
PCTTREAT	=	Percentage of acres treated with a specific pesticide.
PCTSOIL	=	Percentage of state-watershed combination in a specific resource polygon.
k	=	12 irrigated crops and 12 nonirrigated crops.
p	=	Pesticides.
r	=	Approximately 4,700 resource polygons.
t	=	Years from 1960 through 1997.

Separate pesticide risk indicators were constructed for pesticides in leachate and pesticides dissolved in runoff. In addition, the mass loss of pesticides (in leachate, dissolved runoff, and adsorbed runoff) was estimated using a similar algorithm. Databases were constructed by aggregating to three levels: 1) the national level, 2) the state level, and 3) watersheds at the 6-digit Hydrologic Unit level. Twelve pesticide-related time series were produced for each level of aggregation:

- Pounds of pesticides applied
- Leaching mass loss, pounds
- Dissolved runoff mass loss, pounds
- Adsorbed runoff mass loss, pounds
- Pesticide leaching risk index for protection of drinking water, TEUs
- Pesticide runoff risk index for protection of drinking water, TEUs
- Pesticide leaching risk index for protection of fish, TEUs
- Pesticide runoff risk index for protection of fish, TEUs
- Pesticide leaching risk index for protection of algae, TEUs
- Pesticide runoff risk index for protection of algae, TEUs
- Pesticide leaching risk index for protection of crustaceans, TEUs
- Pesticide runoff risk index for protection of crustaceans, TEUs

The leaching risk index for protection of drinking water is a proxy for risk associated with drinking water taken from shallow wells. The runoff risk index for protection of drinking water is a proxy for risk associated with drinking water taken from surface water supplies without land use controls in the upstream watershed. This indicator would not be appropriate for the New York water supply, for example, because much of the source area for that watershed is protected. Some of the other surface water supplies in the country have similar protections. The runoff risk indices for protection of fish, algae, and crustaceans are proxies for risk associated with exposure in surface water during times of the year when runoff is the predominant source of flow. The leaching risk indices for protection of fish, algae, and crustaceans are proxies for risk associated with exposure during times of the year when ground water is the predominant source of flow (such as mid-winter low-flow conditions) or in waterbodies that are predominantly fed by ground water sources (spring-fed ponds, lakes, and streams). Both estimates of risk for fish, algae, and crustaceans are necessary because most waterbodies have both a ground water and a surface water component. For example, USGS reports that in a sample of 54 streams from throughout the United States, ground water contributions ranged from 14 percent to 90 percent, with a median of 55 percent (Winter et al., 1998).

The Analytical Framework

The basis for the analytical framework underlying these estimates is the National Resources Inventory (NRI). The NRI is a national survey of private land use that is based on about 800,000 sample points, 300,000 of which are on cropland (Kellogg et al., 1994). At each NRI sample point, information is collected on nearly 200 attributes, including land use and cover, cropping history, conservation practices, potential cropland, highly eroding land, water and wind erosion estimates, wetlands, wildlife habitat, vegetative cover conditions, and irrigation. The NRI is linked to a national soils database that includes information on soil texture, hydrologic group, and organic matter content, which were the soil characteristics used to define the 120 soil groups for which pesticide loss estimates were simulated using GLEAMS. Using this soil information, each NRI cropland sample point was assigned a soil ID corresponding to one of the 120 soil groups. Similarly, each NRI cropland sample point was assigned one of the 55 climate stations based on geographic proximity. The pesticide use databases provided information on the suite of pesticides used for each crop and state. Linking these data sources together, a list of pesticides, as well as estimates of pesticide loss (mass loss and concentration) relative to application rate, was imputed onto each NRI cropland sample point.

The framework was constructed by identifying resource polygons as the intersection of state, watershed (6-digit level), soil ID, and climate station. To reduce the number of resource polygons to a manageable number, very small polygons were excluded. This procedure was executed using NRI data for both 1982 and 1992. It resulted in a total of 4,891 resource polygons for 1982 and 4,459 for 1992. State-watershed combinations numbered 544 for 1982 and 517 for 1992. The analytical framework based on 1982 was used for calculating environmental indicators for 1960 through 1985, whereas the analytical framework based on 1992 was used for 1986 through 1997.

The percentage of a state-watershed combination that corresponded to each soil-climate combination within a state-watershed combination was estimated using NRI acreage of all 12 crops combined (PCTSOIL in the above algorithm). This percentage was used in the algorithm to distribute acres by crop for each state-watershed combination to the soil-climate groups.

Land Use Time Series

Data by county on acres planted by county were used to capture changes in cropping patterns over space and time. County data were obtained from electronic files published by the National Agriculture Statistics Service (NASS). These data generally cover 1972 to the present, but not all years are included for each of the 12 crops. Gaps in the record were filled one of two ways: 1) state totals were distributed to counties using county shares from a nearby year, or 2) state totals were distributed to counties using county shares derived from the Census of Agriculture. State totals were obtained from NASS publications. The latter approach was used for all states and crops for 1960-71, and for potatoes through 1982 and sugarbeets through 1976. For these estimates, the Census of Agriculture county data were used as follows: 1959 county shares used to estimate acres planted for 1960-61, 1964 county shares used for 1962-66, 1969 county shares used for 1967-71, 1974 county shares used for 1972-76, 1978 county shares used for 1977-80, and 1982 county shares used for 1981-84. Table 9.2 summarizes the number of states with county estimates of acres planted for each crop and the source of the data.

County data on acres planted were converted to state-watershed combinations using conversion weights derived from the NRI, which provides acreage estimates by watershed and county for each of the 12 crops. Separate weights were developed for 1982 and 1992. The 1982 weights were used for years

Table 9.2. Source of County Estimates of Acres Planted for 12 Crops, 1960-97

	Published County NASS data		Estimates derived from state data and county shares	
	Years	Number of states	Years	Number of states
Barley	1972-97	36	1960-71	38
			1987-97	2
Corn	1972-97	41	1960-71	47
			1972-97	5
Cotton	1972-97	18	1960-71	18
Oats	1973-97	32	1960-72	43
			1973-86	20
			1987-97	2
Peanuts	1972-97	8	1960-71	10
			1972-97	2
Potatoes	1983-96	15	1960-82	48
			1983-97	12
			1997	36
Rice	1972-97	6	1960-71	6
Sorghum	1972-97	23	1960-71	23
Soybeans	1972-97	29	1960-71	30
			1972-97	1
Sugarbeets	1977-97	15	1960-76	16
Tobacco	1972-97	15	1960-71	18
	1972-85	1	1986-97	1
Wheat	1972-97	42	1960-71	42

prior to 1986, and the 1992 weights were used for 1986-97. Weights were crop specific; that is, county corn acres planted were converted to state-watershed acres planted using county-watershed proportions derived from NRI corn acreage estimates. In some counties, however, the NRI did not find any acres, even though acres (usually a small number) were reported in the Census of Agriculture or in the NASS county acreage database. This gap occurred more frequently in the early years of the series, where the 1982 NRI did not represent spatial distribution of crops as well as in the later years of the series. In these cases, a conversion weight based on the sum of all 12 crops was used.

Acres-planted data included both irrigated and dryland acreage. The proportion of each crop in each state-watershed combination that was irrigated was estimated using the NRI. These proportions were multiplied by the crop acreage to obtain separate estimates of irrigated and non-irrigated acres for each of the 12 crops in each state-watershed combination.

Pesticide Use Time Series

An annual time series of application rates and percent acres treated for each crop and each state that had acres-planted data was constructed for 1960 through 1997 using all sources of information available, summarized in table 9.3. The primary sources were the Doane farm panel surveys and the USDA National Agricultural Statistics Service (NASS) pesticide-use surveys. For 1990 through 1997, two separate sets of estimates of application rates and percent acres treated were constructed, one with Doane data and another with NASS data, so that results from the two sources could be compared. For 1960 through 1989, a single set of estimates was constructed from published reports of a series of annual surveys by Doane (1970-89), intermittent surveys conducted by Doane and NASS, and a small number of surveys from other sources.

The Doane Pesticide Profile Study is an annual survey of farmers conducted by Doane Marketing Research, Incorporated on behalf of the pesticide industry (Doanes, a). The Doane Pesticide Profile Study includes all crops and all states for 1987 to the present. Sample sizes ranged from 12,252 farms in 1987 to 18,208 farms in 1991. The database is electronic, and estimates of application rates and percent acres treated were obtained by querying the database.

USDA NASS conducted annual pesticide use surveys for selected crops in selected states from 1990 to the present. Sample sizes were comparable to those in the Doane Pesticide Profile Study when compared on a crop-by-crop basis. These data were used to construct a second set of estimates of application rates and percent acres treated for 1990 through 1997 for the selected states and crops.

Table 9.3. Pesticide-use Surveys Used to Construct the Pesticide-use Database for 1960-97

Survey	Source	Crop	Years covered	States or regions covered
Doane Pesticide Profile Study	Doane	All 12 crops	1987-97	All 48 states
Agricultural Chemical Usage: 1990 Field Crops Summary	NASS	Corn	1990	CO, IL, IN, IA, KS, KY, MI, MN, MO, NE, NY, NC, ND, OH, PA, SC, SD, VA, WV, WI
		Cotton	1990	AZ, AR, CA, LA, MS, TX
		Potatoes	1990	CO, ID, ME, MI, MN, NY, ND, OR, PA, WA, WI
		Rice	1990	AR, LA
		Soybeans	1990	AL, AR, DE, FL, GA, ILL, IN, IA, KS, KY, LA, MD, MI, MN, MS, MO, NE, NJ, NC, ND, OH, OK, PA, SC, SD, TN, TX, VA, WI
Agricultural Chemical Usage: 1991 Field Crops Summary	NASS	Corn	1991	GA, IL, IN, IA, KS, KY, MI, MN, MO, NE, NC, OH, PA, SC, SD, TX, WI
		Cotton	1991	AZ, AR, CA, LA, MS, TX
		Peanuts	1991	GA, TX
		Potatoes	1991	CO, ID, ME, MI, MN, NY, ND, OR, PA, WA, WI
		Rice	1991	AR, LA
		Soybeans	1991	AR, GA, IL, IN, IA, KS, KY, LA, MN, MS, MO, NE, NC, OH, SD, TN
		Wheat	1991	CO, ID, KS, MN, MT, NE, ND, OK, OR, SD, TX, WA
Agricultural Chemical Usage: 1992 Field Crops Summary	NASS	Corn	1992	GA, IL, IN, IA, KS, KY, MI, MN, MO, NE, NC, OH, PA, SC, SD, TX,WI
		Cotton	1992	AZ, AR, CA, LA, MS, TX
		Potatoes	1992	CO, ID, ME, MI, MN, NY, ND, OR, PA, WA, WI
		Rice	1992	AR, LA
		Soybeans	1992	AR, GA, IL, IN, IA, KS, KY, LA, MN, MS, MO, NE, NC, OH, SD, TN
		Wheat	1992	AR, CO, ID, KS, MT, NE, OK, OR, SD, TX, WA
Agricultural Chemical Usage: 1993 Field Crops Summary	NASS	Corn	1993	GA, IL, IN, IA, KS, KY, MI, MN, MO, NE, NC, OH, PA, SD, TX, WI
		Cotton	1993	AZ, AR, CA, LA, MS, TX
		Potatoes	1993	CO, ID, ME, MI, MN, NY, ND, OR, PA, WA, WI
		Soybeans	1993	AR, GA, IL, IN, IA, KS, KY, LA, MN, MS, MO, NE, NC, OH, SD, TN
		Wheat	1993	CO, ID, IL, KS, MN, MT, NE, ND, OH, OK, OR, SD, TX, WA

Note: USDA Regions are as follows: NE=Northeast, LK=Lake States, CB=Corn Belt, NP=Northern Plains, AP=Appalachian, SE=Southeast, DT=Delta States, SP=Southern Plains, MT=Mountain, and PA=Pacific.

Table 9.3, *continued*

Survey	Source	Crop	Years covered	States or regions coverec
Doane Pesticide Profile Stud				
Agricultural Chemical Usage: 1994 Field Crops Summary	NASS	Corn	1994	IL, IN, IA, MI, MN, MO, NE, OH, SD, WI
		Cotton	1994	AZ, AR, CA, LA, MS, TX
		Potatoes	1994	CO, ID, ME, MI, MN, NY, ND, OR, PA, WA, WI
		Soybeans	1994	AR, DE, IL, IN, IA, MN, MO, NE, OH
		Wheat	1994	CO, ID, IL, KS, MN, MT, NE, ND, OH, OK, OR, SD, TX, WA
Agricultural Chemical Usage: 1995 Field Crops Summary	NASS	Corn	1995	DE, GA, IL, IN, IA, KS, KY, MI, MN, MO, NE, NC, OH, PA, SD, TX, WI
		Cotton	1995	AZ, AR, CA, LA, MS, TX
		Potatoes	1995	CO, ID, ME, MI, MN, NY, ND, OR, PA, WA, WI
		Soybeans	1995	AR, GA, IL, IN, IA, KY, LA, MN, MS, MO, NE, NC, OH, TN
		Wheat	1995	CO, ID, IL, KS, MN, MO, MT, NE, ND, OH, OK, OR, SD, TX, WA
Agricultural Chemical Usage: 1996 Field Crops Summary	NASS	Corn	1996	IL, IN, IA, KS, KY, MI, MN, MO, NE, NC, OH, PA, SC, SD, TX, WI
		Cotton	1996	AZ, AR, CA, GA, LA, MS, TN, TX
		Potatoes	1996	ID, ME, WA
		Soybeans	1996	AR, IL, IN, IA, LA, MN, MS, MO, NE, OH, TN, WI
		Wheat	1996	CO, ID, KS, MN, MT, NE, ND, OK, OR, SD, TX, WA
Agricultural Chemical Usage: 1997 Field Crops Summary	NASS	Corn	1997	IL, IN, IA, MI, MN, MO, NE, OH, SD, WI
		Cotton	1997	AL, AZ, AR, CA, GA, LA, MS, MO, NC, SC, TN, TX
		Potatoes	1997	ID, ME, MN, ND, OR, WA, WI
		Soybeans	1997	AR, DE, IL, IN, IA, KS, KY, LA, MI, MN, MS, MO, NE, NC, OH, PA, SD, TN, WI
		Wheat	1997	CO, ID, IL, KS, MN, MT, NE, ND, OK, OR, PA, SD, TX, WA
Doane Countrywide Farm Panel Survey	Doane	Corn	1970-86	Doane regions 1, 2, 3, 4, 5; IL, IN, IA, MN, MO, NE, OH, WI, SD, MI
		Soybeans	1970-84, 86	Doane regions 2, 3, 4, 5; IL, IN, IA, MN, MO, NE, OH, WI
		Cotton	1970-84, 86	Doane regions 2, 4, 6, 7
		Wheat	1973-86	Doane regions 5, 6, 7
		Sorghum	1973-86	Doane regions 4, 5, 7
		Rice	1975, 79, 81	CA, TX-AR-LA-MS, other
		Barley	1976	Doane regions 5, 6
Pesticide Use on Cotton	Battelle	Cotton	1985	National
Pesticide Use on Soybeans	Battelle	Soybeans	1985	National

Table 9.3, *continued*

Survey	Source	Crop	Years covered	States or regions covered
Specialty Crop Pesticide Study 1984	Maritz	Rice	1984	CA, TX-AR-LA-MS, other
Specialty Crop Pesticide Study 1983	Doane	Peanuts	1983	TX-OK, AL-GA, VA-NC, other
		Potatoes	1983	CA-ID-WA-OR, ND-MN-WI, NORTHEAST-MI, other
		Sugarbeets	1983	CA, MN-ND, CO-NE, other
		Tobacco	1983	VA-NC-SC-GA-FL, AR-LA-AL-MS-TN-KY, other
Specialty Crop Pesticide Study 1977	Doane	Peanuts	1977	TX-OK, AL-GA, VA-NC, other
		Potatoes	1977	CA-ID-WA-OR, ND-MN-WI, NORTHEAST-MI, other
		Sugarbeets	1977	CA, MN-ND, CO-NE, other
		Tobacco	1977	VA-NC-SC-GA-FL, AR-LA-AL-MS-TN-KY, other
Specialty Crop Pesticide Study 1975	Doane	Peanuts	1975	TX-OK, AL-GA, VA-NC, other
		Potatoes	1975	CA-ID-WA-OR, ND-MN-WI, NORTHEAST-MI, other
		Sugarbeets	1975	CA, MN-ND, CO-NE, other
		Tobacco	1975	VA-NC-SC-GA-FL, AR-LA-AL-MS-TN-KY, other
Specialty Crop Pesticide Study 1973	Doane	Peanuts	1973	TX-OK, AL-GA, VA-NC, other
		Potatoes	1973	CA-ID-WA-OR, ND-MN-WI, NORTHEAST-MI, other
		Sugarbeets	1973	CA, MN-ND, CO-NE, other
		Tobacco	1973	VA-NC-SC-GA-FL, AR-LA-AL-MS-TN-KY, other
Farmer's use of pesticides, Unpublished spreadsheets	NASS	Corn	1982	MD, NY, PA, KY, NC, TN, VA, AL, FL, GA, SC, AR, LA, MS, IL, IA, IN, MO, OH, MI, MN, WI, KS, NE, SD, ND, OK, TX, MT, WA, ID, AZ
		Soybeans	1982	MN, IA, MO, IL, IN, OH, KY, AR, LA, MS, MD, GA, SC, NC, VA, FL, MI, WI, OK, TX
		Cotton	1982	AL, GA, SC, AR, LA, MS, TN, OK, TX, MO, NC, AZ
		Sorghum	1982	KY, NC, TN, VA, AL, GA, SC, AR, LA, MS, IL, MO, KS, NE, SD, OK, TX, AZ
		Wheat	1982	MD, KY, NC, TN, VA, AL, GA, SC, AR, LA, MS, IL, IA, MO, OH, MI, MN, WI, KS, NE, ND, SD, OK, TX, ID, MT, WA, AZ
		Rice	1982	AR, LA, MS, MO, TX
		Peanuts	1982	NC, VA, AL, GA, SC, TX
		Barley	1982	NC, VA, SC, MI, MN, WI, ND, SD, ID, MT, WA, AZ
		Tobacco	1982	PA, KY, NC, TN, VA, FL, GA, SC, OH, IN
		Oats	1982	NY, PA, NC, SC, AR, OH, MI, MN, WI, ND, SD, OK, TX, ID, MT, WA

Table 9.3, *continued*

Survey	Source	Crop	Years covered	States or regions covered
Pesticide Use on Corn	NASS	Corn	1980	IL, IN, IA, MO, OH, MI, MN, WI, KS, NE, SD, TX, KY, PA, NC, GA
Pesticide Use on Soybeans	NASS	Soybeans	1980	NE, KS, MN, IA, MO, IL, IN, OH, KY, TN, AR, LA, MS, AL, GA, SC, NC
Pesticide Use on Sorghum	NASS	Sorghum	1980	KS, MO, NE, OK, SD, TX
Pesticide Use on Cotton	NASS	Cotton	1979	AL, GA, SC, AR, LA, MS, TN, OK, TX
Pesticide Use on Potatoes	NASS	Potatoes	1979	MA, NY, PA, ND, MN, NJ, MI, WA, OR, ID, CO
Farmer's Use of Pesticides, 1976	NASS	Corn	1976	National; USDA regions NE, LK, CB, NP, AP, SE, DT, SP, MT, PA
		Soybeans	1976	National; USDA regions NE, LK, CB, NP, AP, SE, DT, SP, PA
		Cotton	1976	National; USDA regions AP, SE, DT, SP, MT, PA
		Sorghum	1976	National; USDA regions NE, LK, CB, NP, AP, SE, DT, SP, MT, PA
		Wheat	1976	National; USDA regions NE, LK, CB, NP, AP, SE, DT, SP, MT, PA
		Rice	1976	National; USDA regions DT, SP, PA
		Peanuts	1976	National; USDA regions AP, SE, DT, SP, MT
		Tobacco	1976	National; USDA regions NE, LK, CB, AP, SE
		Barley	1976	National; USDA regions NE, LK, CB, NP, AP, SE, SP, MT, PA
Farmer's Use of Pesticides, 1971	NASS	Corn	1971	National; USDA regions NE, LK, CB, NP, AP, SE, DT, SP, MT, PA
		Soybeans	1971	National; USDA regions NE, LK, CB, NP, AP, SE, DT, SP
		Cotton	1971	National; USDA regions CB, AP, SE, DT, SP, MT, PA
		Sorghum	1971	National; USDA regions NE, CB, NP, AP, SE, DT, SP, MT, PA
		Wheat	1971	National; USDA regions NE, LK, NP, AP, SE, SP, MT, PA
		Rice	1971	National; USDA regions DT, SP, PA
		Peanuts	1971	National; USDA regions AP, SE, DT, SP, MT
		Potatoes	1971	National; USDA regions NE, LK, NP, SE, SP, MT, PA
		Tobacco	1971	National; USDA regions NE, CB, AP, SE
		Barley	1971	National; USDA regions NE, LK, CB, NP, SE, SP, MT, PA
		Sugarbeets	1971	National; USDA regions LK, CB, NP, MT, PA
Doane Summary Herbicide Report 1972	Doane	Corn	1961-72	National
		Soybeans	1961-72	National
		Cotton	1961-72	National
		Sorghum	1961-72	National
		Wheat	1961-72	National
		Barley	1961-72	National

Table 9.3, *continued*

Survey	Source	Crop	Years covered	States or regions covered
Doane Summary Insecticide Report 1972	Doane	Corn	1961-72	National
		Soybeans	1961-72	National
		Cotton	1961-72	National
		Sorghum	1961-72	National
		Wheat	1961-72	National
		Barley	1961-72	National
Quantities of pesticides used by farmers, 1966	NASS	Corn	1966	National; USDA regions NE, LK, CB, NP, AP, SE, DT, SP, MT, PA
		Soybeans	1966	National; USDA regions NE, LK, CB, NP, AP, SE, DT, SP
		Cotton	1966	National; USDA regions CB, AP, SE, DT, SP, MT, PA
		Sorghum	1966	National; USDA regions NE, CB, NP, AP, SE, DT, SP, MT, PA
		Wheat	1966	National; USDA regions LK, CB, NP, AP, SE, SP, MT, PA
		Rice	1966	National; USDA regions DT, SP
		Peanuts	1966	National; USDA regions AP, SE, SP, MT
		Potatoes	1966	National; USDA regions NE, LK, CB, SE, MT, PA
		Tobacco	1966	National; USDA regions NE, LK, CB, AP, SE
		Barley	1966	National; USDA regions NE, LK, CB, NP, AP, SE, SP, MT, PA
		Sugarbeets	1966	National; USDA regions LK, CB, NP, MT, PA
Quantities of pesticides used by farmers, 1964	NASS	Corn	1964	National; USDA regions NE, LK, CB, NP, AP, SE, DT, SP, MT, PA
		Soybeans	1964	National; USDA regions NE, LK, CB, NP, AP, SE, DT
		Cotton	1964	National; USDA regions CB, AP, SE, DT, SP, PA
		Sorghum	1964	National; USDA regions CB, NP, AP, SE, SP, MT
		Wheat	1964	National; USDA regions LK, CB, NP, AP, SP, MT, PA
		Rice	1964	National; USDA regions CB, DT, SP
		Peanuts	1964	National; USDA regions AP, SE, SP, MT
		Potatoes	1964	National; USDA regions NE, LK, CB, SE, MT, PA
		Tobacco	1964	National; USDA regions NE, LK, CB, AP, SE
		Sugarbeets	1964	National; USDA regions LK, CB, NP
		Barley	1964	National; USDA regions NE, LK, CB, NP, AP, SE, DT, SP, MT, PA
Extent of spraying and dusting on farms, 1958 with comparisons (1962)	NASS	Corn	1958	National; USDA regions NE, LK, CB, NP, AP, SE, DT, SP, MT, PA
		Cotton	1958	National; USDA regions CB, AP, SE, DT, SP, PA
		Potatoes	1958	National; USDA regions NE, LK, CB, SE, MT, PA
		Tobacco	1958	National; USDA regions NE, LK, CB, AP, SE

The Doane Countrywide Farm Panel Surveys for insecticides and herbicides, precursors to the Doane Pesticide Profile Study, were available (although not in electronic form) annually for most years from 1970 through 1986 for corn, soybeans, cotton, wheat, and sorghum (Doanes, b). Results were reported for the 7 Doane regions, shown below, and for most states in regions 3 and 5 for corn and soybeans. State estimates from the survey were used for corn and soybeans in regions 3 and 5. State estimates for cotton, wheat, sorghum, corn, and soybeans in regions other than 3 and 5 were obtained by assigning the regional estimates to each of the states within each region, implicitly assuming homogeneity within the regions. These estimates, particularly application rate estimates, were supplemented with additional data from NASS surveys and surveys by the Batelle Corporation.

Doane Region	States	Crops
1	Maine, Connecticut, Massachusetts, New Hampshire, New Jersey, New York, Pennsylvania, Rhode Island, Vermont	Corn
2	Delaware, Florida, Georgia, Maryland, North Carolina, South Carolina, Virginia, West Virginia	Corn, soybeans, cotton
3	Illinois, Indiana, Ohio, Michigan, Wisconsin	Corn, soybeans
4	Alabama, Arkansas, Kentucky, Louisiana, Mississippi, Tennessee	Corn, soybeans, cotton, sorghum
5	Iowa, Minnesota, Missouri, Nebraska, Kansas, North Dakota, South Dakota	Corn, soybeans, wheat, sorghum
6	Arizona, California, Colorado, Idaho, Montana, Nevada, New Mexico, Oregon, Utah, Washington, Wyoming	Wheat, cotton
7	Texas, Oklahoma	Wheat, cotton, sorghum

Pesticide use estimates for these 5 crops for the early years not covered in the Doane Countrywide Panel Surveys were largely constructed using information in three USDA pesticide use surveys for 1964, 1966, and 1971 for each of the 10 U.S. production regions and national use estimates from Doane for most years. The national Doane data together with regional shares from 1970-73 were used to generate time series for acres treated by region. Application rates were taken from all sources.

These procedures and databases provided estimates for most states, years, and pesticides for these five crops, but a few additional estimates were needed to complete the series. Missing data on application rates were filled using application rates for nearby years or nearby states. ccasionally it was possible to derive percent acres-treated data for a state (or groups of states) or pesticide using the acres-treated residuals derived from regional or national data. Percent acres-treated data for some missing years were constructed using regional

Table 9.4. Pesticides Used to Estimate Environmental Risk Indicators

A "1" indicates the pesticide was included in the risk estimate for one or more of the 12 crops during the indicated time interval; a "0" indicates the pesticide was not included.

Pesticide (active ingredient name)	1960-65	1966-71	1972-77	1978-83	1984-90	1991-97
2,4,5-T	1	1	1	1	1	0
2,4-D	1	1	1	1	1	1
2,4-DB	1	1	1	1	1	1
Acephate	0	1	1	1	1	1
Acifluorfen	0	0	1	1	1	1
Alachlor	0	1	1	1	1	1
Aldicarb	1	1	1	1	1	1
Aldrin	1	1	1	1	1	0
Ametryn	0	0	0	0	1	1
Amitraz	0	0	0	0	1	1
Amitrole	0	0	0	0	1	0
Anilazine	0	0	0	0	1	1
Arsenic Acid	0	0	0	0	0	1
Atrazine	1	1	1	1	1	1
Azinphos-Methyl	1	1	1	1	1	1
Benomyl	0	1	1	1	1	1
Bensulfuron	0	1	1	0	1	1
Bensulide	0	0	0	0	1	1
Bentazone	0	0	1	1	1	1
Bifenox	0	0	1	1	1	1
Bifenthrin	0	0	0	0	1	1
Bromacil	0	0	0	0	1	1
Bromoxynil	1	1	1	1	1	1
Butylate	1	1	1	1	1	1
Captan	1	1	1	1	1	1
Carbaryl	1	1	1	1	1	1
Carbofuran	1	1	1	1	1	1
Chloramben	1	1	1	1	1	1
Chlordane	1	1	1	1	1	0
Chlordimeform	0	1	1	1	1	1
Chloridazon	0	0	0	0	1	1
Chlorimuron	0	0	0	0	1	1
Chlorobenzilate	0	0	0	0	1	0
Chloropicrin	0	0	0	0	1	1
Chlorothalonil	0	1	1	1	1	1
Chloroxuron	0	0	0	0	0	1
Chlorpropham	1	1	1	1	1	1
Chlorpyrifos	1	1	1	1	1	1
Chlorsulfuron	0	0	1	1	1	1
Clomazone	0	0	0	0	1	1
Clopyralid	0	0	0	0	1	1
Cyanazine	0	1	1	1	1	1
Cycloate	0	1	1	1	1	1
Cyfluthrin	0	0	0	0	1	1
Cypermethrin	0	0	1	1	1	1

Table 9.4, *continued*

Pesticide (active ingredient name)	1960-65	1966-71	1972-77	1978-83	1984-90	1991-97
Cyromazine	0	0	0	0	0	1
Dalapon	1	1	1	1	1	1
Daminozide	0	0	0	0	1	1
DCPA	0	0	0	0	1	1
DDT	1	1	1	0	0	0
Desmedipham	0	0	1	1	1	1
Diazinon	1	1	1	1	1	1
Dicamba	1	1	1	1	1	1
Dichlobenil	0	0	0	0	0	1
Dichloropropene	1	1	1	1	1	1
Dichlorprop	0	0	0	0	1	1
Diclofop	0	0	1	1	1	1
Dicloran	0	0	0	0	1	1
Dicofol	1	1	1	1	1	1
Dicrotophos	1	1	1	1	1	1
Difenzoquat	0	0	1	1	1	1
Diflubenzuron	0	0	0	0	1	1
Dimethipin	0	0	0	0	1	1
Dimethoate	1	1	1	1	1	1
Dinocap	0	0	0	0	1	1
Dinoseb	1	1	1	1	1	1
Diphenamid	0	1	1	1	1	1
Dipropetryn	0	0	0	0	1	1
Diquat	0	1	1	1	1	1
Disulfoton	1	1	1	1	1	1
Diuron	1	1	1	1	1	1
Dodine	0	0	0	0	1	0
Endosulfan	1	1	1	1	1	1
Endothal	0	0	0	0	1	1
EPN	1	1	1	1	1	1
EPTC	1	1	1	1	1	1
Esfenvalerate	0	0	0	0	1	1
Ethalfluralin	0	0	0	1	1	1
Ethephon	0	0	0	0	1	1
Ethion	1	1	1	0	1	1
Ethofumesate	0	0	1	1	1	1
Ethoprophos	0	1	1	1	1	1
Etridiazole	0	0	0	1	1	1
Fenamiphos	0	0	0	0	1	1
Fenoxaprop	0	0	0	0	1	1
Fensulfothion	1	1	1	1	1	1
Fenvalerate	0	0	1	1	1	1
Ferbam	0	0	0	0	1	1
Fluazifop	0	0	0	1	1	1
Flucythrinate	0	0	0	1	1	1
Flumetralin	0	0	0	0	1	1
Fluometuron	1	1	1	1	1	1

Table 9.4, *continued*

Pesticide (active ingredient name)	1960-65	1966-71	1972-77	1978-83	1984-90	1991-97
Fluridone	0	0	0	0	0	1
Fluvalinate-Tau	0	0	0	0	0	1
Fomesafen	0	0	0	0	1	1
Fonofos	1	1	1	1	1	1
Formetanate	0	0	0	0	1	1
Fosetyl	0	0	0	0	0	1
Glufosinate	0	0	0	0	0	1
Glyphosate	0	0	1	1	1	1
Hexazinone	0	0	0	0	1	1
Hexythiazox	0	0	0	0	0	1
Imazapyr	0	0	0	0	1	1
Imazaquin	0	0	0	0	1	1
Iprodione	0	0	0	0	1	1
Isofenphos	0	0	0	1	0	1
Isopropalin	0	0	1	1	1	1
Lactofen	0	0	0	0	1	1
Lambda-Cyhalothirin	0	0	0	0	1	1
Lindane	1	1	1	1	1	1
Linuron	1	1	1	1	1	1
Malathion	1	1	1	1	1	1
Maleic Hydrazide	1	1	1	1	1	1
Mancozeb	1	1	1	1	1	1
Maneb	1	1	1	1	1	1
MCPA	1	1	1	1	1	1
MCPB	0	0	0	0	1	1
Mepiquat	0	0	0	0	1	1
Metalaxyl	1	1	1	1	1	1
Metaldehyde	0	0	0	0	1	1
Metam	1	1	1	1	1	1
Methamidophos	0	1	1	1	1	1
Methazole	0	0	0	0	1	1
Methidathion	0	0	1	1	1	1
Methiocarb	0	0	0	1	1	1
Methomyl	0	1	1	1	1	1
Methoxychlor	0	0	1	1	1	1
Methyl Bromide	0	0	0	0	1	1
Metiram	1	1	1	1	1	1
Metolachlor-S	0	0	1	1	1	1
Metribuzin	0	0	1	1	1	1
Metsulfuron	0	0	0	0	1	1
Mevinphos	0	0	0	0	1	1
Molinate	1	1	1	1	1	1
Monocrotophos	1	1	1	1	1	1
MSMA	1	1	1	1	1	1
Naled	0	0	0	0	1	1
Napropamide	0	0	1	1	1	1
Naptalam	1	1	1	1	1	1

Table 9.4, *continued*

Pesticide (active ingredient name)	1960-65	1966-71	1972-77	1978-83	1984-90	1991-97
Norflurazon	0	0	1	1	1	1
Oryzalin	0	0	1	1	1	1
Oxadiazon	0	0	0	0	0	1
Oxamyl	0	0	1	1	1	1
Oxydemeton-Methyl	0	0	0	0	1	1
Oxyfluorfen	0	0	0	0	1	1
Paraquat	1	1	1	1	1	1
Parathion	1	1	1	1	1	1
Pebulate	1	1	1	1	1	1
Pendimethalin	0	0	1	1	1	1
Permethrin	1	1	1	1	1	1
Phenmedipham	0	0	0	0	1	1
Phorate	1	1	1	1	1	1
Phosmet	1	1	1	1	1	1
Phosphamidon	1	1	1	1	1	1
Picloram	0	0	0	0	1	1
Profenofos	0	0	1	1	1	1
Profluralin	1	1	1	1	1	1
Prometon	0	0	0	0	1	1
Prometryn	0	1	1	1	1	1
Propachlor	1	1	1	1	1	1
Propamocarb HCL	0	0	0	0	0	1
Propanil	1	1	1	1	1	1
Propargite	1	1	1	1	1	1
Propazine	1	1	1	1	1	1
Propiconazole	0	0	0	0	1	1
Propoxur	0	0	0	0	0	1
Quizalofop	0	0	0	0	1	1
Sethoxydim	0	0	0	1	1	1
Simazine	1	1	1	1	1	1
Sulprofos	0	0	1	1	1	1
Tebuthiuron	0	0	0	0	1	1
Terbacil	0	0	0	0	0	1
Terbufos	0	1	1	1	1	1
Terbutryn	0	0	1	1	1	1
Thiabendazole	1	1	1	1	1	1
Thidiazuron	0	0	0	0	1	1
Thifensulfuron	0	0	0	0	1	1
Thiobencarb	0	0	0	1	1	1
Thiodicarb	0	0	0	0	1	1
Thiophanate-Methyl	0	0	0	0	1	1
Thiram	0	0	0	0	1	1
Toxaphene	1	1	1	1	1	0
Tralomethrin	0	0	0	0	1	1
Tri-Allate	1	1	1	1	1	1
Triadimefon	0	0	0	1	1	1
Tribufos	0	0	0	0	1	1

Table 9.4, *continued*

Pesticide (active ingredient name)	1960-65	1966-71	1972-77	1978-83	1984-90	1991-97
Trichlorfon	1	1	1	1	1	1
Triclopyr	0	0	0	0	1	1
Tridiphane	0	0	0	0	1	1
Trifluralin	1	1	1	1	1	1
Triforine	0	0	0	0	1	1
Trimethacarb	0	0	0	0	1	1
Vernolate	1	1	1	1	1	1
Number of pesticides	67	82	106	112	178	185

shares from nearby years to distribute national totals of acres treated. In cases where acres-planted data indicated that crops were grown in states not covered by the pesticide-use surveys, pesticide-use data from a nearby state or region was used.

Pesticide-use data for the remaining 7 crops for 1960-86 was comparatively sparse, and considerable extrapolation was required to construct complete series. Very little state-specific data was available, so the focus was on constructing regional time series. State estimates were made by assigning the regional estimates to each of the states within each region. Missing data for application rates were filled by extending data from survey years to years without data. In cases where only national data on acres treated were available, acres treated were distributed to regions using regional shares from survey years and percent acres-treated estimates were calculated using acres-planted data. Pesticide-use data for small grains and wheat were used for oats and sometimes for barley. Most other missing data were filled using procedures already presented for the first 5 crops. In all cases, the series were constructed using all the information available and using the most precise extrapolation procedures possible.

Not all pesticides reported in the surveys could be used to estimate the environmental risk indicators because information on pesticide loss or water quality thresholds was not available. The 194 pesticides included in the estimates are summarized by 6-7 year time intervals in table 9.4. Most of the pesticides not included were very minor. The few pesticides with significant or frequent use that were not included were: acetochlor, clethodim, dimethenamid, DSMA, flumetsulam, halosulfuron, heptachlor, imazethapyr, nicosulfuron, primisulfuron, prosulfuron, tefluthrin, and zetamethain.

Water Quality Thresholds

The EPA Office of Water sets drinking water standards for some pesticides. A Health Advisory (HA) is the maximum concentration of a chemical in drinking water that is not expected to cause any adverse noncarcinogen effect over a lifetime exposure with a margin of safety. A Maximum Contaminant Level (MCL) is the maximum permissible pesticide concentration allowed in a public water source. HAs and MCLs were used as water-quality thresholds to calculate the environmental-risk indicators for drinking water. Many pesticides, however, do not have HAs or MCLs. For pesticides categorized by the EPA Office of Water as noncarcinogens or only "possible" carcinogens, "safe" thresholds were estimated from EPA published Reference Dose (RFD) values using procedures similar to those used to derive HAs. For pesticides categorized by the EPA Office of Water as probable or known carcinogens, EPA-published data on cancer slopes were used to estimate the CHCL (Chronic Human Carcinogen Level), which produces concentrations comparable to the MCL.

Thresholds for fish, algae, and crustaceans were estimated from published toxicity data. Since so little data on long-term exposures are available, Maximum Acceptable Toxicant Concentrations (MATCs) were calculated from 96-hour LC50's using the method of Barnthouse, Suter, and Rosen (1990). Most 96-hour LC50's used in this study were taken from EPA's Office of Pesticide Toxicity Database (EPA, 1997). MATCs are equivalent to the maximum concentration permitting survival for chronic exposures. When toxicity data for more than one species were available, the "safe" threshold for the most sensitive species was used.

The estimated water-quality thresholds are shown in table 9.5 for the pesticides included in this study. Thresholds are constant over space and time, but the suite of pesticides included in the indicator differs depending on the target group. For example, more toxicity results were available for humans than for the other groups. The potential risk indicators are thus constrained to represent temporal and spatial changes only for the suite of pesticides for which toxicity data were available for each target group. Additional information on toxicity data and derivation methods can be found in Plotkin et al.

Combining Risk Indicators Based on Alternative Pesticide Use Sources

The separate pesticide use databases for NASS 1990-97 and Doane 1990-97 allowed separate risk indicators to be estimated for pesticides that were common to both surveys. In the majority of cases, risk indicators did not differ by more than about 20 percent when aggregated to the state level and compared by pesticide. When aggregated over pesticides, there was even less difference. For the risk indicators presented in this paper, the two estimates were combined. For each pesticide in each resource polygon, the average of the two risk scores was calculated wherever pesticide use estimates from both surveys existed (see table 9.3 for NASS coverage).

Table 9.5. Water Quality Thresholds (concentrations in parts per billion) **for Pesticides Included in the Potential Risk Indicators**

Pesticide (active ingredient name)	Humans	Fish	Algae	Crustaceans
2,4-D	70	4247	19200	79000
2,4-DB	70	578.46	NA	NA
2,4,5-T	70	125.89	NA	NA
AC 263,499	NA	NA	NA	NA
Acephate	2.8	2725.52	NA	150
Acifluorfen	3.27	2841.12	NA	NA
Alachlor	2	26.2	0.35	230
Aldicarb	7	3.75	NA	NA
Aldrin	0.02	0.18	NA	NA
Ametryn	60	452.54	1.14	320
Aminochlor	NA	NA	NA	NA
Aminocarb	NA	NA	NA	NA
Amitraz	1.75	38.43	NA	1.1
Amitrole	0.31	12422.38	NA	NA
Ancymidol	NA	19952.62	68	NA
Anilazine	2.8	14.48	NA	58
Arsenic Acid	50	4918.92	NA	20
Asulam Na Salt	252	36926.66	20	NA
Atrazine	3	658.46	0.02	60
Azinphos-Methyl	10.5	0.02	NA	0.25
Bendiocarb	35	54.87	NA	0.74
Benefin	NA	NA	NA	NA
Benomyl	17.5	0.97	1100	13
Bensulfuron	1400	12002.45	400	17000
Bensulide	46.2	66.53	NA	NA
Bentazone	200	40423.04	900	NA
BHC	NA	NA	NA	NA
Bifenox	1050	19.09	NA	NA
Bifenthrin	10.5	0.01	NA	0
Bromacil	90	6485.34	1.1	NA
Bromoxynil	140	3.85	9.3	2.6
Butylate	350	22.62	NA	NA
Captan	289.26	2.29	80	NA
Carbaryl	700	27.4	370	3300
Carbofuran	40	2.95	NA	0.4
Chloramben	100	1584.89	NA	NA
Chlordane	2	0.49	NA	NA
Chlordimeform	0.27	2150.96	NA	NA
Chlorimuron	140	1308.3	NA	NA
Chlorobenzilate	140	81.04	NA	NA
Chloroneb	91	530.91	NA	NA
Chloropicrin	NA	1.38	NA	NA
Chlorothalonil	45.45	2.3	50	0.83
Chloroxuron	NA	49.75	NA	NA
Chlorpropham	350	424.63	NA	NA

NA denotes data not available.

Table 9.5, *continued*

Pesticide (active ingredient name)	Humans	Fish	Algae	Crustaceans
Chlorsulfuron	350	9308.29	NA	20000
Clomazone	301	3210.92	990	2200
Clopyralid	3500	20721.92	NA	NA
Chlorpyrifos	20	0.02	NA	0
Cyanazine	1	1411.46	1.2	NA
Cycloate	35	658.46	NA	NA
Cyfluthrin	175	0.02	NA	0
Cypermethrin	7	0.04	NA	0
Cyromazine	52.5	17313.7	NA	310
Dalapon	200	21052.81	NA	NA
Dalapon NA Salt	200	21052.81	NA	NA
Daminozide	40.28	31007.73	NA	NA
DBCP	NA	NA	NA	NA
Dicloran	175	66.53	NA	NA
DCPA	7	1003.46	3090	NA
DDT	1.03	0.1	NA	NA
Desmedipham	280	225.68	30	NA
Diazinon	0.6	8.91	NA	0.17
Dicamba	200	4918.92	5	NA
Dichlobenil	9.1	728	NA	1000
Dichloropropene	2	137.02	NA	NA
Dichlorprop	35	182.28	NA	NA
Diclofop	1.4	13.35	NA	64
Dicofol	0.84	4.97	NA	19
Dicrotophos	0.7	953.4	1000	3
Dieldrin	0.02	0.08	NA	NA
Diethatyl	NA	NA	NA	NA
Difenzoquat	1400	8594.09	NA	NA
Diflubenzuron	140	10751.33	190	0.05
Dimethipin	14	2988.55	800	610
Dimethoate	0.35	903.58	NA	40
Dinocap	28	1.06	NA	NA
Dinoseb	7	2.86	NA	NA
Dinoseb Salts	NA	NA	NA	NA
Diphenamid	200	5	NA	NA
Dipropetryn	NA	211.12	NA	NA
Diquat	35	2276.77	NA	170
Disulfoton	0.3	0.64	NA	0.04
Diuron	10	86.37	0.44	6
DNOC Na Salt	NA	6.33	NA	NA
Dodine	28	106.65	NA	NA
Endosulfan	42	0	NA	2.7
Endothal	100	240.33	6.5	2200
EPN	0.07	11.11	NA	NA
EPTC	175	1848.29	860	NA
Ethion	3.5	4.56	NA	0.03
Esfenvalerate	140	0	NA	NA
Ethalfluralin	28	2.86	10.9	23.7

Table 9.5, *continued*

Pesticide (active ingredient name)	Humans	Fish	Algae	Crustaceans
Ethephon	126	26628.09	100	17000
Ethofumesate	2800	58.73	2700	250
Ethoprophos	0.7	0.48	NA	0.36
Etridiazole	4.86	369.29	NA	NA
Fenac	NA	NA	NA	NA
Fenamiphos	2	0.33	NA	0.12
Fenarimol	455	112.12	NA	113
Fenbutatin Oxide	350	0.1	NA	NA
Fenitrothion	NA	NA	NA	NA
Fenoxaprop	17.5	34.71	27	NA
Fenoxycarb	NA	135.62	NA	0
Fensulfothion	1.75	6.97	NA	NA
Fenthion	NA	NA	NA	NA
Fenvalerate	175	0.01	NA	0.03
Ferbam	140	437.02	NA	NA
Fluazifop	70	135.62	NA	NA
Flucythrinate	140	0.02	NA	NA
Flumetralin	NA	1.51	16	8.8
Fluometuron	90	77.05	107	NA
Fluridone	560	618.34	NA	NA
Fluvalinate-Tau	70	0.06	NA	0.02
Fomesafen	1.75	164346.3	NA	690
Fonofos	10	0.52	NA	0.05
Formetanate	14	375.4	NA	NA
Fosamine	70	85896.5	15000	NA
Fosetyl	2100	14710.93	320	NA
Glufosinate	140	1984.86	2500	32000
Glyphosate	700	19295.18	240	50000
Hexazinone	200	30413.97	4	20000
Hexythiazox	17.5	62.62	NA	6.1
Hydramethylnon	0.21	8.91	NA	NA
Imazamethabenz-N	NA	NA	NA	NA
Imazamethabenz	NA	NA	NA	NA
Imazapyr	NA	19952.62	10	97100
Imazapyr Amine	NA	NA	NA	NA
Imazaquin	1750	61925.57	NA	NA
Iprodione	7.97	430.81	130	170
Isazofos	0.14	0.62	NA	NA
Isofenphos	3.5	168.01	NA	0.05
Isopropalin	105	98.9	2	NA
Lactofen	2.06	284.74	NA	NA
Lambda-Cyhalothirin	7	0.01	NA	0
Lindane	0.2	0.11	NA	54
Linuron	5.6	110.75	5.35	130
Malathion	200	0.29	NA	0.06
Maleic Hydrazide	4000	19952.62	9840	NA
Mancozeb	5.83	53.58	NA	7

Table 9.5, *continued*

Pesticide (active ingredient name)	Humans	Fish	Algae	Crustaceans
Maneb	5.83	0	NA	NA
MCPA	10	17552.14	6	NA
MCPA Ester	NA	NA	NA	NA
MCPB	70	626.35	50	NA
Mecoprop	NA	NA	NA	NA
Mepiquat	4200	415453.9	200	6250
Metalaxyl	518	3099.56	56000	1200
Metaldehyde	NA	1121.13	NA	NA
Methamidophos	7	165.17	NA	NA
Metam	1.77	60.02	NA	NA
Methanearsonic Acid	NA	NA	NA	NA
Methazole	NA	592.78	NA	NA
Methidathion	1.05	0.15	NA	0.02
Methiocarb	35	0.05	NA	0.1
Methomyl	200	33.48	NA	0.4
Methoxychlor	40	0.11	NA	NA
Methyl Bromide	10	1760.08	NA	NA
M-Isothiocyanate	NA	9.34	NA	NA
Parathion	2	5.6	NA	0.18
Metiram	2.1	33460.31	NA	NA
Metolachlor	70	562.56	0.7	354
Metribuzin	100	7683.69	2.33	1290
Metsulfuron	1750	31167.36	0.16	5100
Mevinphos	1.75	0.96	NA	NA
Molinate	1.4	124.51	170	380
Monocrotophos	0.35	728	NA	NA
MSMA	70	1936.87	NA	NA
NAA Amide	NA	NA	NA	NA
Napthaleneacetic	NA	NA	NA	NA
Naled	14	8.58	2	0.05
Napropamide	700	1480.61	1200	NA
Naptalam	371	14774.85	NA	NA
Nitrapyrin	NA	NA	NA	NA
Norflurazon	14	1256.98	3.2	1000
Oryzalin	84	403.02	5.48	NA
Oxadiazon	3.5	109.38	5.6	30
Oxamyl	200	360.14	NA	1000
Oxycarboxin	NA	3378.61	NA	NA
Oxydemeton-Methyl	3.5	89.06	NA	27
Oxyfluorfen	2.1	21.44	NA	NA
Oxythloquinox	NA	NA	NA	NA
Paraquat	30	2115.14	0.22	NA
Parathon (Ethyl)	0.23	1.52	NA	0
PCNB	2.1	10	NA	18
Pebulate	49	1138.03	76	NA
Pendimethalin	70	14.25	0.7	14
Permethrin	35	0.05	NA	0.02

Table 9.5, *continued*

Pesticide (active ingredient name)	Humans	Fish	Algae	Crustaceans
Petroleum Oil	NA	NA	NA	NA
Phenmedipham	1750	183.71	NA	NA
Phenthoate	NA	NA	NA	NA
Phorate	3.5	0.06	NA	0.01
Phosalone	17.5	10	NA	NA
Phosmet	7	1.89	NA	0.37
Phosphamidon	0.14	483.76	NA	2.2
Picloram	500	182.28	970	11800
Piperalin	NA	94.44	NA	NA
Prochloraz	NA	NA	NA	NA
Profenofos	0.35	0.6	NA	0.2
Profluralin	42	1.24	NA	NA
Prometon	100	2566.61	32	3450
Prometryn	280	406.11	0.3	1000
Pronamide	NA	NA	NA	NA
Propachlor	90	17.93	NA	NA
Propamocarb HCL	700	19251.3	22400	NA
Propanil	35	314.7	6	86
Propargite	20.47	2.76	NA	9
Propazine	10	2933.19	NA	NA
Propham (IPC)	NA	NA	NA	NA
Propiconazole	9.1	102.56	18	310
Propoxur	3	168.01	NA	23
Chloridazon	NA	NA	60	NA
Quizalofop	62.93	53.58	98	NA
Sethoxydim	630	153.85	NA	NA
Siduron	NA	2115.13	170	NA
Silvex	NA	NA	NA	NA
Simazine	4	739.38	30	2500
Sulfometuron	NA	NA	NA	NA
Sulprofos	21	125.89	26300	0.02
Tebuthiuron	500	21273.55	13	21800
Terbacil	90	8533.16	4.4	NA
Terbufos	0.9	0.05	NA	0.03
Terbutryn	0.7	101.2	NA	NA
Thiabendazole	700	66.53	NA	42
Thidiazuron	NA	3210.92	NA	100
Thifensulfuron	91	19952.62	NA	NA
Thiobencarb	70	21.44	13	1
Thiodicarb	NA	54.87	1600	9
Thiophanate-Methyl	560	2.66	110	NA
Thiram	56	0.54	NA	NA
Toxaphene	3	0.05	NA	NA
Tralomethrin	52.5	0.11	NA	0
Triadimefon	28	594.37	100	52
Tri-Allate	9.1	153.85	NA	13
Tribufos	0.7	168.01	118	NA

Table 9.5, *continued*

Pesticide (active ingredient name)	Humans	Fish	Algae	Crustaceans
Trichlorfon	14	25	NA	0.01
Triclopyr Amine	NA	NA	NA	NA
Triclopyr	350	90.4	NA	NA
Tridiphane	2.1	42.17	NA	NA
Trifluralin	5	0.66	4.6	2.4
Triforine	175	3659.82	NA	NA
Trimethacarb	17.5	125.89	NA	NA
Triphenyltin Hyd	0.13	1.2	NA	NA
Vernolate	7	344.93	230	NA

Trends in the Potential for Environmental Risk

Pounds of pesticides applied to the 12 crops included in the study steadily increased throughout the 1960s and 1970s to peak in the early 1980s at about 560 million pounds, and then fell to about 450-500 million pounds throughout the 1990s (figure 9.1). Pesticide loss estimates generally followed a similar trend (figure 9.2). As shown in the table below, the highest average percent losses are for dissolved runoff and the average lowest are for leaching for most time periods; however, percent leaching losses were much higher and percent runoff losses lower in the 1960s than in later years. The sum of loss estimates for both leaching and runoff ranged from about 4 to 5.5 percent of the amount applied. However, this range represents an upper bound on total loss because loss results correspond to the 95th percentile, which is a nearly "worst" case.

95th percentile loss as a percent of amount applied	1960s	1970s	1980s	1990s
Leaching	1.5%	1.0%	0.5%	0.5%
Dissolved Runoff	1.9%	3.4%	3.5%	3.1%
Adsorbed Runoff	1.0%	1.1%	1.2%	1.5%
Sum	4.4%	5.5%	5.2%	5.1%

Corn was generally, but not always, the dominant crop associated with high potential risk scores at the national level. Corn was dominant for all time periods for runoff and leaching risk for protection of drinking water, runoff and leaching risk for protection of algae, and all but the early 1960s for runoff and leaching risk for protection of crustaceans (table 9.6). Cotton, potatoes, soybeans, tobacco, and sorghum were frequently the second or third most important crop in determining the national potential risk scores for these 6 indicators. For fish-risk indicators, however, several crops were dominant at one

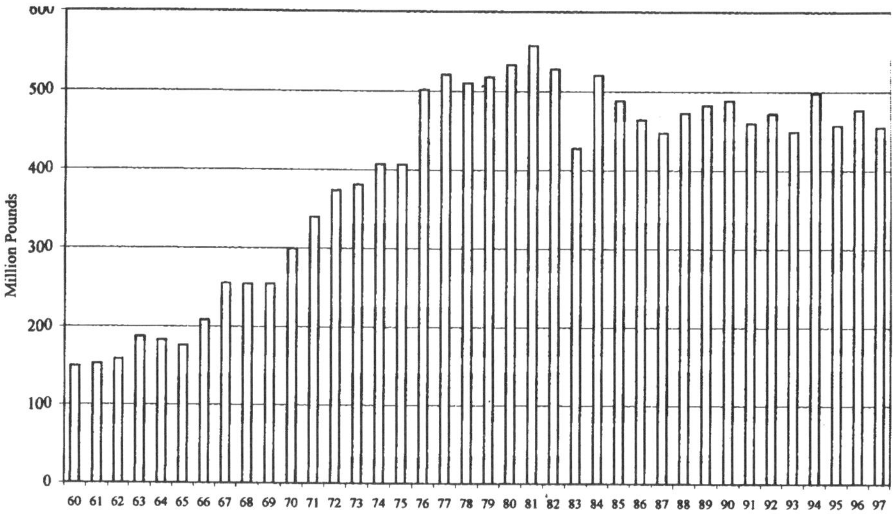

Figure 9.1. Trend in quantity of pesticides applied to 12 crops: corn, soybeans, cotton, wheat, sorghum, potatoes, barley, rice, tobacco, sugarbeets, peanuts, and oats.

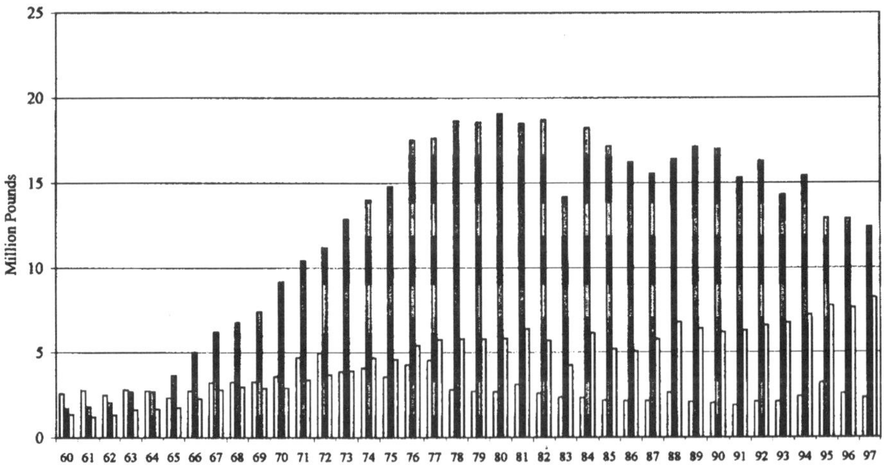

Figure 9.2. Trends in quantities of pesticides lost from farm fields for 12 crops. (left column: leaching; middle: dissolved runoff; right: adsorbed runoff.)

Table 9.6. Percentage of Pesticide Risk Indicator (TEUs) by Crop

	1960-65	1966-71	1972-77	1978-83	1984-90	1991-97
	Pesticide leaching index for protection of drinking water					
Barley	0.2%	0.0%	0.0%	0.0%	0.0%	0.0%
Corn	62.5%	75.6%	78.5%	72.6%	61.4%	41.6%
Cotton	3.4%	5.9%	3.5%	2.0%	6.9%	12.9%
Oats	0.0%	0.0%	0.0%	0.0%	0.0%	0.0%
Peanuts	7.8%	2.7%	2.5%	5.9%	4.2%	8.4%
Potatoes	19.8%	7.8%	3.7%	4.1%	6.2%	5.1%
Rice	0.4%	0.1%	0.4%	0.5%	0.4%	0.9%
Sugarbeets	0.1%	0.2%	0.1%	0.0%	2.4%	5.4%
Sorghum	2.7%	7.2%	6.3%	7.6%	7.5%	5.0%
Soybeans	0.0%	0.1%	0.5%	1.9%	2.1%	3.8%
Tobacco	3.0%	0.2%	4.4%	5.3%	8.8%	16.7%
Wheat	0.1%	0.1%	0.1%	0.2%	0.2%	0.2%
	100.0%	100.0%	100.0%	100.0%	100.0%	100.0%
	Pesticide runoff index for protection of drinking water					
Barley	0.4%	0.5%	0.0%	0.0%	0.0%	0.1%
Corn	92.1%	91.9%	81.7%	75.1%	80.8%	79.9%
Cotton	2.1%	2.6%	5.3%	3.3%	3.5%	6.9%
Oats	0.0%	0.0%	0.1%	0.0%	0.0%	0.0%
Peanuts	0.2%	0.2%	0.3%	0.3%	0.3%	0.1%
Potatoes	0.1%	0.4%	0.3%	0.2%	0.6%	1.7%
Rice	0.1%	0.1%	0.5%	1.3%	1.3%	2.0%
Sugarbeets	0.3%	0.1%	0.0%	0.0%	0.0%	0.2%
Sorghum	0.9%	1.3%	2.7%	2.8%	2.6%	2.7%
Soybeans	0.5%	1.4%	7.1%	15.3%	9.6%	5.8%
Tobacco	3.2%	1.4%	1.3%	0.9%	0.3%	0.1%
Wheat	0.0%	0.2%	0.6%	0.7%	0.8%	0.5%
	100.0%	100.0%	100.0%	100.0%	100.0%	100.0%
	Pesticide leaching index for protection of fish					
Barley	0.1%	0.6%	0.1%	0.1%	0.1%	0.0%
Corn	10.2%	66.6%	73.4%	65.5%	42.9%	22.4%
Cotton	2.9%	4.7%	2.9%	2.4%	7.5%	23.0%
Oats	0.0%	0.0%	0.0%	0.0%	0.0%	0.0%
Peanuts	49.1%	18.1%	7.4%	6.3%	15.9%	17.2%
Potatoes	5.2%	4.2%	1.6%	1.8%	2.1%	4.2%
Rice	0.0%	0.1%	5.6%	3.6%	2.6%	3.1%
Sugarbeets	0.9%	2.8%	0.3%	0.1%	0.5%	1.1%
Sorghum	0.0%	0.0%	0.3%	4.4%	3.3%	1.7%
Soybeans	0.4%	2.0%	0.7%	4.7%	4.3%	0.8%
Tobacco	31.2%	0.8%	7.5%	11.0%	20.4%	26.3%
Wheat	0.0%	0.1%	0.1%	0.1%	0.5%	0.1%
	100.0%	100.0%	100.0%	100.0%	100.0%	100.0%
	Pesticide runoff index for protection of fish					
Barley	0.1%	0.4%	0.0%	0.0%	0.5%	0.1%
Corn	10.7%	14.4%	14.5%	15.8%	15.6%	12.4%
Cotton	17.0%	18.9%	38.2%	12.2%	17.7%	42.4%
Oats	0.0%	0.0%	0.1%	0.0%	0.1%	0.0%
Peanuts	0.9%	0.8%	6.6%	27.5%	4.7%	2.6%

Table 9.6, *continued*

	1960-65	1966-71	1972-77	1978-83	1984-90	1991-97
	Pesticide runoff index for protection of fish, *continued*					
Potatoes	68.6%	59.3%	27.7%	22.9%	39.4%	26.2%
Rice	0.0%	0.1%	0.2%	0.5%	0.3%	0.5%
Sugarbeets	0.2%	0.2%	0.3%	1.6%	3.6%	3.1%
Sorghum	0.4%	0.4%	0.8%	0.7%	0.6%	0.8%
Soybeans	0.7%	4.6%	9.8%	15.2%	10.6%	7.2%
Tobacco	1.2%	0.6%	0.8%	1.8%	0.6%	2.1%
Wheat	0.0%	0.2%	1.1%	1.9%	6.2%	2.6%
	100.0%	100.0%	100.0%	100.0%	100.0%	100.0%
	Pesticide leaching index for protection of algae					
Barley	0.0%	0.0%	0.0%	0.0%	0.0%	0.0%
Corn	96.1%	92.9%	94.0%	91.6%	89.4%	89.2%
Cotton	0.1%	0.1%	0.0%	0.0%	0.1%	0.1%
Oats	0.0%	0.0%	0.0%	0.0%	0.0%	0.0%
Peanuts	0.0%	0.0%	0.0%	0.0%	0.2%	0.2%
Potatoes	0.0%	0.0%	0.0%	0.0%	0.0%	0.0%
Rice	0.0%	0.0%	0.0%	0.0%	0.0%	0.0%
Sugarbeets	0.0%	0.0%	0.0%	0.0%	0.0%	0.0%
Sorghum	3.7%	7.1%	5.9%	8.1%	9.9%	9.9%
Soybeans	0.0%	0.0%	0.0%	0.2%	0.4%	0.3%
Tobacco	0.0%	0.0%	0.0%	0.0%	0.0%	0.0%
Wheat	0.0%	0.0%	0.0%	0.0%	0.1%	0.2%
	100.0%	100.0%	100.0%	100.0%	100.0%	100.0%
	Pesticide runoff index for protection of algae					
Barley	0.0%	0.0%	0.0%	0.0%	0.0%	0.0%
Corn	96.7%	94.7%	94.3%	92.4%	92.2%	91.9%
Cotton	0.4%	0.2%	0.2%	0.2%	0.2%	0.4%
Oats	0.1%	0.0%	0.0%	0.0%	0.0%	0.0%
Peanuts	0.0%	0.0%	0.0%	0.0%	0.0%	0.0%
Potatoes	0.0%	0.0%	0.0%	0.0%	0.0%	0.0%
Rice	0.0%	0.0%	0.0%	0.0%	0.0%	0.0%
Sugarbeets	0.0%	0.0%	0.0%	0.0%	0.0%	0.0%
Sorghum	2.7%	4.8%	4.7%	5.9%	6.1%	6.5%
Soybeans	0.1%	0.2%	0.8%	1.5%	1.3%	0.9%
Tobacco	0.0%	0.0%	0.0%	0.0%	0.0%	0.0%
Wheat	0.2%	0.0%	0.0%	0.0%	0.1%	0.3%
	100.0%	100.0%	100.0%	100.0%	100.0%	100.0%
	Pesticide leaching index for protection of crustaceans					
Barley	0.0%	0.6%	0.1%	0.1%	0.1%	0.0%
Corn	0.4%	57.9%	80.5%	74.2%	64.5%	49.2%
Cotton	95.7%	35.9%	2.3%	0.7%	2.5%	8.5%
Oats	0.0%	0.0%	0.0%	0.0%	0.0%	0.0%
Peanuts	1.3%	1.1%	5.6%	5.9%	7.8%	8.2%
Potatoes	0.7%	0.8%	0.3%	0.5%	0.9%	3.6%
Rice	0.0%	0.1%	6.1%	4.1%	3.9%	7.0%
Sugarbeets	1.7%	1.6%	0.2%	0.1%	0.3%	0.8%
Sorghum	0.0%	0.1%	0.4%	5.3%	5.2%	3.6%

continued

Table 9.6, *continued*

	1960-65	1966-71	1972-77	1978-83	1984-90	1991-97
Pesticide leaching index for protection of crustaceans, *continued*						
Soybeans	0.1%	1.8%	1.2%	4.1%	3.4%	0.9%
Tobacco	0.0%	0.0%	3.0%	5.1%	10.8%	17.8%
Wheat	0.0%	0.1%	0.1%	0.1%	0.6%	0.3%
	100.0%	100.0%	100.0%	100.0%	100.0%	100.0%
Pesticide runoff index for protection of crustaceans						
Barley	0.6%	6.3%	0.2%	0.1%	0.2%	0.1%
Corn	30.5%	44.9%	58.5%	66.3%	56.7%	46.7%
Cotton	33.4%	15.0%	15.5%	11.2%	23.7%	31.3%
Oats	0.2%	0.6%	0.6%	0.2%	0.0%	0.0%
Peanuts	5.0%	3.2%	2.1%	2.4%	3.5%	6.8%
Potatoes	5.0%	6.1%	2.9%	1.8%	2.4%	4.3%
Rice	0.3%	0.3%	0.5%	1.3%	0.7%	0.8%
Sugarbeets	3.7%	2.0%	1.1%	1.0%	0.8%	1.4%
Sorghum	17.8%	8.5%	9.1%	4.6%	2.7%	1.7%
Soybeans	1.9%	8.3%	4.7%	6.9%	4.8%	2.6%
Tobacco	1.3%	0.6%	0.7%	0.9%	1.5%	2.1%
Wheat	0.4%	4.3%	4.2%	3.3%	2.9%	2.0%
	100.0%	100.0%	100.0%	100.0%	100.0%	100.0%

time or another: corn, peanuts, tobacco, cotton, and potatoes. Relationships between potential risk scores and crops at the state and regional level will often be quite different than those shown for the nation, however, because no single region or state has the same crop mix as the entire nation.

The 8 pesticide-risk indicators varied markedly over both space and time. The spatial distributions for the two indicators associated with drinking water are shown in maps 9.1 and 9.2. The runoff risk indicator is greater in the Midwest states and the Mississippi embayment region than in other parts of the country. The leaching risk indicator is high in some of these same areas, but is also high in the Southeast and Mid-Atlantic states.

National aggregates demonstrate the general temporal trends. For all four target groups, pesticide risk in runoff was much greater than pesticide risk in leachate. The pesticide runoff indicator for protection of drinking water showed a dramatic reduction in risk from the 1960s and early 1970s (figure 9.3). Leaching risk for drinking water, however, had highest scores in 1996 and 1997, and the series appears to be trending upward (figure 9.4). For fish, there was little overall trend in the runoff indicator, but leaching risk in 1996-97 was less than half the level in 1976-78. Potential risk levels for algae were, overall, more than 25 times greater than risk levels for the other three groups, reflecting algae's sensitivity to herbicides. Pesticide risk for algae showed modest reductions in recent years from the high levels in the late 1970s for both runoff and leaching. Crustacean risk had trends generally similar to those of fish.

Pesticide Runoff Risk For Protection Of Drinking Water, 1996 - 1997 Average

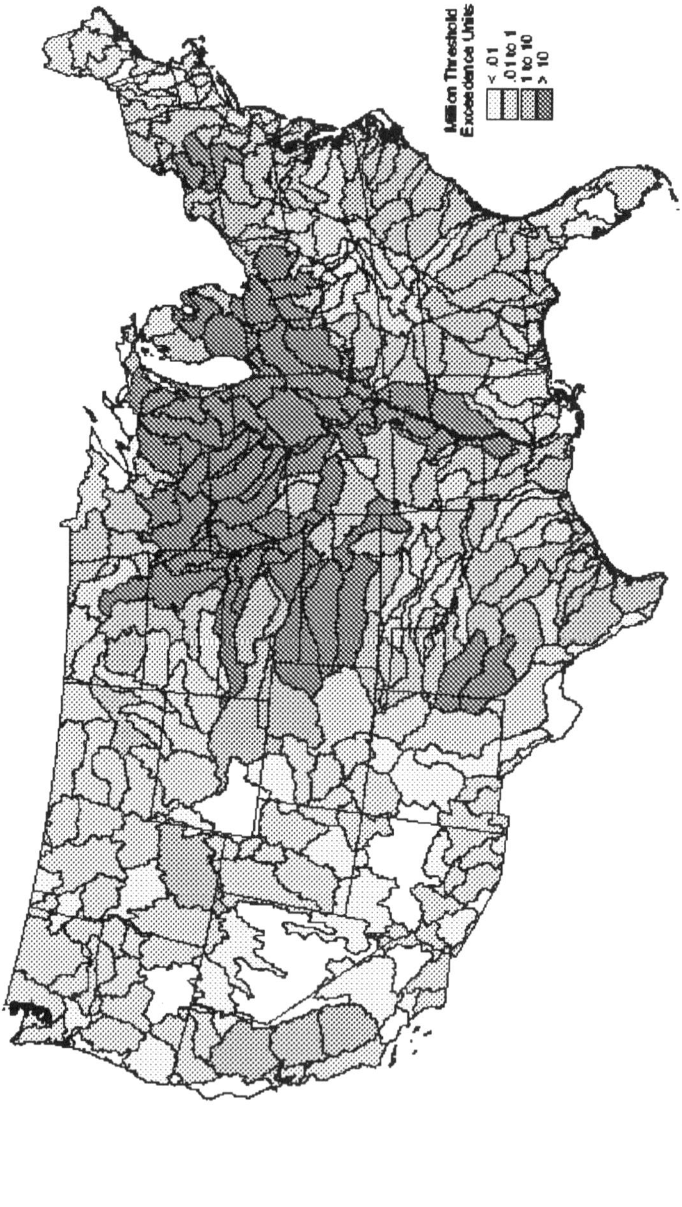

Map 9.1

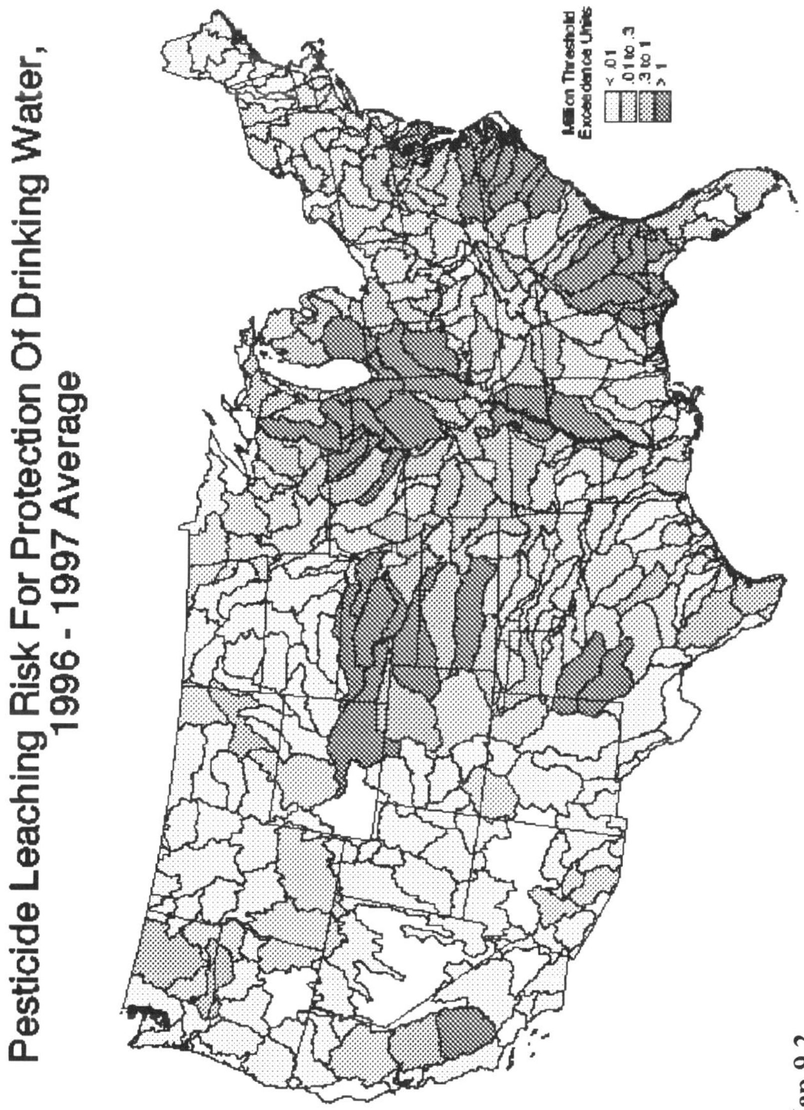
Pesticide Leaching Risk For Protection Of Drinking Water,
1996 - 1997 Average
Million Threshold Exceedance Units
< .01
.01 to .3
.3 to 1
> 1

Map 9.2

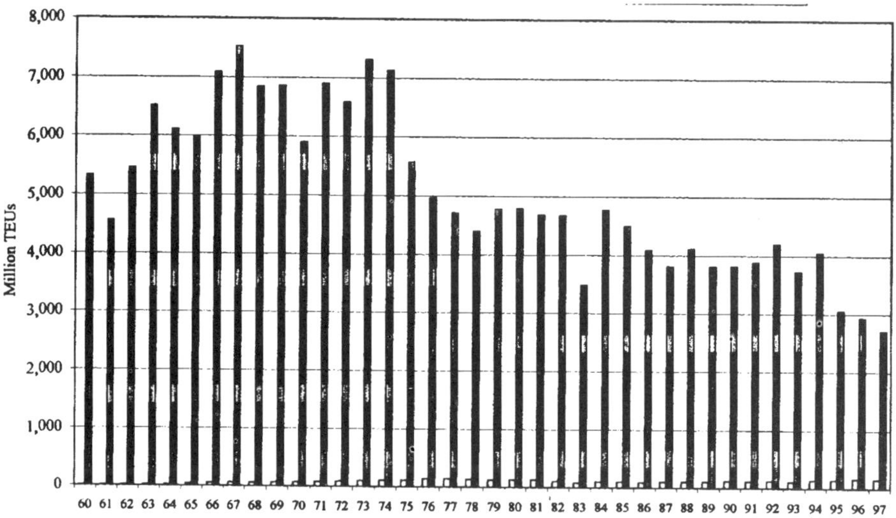

Figure 9.3. Pesticide risk indicators for protection of drinking water (left column: leaching; right: dissolved runoff).

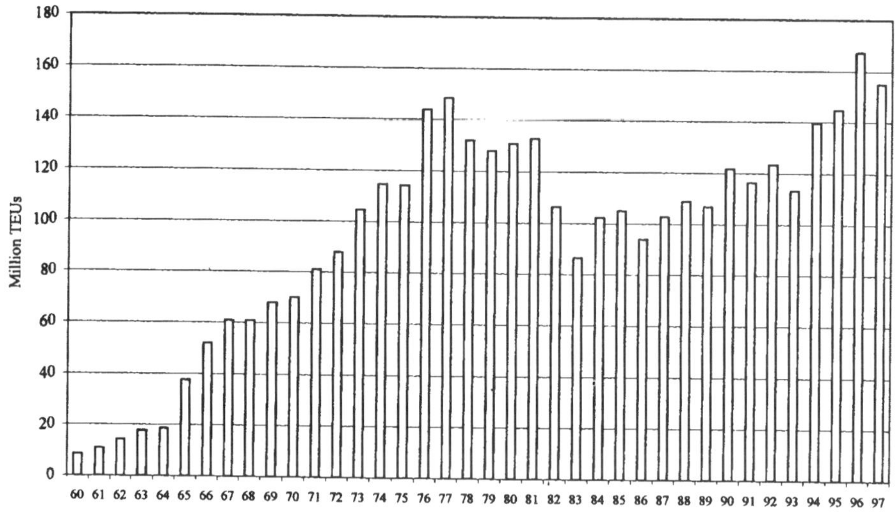

Figure 9.4. Pesticide leaching risk indicator for protection of drinking water.

Overall, it appears that pesticide risk for all groups initially increased in the early 1960s to high levels in the 1970s, and then either decreased or remained about constant through the 1980s and 1990s.

These national-level trends, however, obfuscate the underlying spatial-temporal diversity. In spite of the sharp fall in the national indicator for runoff risk for drinking water (decrease of 60 percent from 1973-74 to 1996-97, ending up at a level far below levels in the 1960s), for example, there are areas of the country where that risk is increasing, as well as areas where the potential risk has fallen more dramatically. Runoff risk for protection of drinking water in Louisiana generally increased throughout the time period. Runoff risk in Alabama and Mississippi increased to a peak in 1976-77, followed by a decline through 1988, and then increased again through most of the 1990s. In Wisconsin, runoff risk increased steadily through the 1960s and 1970s and then remained relatively high except for the last 3 years in the series. New York and Pennsylvania had trends similar to Wisconsin's, but the downward trend began earlier. Runoff risk in Michigan and Minnesota also increased steadily through the 1960s and 1970s, but then mostly decreased through the 1980s and 1990s to levels similar to those in the 1960s. Texas showed a jump in runoff risk in the 1970s, but otherwise showed little trend. Runoff risk in four Northern Plains states fluctuated throughout the period, but showed little or no overall trends. Runoff risk in Iowa, which had the highest risk score of any state throughout most of the time period, doubled between 1960 and 1973-74, and then gradually fell (with some fluctuation) to levels below those at the beginning of the series. Midwest states (Ohio, Illinois, Indiana, and Missouri) had trends similar to the national trends, except that the percent reduction was greater (75-90 percent decrease from 1973-74 to 1996-97 for Ohio, Indiana, and Missouri). The Southeast showed trends similar to those of the Midwest, but the decreases in risk occurred slightly earlier and generally were sharper. Runoff risk scores for these 25 states comprised 96-97 percent of the total risk score for all 48 states.

The other 7 pesticide risk indicators similarly show this kind of diversity in temporal trends at the state and regional levels. Thus, the national level trends are useful as overall trends for the agriculture sector, but do not necessarily represent specific production areas.

The indices of potential risk change over space and time because of: 1) changes in acres planted, 2) changes in percent acres treated, 3) changes in the suite of pesticides used, and 4) changes in application rates. In addition, the changes in these factors for the more mobile and/or persistent pesticides and on the more vulnerable acres (high intrinsic potential for pesticide loss) influence the indices the most. Sometimes these factors work together to increase/decrease pesticide risk, and sometimes they offset each other. Changes in acres planted and changes in application rates would both generally be expected to be gradual; however, these changes can introduce sharp fluctuations when relative prices shift quickly or when government programs are introduced. The Payment in Kind (PIK) program in 1983, for example, sharply reduced acreage in field crops in most areas of the country in 1983, but levels returned in 1984 when the program was discontinued. Sharp changes in the indicators would also occur during times when the relative price of corn and soybeans changes, since corn generally has had a higher risk score than soybeans have. The introduction and rapid adoption of new pesticides or the eradication of a pest problem (such as the boll weevil in cotton production) can also result in sharp changes in the risk indicators in specific areas. Fluctuations in the state-level time series for the pesticide-risk indicators can be tracked back to one or more of these factors.

Notes

1. Preliminary results have previously been reported in: R. L. Kellogg, R. Nehring, and A. Grube. "Spatial and Temporal Trends in the Potential for Agrochemicals to Leach." Poster presented at the 50th annual meeting of the Soil and Water Conservation Society, Des Moines Iowa, August 6-9, 1995; R. L. Kellogg, and D. W. Goss. "Development of Environmental Indicators for Use in Macroeconomic Models." *Journal of Agricultural and Applied Economics*, 29, (1) (July 1997):77-86; R. L. Kellogg, R. Nehring, A. Grube, S. Plotkin, D.W. Goss, and S. Wallace. "Trends in the Potential for Environmental Risk from Pesticide Loss from Farm Fields," poster presented at a Soil and Water Conservation Society Conference on "The State of North America's Private Land," January 19-21, 1999, Chicago, Illinois.

2. To measure actual risk, the edge-of-field and bottom-of-rootzone concentration estimates would have to be adjusted using hydrology models to represent in-stream concentrations (or in the case of drinking water taken from ground water sources, aquifer concentrations). For human risk, additional adjustments would be needed to account for sources of drinking water and population. These adjustments are beyond the scope of this study.

References

Aspelin, A. L. 1997. "Pesticides Industry Sales and Usage: 1994 and 1995 Market Estimates." Office of Pesticide Programs, Environmental Protection Agency, Washington, D.C.

Association of State and Interstate Water Pollution Control Administrators. 1985. *America's Clean Water: The States' Nonpoint Source Assessment 1985*. Washington, D.C.

Baker, D. B., and R. P. Richards. 1995. "Herbicides in Ohio's Drinking Water: Risk Analysis, Reduction, and Communication." Proceedings of the Fourth National Conference on Pesticides: New Directions in Pesticide Research, Development, and Policy, Virginia Water Resources Center, VPISU, Blacksburg, Va.

Baker, J., E. Adcock, and G. Miller.1992. Understanding and reducing pesticide losses. Cooperative Extension Service, Iowa State University, Ames, September.

Barbash, J. E., G. P. Thelin, D. W. Kolpin, and R. J. Gilliom. 1999. Distribution of Major Herbicides in Ground Water of the United States. USGS Water Resources Investigations Report 98-4245, U.S. Government Printing Office.

Barnthouse, L. W., G. W. Suter, and A. E. Rosen. 1990. "Risk of Toxic Contaminants to Exploited Fish Populations: Influence of Life History, Data Uncertainty and Exploitation Intensity." *Environmental Toxicology and Chemistry,* 9:297-311.

Doanes Countrywide Farm Panel Surveys. Doane Agricultural Service Inc. Unpublished manuscript.

Doanes (a). "The Doane Pesticide Profile Study: Users Manual." Doane Marketing Research, Inc. Unpublished manuscript.

Doanes (b). "The Doane Herbicide Marketing Study: Manual for Users." Doane Agricultural Service, Inc. Unpublished manuscript.

Goolsby, D. A., and W. A. Battaglin. 1993. "Occurrence, distribution, and transport of agricultural chemicals in surface waters of the Midwestern United States." In *Selected papers on agricultural chemicals in water resources of the Mid-continental United States,* ed. D. A. Goolsby et al. U.S. Geological Survey Pen-file Report 93-418.

Goolsby, D. A., W. A. Battaglin, and E. M. Thurman. 1993. "Occurrence and Transport of Agricultural Chemicals in the Mississippi River Basin." U.S. Geological Survey Circular 1120-C, U. S. Government Printing Office.

Hamilton, P. A., and R. J. Shedlock. 1992. "Are Fertilizers and Pesticides in the Ground Water?: A Case Study of the Delmarva Peninsula." U. S. Geological Survey Circular 1080, U.S. Government Printing Office.

Insect Control Guide, volume 8. 1995. Willoughby, Ohio: Meister Publishing Company.

Kellogg, R. L., G. W. TeSelle, and J. J. Goebel. 1994. "Highlights from the 1992 National Resources Inventory." *Journal of Soil and Water Conservation,* 49:521-27.

Knisel, W. G., ed. 1993. "GLEAMS: Groundwater Loading Effects of Agricultural Management. Systems Version 2.10." Biological & Agricultural Engineering Department. Publication No. 5, University of Georgia-Coastal Plain Experiment Station, Tifton.

Kolpin, Dana W., Michael R. Burkart, and E. Michael Thurman. 1994. "Herbicides and Nitrate in Near-Surface Aquifers in the Midcontinental United States, 1991." USGS Water Supply Paper 2413, U.S. Government Printing Office.

Larsen, S. J., R. J. Gilliam, and P. D. Capel. 1999. "Pesticides in Streams of the United States — Initial Results from the National Water-Quality Assessment Program." U.S. Geological Survey, Water Resources Investigations Report 98-4222, U.S. Government Printing Office.

Monsanto Agricultural Company. 1990. *The National Alachlor Well Water Survey (NAWWS): Data Summary.* Monsanto Technical Bulletin, July.

Pimentel, D., et al. 1992. "Environmental and Economic Costs of Pesticide Use." *BioScience,* 42(10)(November):750-60.

Plotkin, S., J. K. Bagdon, E. S. Hesketh, and M. H. Hugo. "Pesticide Short-term and Long-term Toxicity Derivation and Data: Humans, Fish, Crustaceans and Small Aquatic Plants." Unpublished manuscript. USDA NRCS and University of Massachusetts Extension.

USEPA. 1990. *National Pesticide Survey Project Summary,* Fall, Washington, D.C.

USEPA. 1997. *Office of Pesticide Toxicity Database,* Washington, D.C.

U.S. Department of Agriculture. 1992. "Atrazine in Surface Waters: A Report of the Atrazine Task Group to the Working Group on Water Quality." Washington, D.C, April.

U.S. Geological Survey. 1999. "The Quality of Our Nation's Waters — Nutrients and Pesticides." U.S. Geological Survey Circular 1225.

U. S. Geological Survey. 1995. "National Water Quality Assessment Program." *Pesticides in the Atmosphere.* U. S. Department of Interior, Washington, D.C.

Wauchope, R. D., T. M. Butler, A. G. Hornsby, P. W. M. Augustijn-Beckers, and J. P. Burt. 1992. "The SCS/ARS/CES pesticide properties database for environmental decision making." *Environmental Contaminant Toxicology Reviews*, p. 123.

Weed Control Guide, volume 30. 1996. Willoughby, Ohio: Meister Publishing Company.

Willis, G. H., and L. L. McDowell. 1987. "Pesticide persistence of foliage." *Environmental Contaminant Toxicology Reviews*, 100:23-73.

Winter, T. C., J. W. Harvey, O. L. Franke, and W. M. Alley. 1998. "Ground Water and Surface Water: A Single Resource." U.S. Geological Survey Circular 1139, U.S. Geological Survey, Denver, Colorado.

10

The Environmental Performance of the U.S. Agricultural Sector

V. Eldon Ball, Rolf Färe, Shawna Grosskopf, F. Hernandez-Sancho, and Richard F. Nehring

Abstract

This paper introduces an environmental performance index and evaluates the environmental performance of the U.S. agricultural sector. Our motivation for developing an index of environmental performance was to assess improvements in the ability of the agricultural sector to produce marketable agricultural outputs while minimizing environmental damage. Since our index is based on distance functions that are representations of technology, we are in effect measuring "environmental productivity". And since these distance functions explicitly allow us to model joint production of good and bad outputs, and do not require information on shadow prices of "bads," our index has fairly minimal data requirements. The distance function allows us to aggregate across multiple undesirable outputs (again without requiring information on shadow prices), allowing us to provide a stand-alone index of bad outputs as well. Our index of environmental performance is then constructed as the ratio of a quantity index of marketed agricultural outputs to the index of undesirable outputs.

Introduction

Recently, Ball, Färe, Grosskopf, and Nehring (2001) studied the productivity growth of U .S. agriculture in the presence of undesirable (or bad) outputs. Their productivity model accounted for the joint production of good and bad outputs and explicitly modeled expansion of good outputs and contraction of bad outputs.

The purpose of this paper is to introduce an environmental performance index or "green" index and to evaluate the environmental performance of the U.S. agricultural sector. This work extends the bilateral environmental index introduced by Fare, Grosskopf and Hernandez-Sancho (2000) to multilateral comparisons of environmental performance. It also exploits the recently revised and extended data set employed by Ball et al. (1999).

The paper is organized as follows. Section 2 begins with a discussion of how we model the joint production of desirable and undesirable outputs. Then we turn to a discussion of our environmental performance index and the index number properties it satisfies. Since our index is constructed from distance functions, we can exploit the connection to the efficiency measurement literature and employ Data Envelopment Analysis (DEA) to estimate the index. We apply our methods to state-by-year panel data referenced in Chapters 2 and 9 of this volume. This data set includes variables that proxy effects of pesticides and fertilizer on groundwater and surface water quality for the period 1960 to 1996.

THE ENVIRONMENTAL PERFORMANCE INDEX

In this section, we develop our environmental performance index, show that it satisfies a number of desirable index number properties, and demonstrate how it can be calculated. The index is defined as ratios of distance functions, in the tradition of S. Malmquist (1953). Since our goal is to explicitly provide an index of environmental performance that provides a summary of the degree to which we have succeeded in producing maximum good outputs with minimal bad outputs, our distance functions must be modified to account for joint production of good and bad outputs. This index is a generalization of the bilateral index first used by Fare, Grosskopf and Hernandez-Sancho (2000) in their analysis of environmental performance and income in OECD countries.

Some notation is required. We let $x = (x_1, \ldots, x_N) \in \Re^N_+$ denote a vector of inputs, $y = (y_1, \ldots, y_M) \in \Re^M_+$, a vector of desirable outputs, while $b = (b_1, \ldots, b_J) \in \Re^J_+$ refers to the undesirable outputs produced. The production technology T consists of all feasible input and output vectors (x, y, b):

$$T = \{(x, y, b): x \text{ can produce } (y, b)\}, \tag{1}$$

In words, it consists of all desirable and undesirable outputs that can be produced by the given input vector.

To model the condition that good and bad outputs are produced jointly, and that some outputs are undesirable and costly to dispose of, we impose the following conditions on technology T:

T.1 $(x, y, b) \in T$ and $0 \leq \theta \leq 1$ imply $(x, \theta y, \theta b) \in T$

T.2 if $(x, y, b) \in T$ and $b = 0$, then $y = 0$.

Weak Disposability of Outputs is expressed by the first property, and it states that proportional reductions of undesirable and desirable outputs are feasible. On the other hand, it may be infeasible to reduce the undesirable outputs only. If (y, b) is feasible and $b' < b$ then it may be impossible to produce (y, b') using x; that is, $(x, y, b) \in T$ and $(x, y, b') \notin T$. Clearly if undesirable outputs could be disposed of without cost (freely), then this problem would not arise.

The second condition captures the idea that desirable and undesirable outputs are jointly produced. This joint production means that if no bad outputs are produced, then there can be no production of good outputs. Alternatively, if one wishes to produce some good outputs, there will be by-products of bad outputs.

In addition to $T.1$ and $T.2$, we assume that good outputs are freely disposable:

T.3 $(x, y, b) \in T$ and $y' \leq y$ imply $(x, y, b) \in T$

We also impose the standard assumptions found, for example, in Fare and Primont (1995).

The first step in defining our environmental performance index is to introduce a quantity index for the good outputs. To do so, we define a (sub-vector) output distance function in the spirit of Shephard (1970):

$$D_y(x, y, b) = \inf \{\theta : (x, y / \theta b) \in T\} \tag{2}$$

This function is homogeneous of degree one in good outputs, and it inherits its properties from the technology T.

Let y^t and y^{t+1} be two output vectors. We define the output quantity index for these as:

$$Q_y^t(x^t, y^t, \mathrm{b}^{t+1}) = D_y^t(x^t, y^{t+1}, b^t) / D_y^t(x^t, y^t, b^t) \tag{3}$$

Note that if only one good output is produced, then the index takes the following simple form:

$$Q_y^t(x^t, y^t, \mathrm{b}^{t+1}) = y^{t+1}/y^t. \tag{4}$$

This result follows from the homogeneity of the distance functions, and it gives the intuitively appealing result that in the single output case, the good output index is just the ratio of outputs in the two periods, t and $t + 1$. As this simple form illustrates, in the definition of our good output quantity index, we have chosen to use the period t inputs, bad outputs, and technology as references.

This quantity index satisfies several properties or "tests" due to Fisher (1922):

- HOMOGENEITY: $Q_y^t(x^t, y^t, b^t, \lambda y^{t+1}) = \lambda\, Q_y^t(x^t, y^t, b^t, \lambda y^{t+1}),\ \lambda > 0.$

This test is met since $D_y^t(x^t, b^t, \lambda y^{t+1}) = \lambda\, D_y^t(x^t, b^t, \lambda y^{t+1})$.

- TIME-REVERSAL: $Q_y^t(x^t, y^t, b^t, y^{t+1}) = Q_y^t(x^t, y^{t+1}, b^t, y^t)$.

To see that the index satisfies this test, use (3),

$$D_y^t(x^t, y^{t+1}, b^t) \,/\, D_y^t(x^t, y^t, b^t) \cdot D_y^t(x^t, y^t, b^t) \,/\, D_y^t(x^t, y^{t+1}, b^t) = 1.$$

- TRANSITIVITY: $Q_y^t(x^t, y^t, b^t, y^{t+1}) Q_y^t(x^t, y^{t+1}, b^t, y^{t+2}) = Q_y^t(x^t, y^t, b^t, y^{t+2})$

To show transitivity, use definition (3)

$$D_y^t(x^t, y^{t+1}, b^t) \,/\, D_y^t(x^t, y^t, b^t) \cdot D_y^t(x^t, y^t, b^t) \,/\, D_y^t(x^t, y^{t+1}, b^t) =$$

$$D_y^t(x^t, y^{t+2}, b^t) \,/\, D_y^t(x^t, y^t, b^t) \;=\; Q_y^t(x^t, y^t, b^t, y^{t+2})$$

- DIMENSIONALITY: $Q_y^t(x^t, \lambda y^t, b^t, \lambda y^{t+1}) = Q_y^t(x^t, y^t, b^t, y^{t+2})$

This test is satisfied due to homogeneity of the distance functions and (3).

Now we turn to the estimation method we use to construct our good output quantity index — namely linear programming techniques borrowed from the DEA literature. We assume that at each period $t = 1, \ldots, T$ we have $k = 1, \ldots, K$ observations of inputs and outputs $(x^{k,t}, y^{k,t}, b^{k,t})$. From this data, we construct the best practice reference technology for each t,

$$\begin{aligned} T^t = \{(x, y, b)\colon\ & \textstyle\sum_{k=1}^{K} z_k^t y_{km}^t \ge y_m^t, \quad m = 1, \ldots, M, \\ & \textstyle\sum_{k=1}^{K} z_k^t b_{kj}^t = b_j^t, \quad j = 1, \ldots, J, \\ & \textstyle\sum_{k=1}^{K} z_k^t x_{kn}^t \le x_n^t, \quad n = 1, \ldots, N, \\ & z_k^t \ge 0, \quad k = 1, \ldots, K\}, \end{aligned} \tag{5}$$

where z_k^t are the intensity variables, which serve to form the technology from convex combinations of the data.[1]

This technology exhibits constant returns to scale

$$\lambda T^t = T^t,\ \lambda > 0,$$

and satisfies weak disposability of outputs. It also satisfies the joint production condition, provided the following is met:

$$\Sigma_{k=1}^{K} b_{kj}^{t} > 0 \quad j = 1, \ldots, J,$$

$$\Sigma_{j=1}^{J} b_{kj}^{t} > 0 \quad k = 1, \ldots, K.$$

The first inequality says that each bad is produced by at least one observation. The second states that each observation produces at least one bad.

The good output quantity index may be calculated for each observation $k' = 1, \ldots, K$ by solving two linear programming problems, namely

$$(D_y^t(x^{k',t}, y^{k',t}, b^{k',t}))^{-1} = \max\ \{\theta : (x^{k',t}, \theta y^{k',t}, b^{k',t}) \in T^t\} \qquad (6)$$

and

$$(D_y^t(x^{k',t}, y^{k',t+1}, b^{k',t}))^{-1} = \max\ \{\theta : (x^{k',t}, \theta y^{k',t+1}, b^{k',t}) \in T^t\}. \qquad (7)$$

The ratio of the solutions to these two problems yields the index for observation k' between t and $t + 1$.[2] It will tell us whether the quantity of good outputs has increased over time for each observation.

Next we introduce an analogous quantity index for undesirable outputs. Of course, our goal is to reduce these outputs to a minimum, rather than seeking to increase them as we did for the good output quantity index. Thus, we scale the "bads," so the distance function we use to construct our quantity index resembles Shephard's input distance function

$$D_b(x, y, b) = \sup\ \{\lambda : (x, y, b/\lambda) \in T\}, \qquad (8)$$

which is homogeneous of degree one in the bad outputs, b. Moreover, it inherits the properties of the technology, T. To evaluate how the quantity of bads changes between periods t and $t + 1$, we define the quantity index for the bad outputs as

$$Q_b^t(x^t, y^t, b^t, b^{t+1}) = D_b^t(x^t, y^t, b^{t+1}) \,/\, D_b^t(x^t, y^t, b^t). \qquad (9)$$

One can show that this quantity index also satisfies the four Fisher tests discussed for the good output quantity index above. Again, we use linear programming to estimate the distance functions used to construct our index. In particular for k' we solve the following two problems:

$$(D_b^t(x^{k',t}, y^{k',t}, b^{k',t}))^{-1} = \min \{\lambda: (x^{k',t}, y^{k',t}, \lambda b^{k',t}) \in T^t\} \tag{10}$$

and

$$(D_b^t(x^{k',t}, y^{k',t}, b^{k',t+1}))^{-1} = \min \{\lambda: (x^{k',t}, y^{k',t}, \lambda b^{k',t+1}) \in T^t\} \tag{11}$$

The ratio of the solutions to these problems yields the index for k' between t and $t + 1$.

We can now define our overall environmental performance index as the ratio of the good and bad quantity indices

$$E^{t,t+1}\left(x^t, y^t, b^t, y^{t+1}, b^{t+1}\right) = \frac{Q_y^t\left(x^t, y^t, b^t, y^{t+1}\right)}{Q_b^t\left(x^t, y^t, b^t, b^{t+1}\right)}. \tag{12}$$

If a single good and a single bad output are produced, then this index becomes

$$E^{t,t+1}\left(x^t, y^t, b^t, y^{t+1}, b^{t+1}\right) = \frac{y^{t+1}/y^t}{b^{t+1}/b^t} = \frac{y^{t+1}/b^{t+1}}{y^t/b^t}, \tag{13}$$

which is simply the ratio of the good to bad output in the two time periods. Finally, one can show that our environmental performance index satisfies the four Fisher index number tests discussed above.

Empirical Illustration: The Case of U.S. Agriculture

In this section, we provide an empirical illustration of the measurement of productivity in the presence of undesirable outputs using a unique data set developed by the U .S. Department of Agriculture's Economic Research Service (ERS) in cooperation with the Natural Resources Conservation Service (NRCS). Before turning to these results, we first provide a brief discussion of the data.

Indices of Desirable Outputs and Inputs

The data used to construct our productivity indices based on desirable outputs and inputs alone are described in Ball et al. (1999) and in Chapter 2 of this volume. We identify two desirable outputs, crops and livestock. The inputs include services of capital, land, labor and intermediate goods. The data are available for 48 contiguous states over the period 1960 to 1996 and are used to construct a state-by-year panel.

As a first step, we construct longitudinal indices of outputs and inputs An index of relative real output between two states is obtained by dividing the nominal output-value ratio for the two states by the corresponding output-price index. We construct multilateral price indices using a method proposed independently by Eltetö and Köves (1964) and Szulc (1964). The "EKS" index is based on the idea that the most appropriate index to use in a comparison between two states is the Fisher Ideal index, which is expressed as the geometric mean of the widely used Laspeyres and Paasche indexes. The Fisher index possesses many of the properties considered desirable in classical index number analysis, and in addition represents exactly a production structure that has attractive properties.

When the number K of states exceeds two, the application of the Fisher index number procedure to the $K(K-1)/2$ possible pairs of states yields a matrix of bilateral indices that may not satisfy Fisher's circularity test. The problem is to obtain results that satisfy the transitivity test and that deviate the least from the bilateral Fisher indices.

Let P_F^{ij} denote the bilateral Fisher index for state i relative to state j. If P_{EKS}^{ij} denotes the multilateral index between states i and j, then the Eltetö and Köves (1964) and Szulc (1964) method suggests that P_{EKS}^{ij} should deviate the least from the bilateral Fisher index P_F^{ij}. Thus P_{EKS}^{ij} should minimize the following distance criterion:

$$\sum_{i=1}^{K}\sum_{j=1}^{K}\left(\ln P_{EKS}^{ij} - \ln P_F^{ij}\left(p^i, p^j, y^i, y^j\right)\right)^2 \tag{14}$$

From the first order conditions for a minimum, it can be shown that the multilateral index with the minimum distance is given by

$$P_{EKS}^{ij} = \left(\prod_{k=1}^{K} P_F(p^i, p^k, y^i, y^k) \cdot P_F(p^k, p^j, y^k, y^j)\right)^{1/K} \tag{15}$$

The EKS index between states i and j may, therefore, be expressed as the geometric mean a geometric mean of K indirect comparisons of i and j through other states.

Using the above formula, we construct indices of relative output prices for the forty-eight contiguous states. The corresponding quantity indices are formed implicitly. Comparisons of levels of capital, land, labor and intermediate inputs require data on relative input prices. Relative price levels of capital input among states are obtained based on relative prices of investment goods, taking into account the flow of capital services per unit of capital stock in each state.

Differences in the relative efficiency of land across states prevent the direct comparison of observed prices. Our estimates of the relative price of land in each state are based on hedonic regressions. For our cross section of states, we estimate the following equation by ordinary least squares:

$$\ln w_i^j = \sum_{i=1}^{K} \delta_i D_i + \sum_{c=1}^{C} \beta_c x_{ic}^j + \varepsilon_{ij}, i = 1,\Lambda ,K, \qquad (16)$$

where w_i^j is the price of land in region j in state i, x_i^j is a vector of land characteristics, D_i is a dummy variable equal to unity for the corresponding state and zero otherwise, and ε_{ij} is a stochastic error term.[3] When the log of price is related to linear state dummy variables as above, a hedonic price index can be calculated from the antilog of the δ_i coefficients.[4]

In constructing EKS multilateral indices of relative labor input, we assume that the relative efficiency of an hour worked is the same for a given type of labor in all forty-eight states. Hours worked and average hourly compensation are cross-classified by sex, age, education, and employment class (employee versus self-employed and unpaid family workers). Since average compensation data are not available for self-employed and unpaid family workers, each self-employed worker is imputed the mean wage of hired workers with the same demographic characteristics.

Fertilizers and pesticides are important intermediate inputs. We construct relative prices of fertilizers and pesticides among states from hedonic regression results. A price index for fertilizer is formed by regressing the prices of single nutrient and multi-grade fertilizers on the proportion of nutrients contained in the fertilizer materials. Price differences among pesticides are assumed due to differences in physical characteristics such as toxicity, persistence in the environment, and leaching potential.

Indices of Undesirable Outputs

Next we turn to the data we use to proxy the environmental impacts arising from chemical use. A detailed discussion of the construction of these series is included in Kellogg et al. (this volume). Here we provide a brief overview. The environmental variables are indicators of risk to human health and to fish from exposure to agricultural pesticides. For each population, we construct indicators of risk from exposure to pesticide runoff (i.e., surface water contamination) and pesticide leaching into groundwater.[5]

Our assessment of risk is based on the extent to which the concentration of a specific pesticide exceeds a water quality threshold. For each of 200 pesticides applied to twelve crops, we estimate the annual concentration at the bottom of the root zone and the edge of the field for 4700 representative soils. The estimated concentrations are compared to water quality thresholds that represent "safe" levels for chronic exposure. When concentration of a specific pesticide exceeds the threshold, a risk indicator is constructed using the concentration-threshold ratio.[6] As such, the series proxy changes over time and across states in the risk from pesticide exposure.

In computing our environmental performance indices, we include the two measures of risk to human health (table 10.1). We then compute a second index that includes measures of risk to fish. We report the results averaged over states for each year in tables 10.2 and 10.3. The results averaged over years by state appear in tables 10.4 and 10.5. Due to the difference in states and years reported for the measures of risk to human health and to fish, the average indices of good output Q_y will not necessarily be identical in tables 10.2 and 10.3 and tables 10.4 and 10.5. We compute the averages for Q_y on the basis of available data for the corresponding index of bad output Q_b.[7]

We begin with the case where pesticide leaching and runoff pose a risk to human health. The first column of table 10.2 displays annual average indices of good outputs. Note that the values exceed unity (indicating increases over the previous year) in all years except 1973, 1987, and 1994. In the second column of table 10.2, we report indices of our two undesirable outputs. This series trends upward through 1974 and declines thereafter. Our index of environmental performance (defined as the ratio of the indices of desirable and undesirable outputs) is given in the last column of table 10.2. That the value of the index is generally less than one before 1975 and greater than one afterwards suggests a general pattern of improvement in the latter sub-period. A similar picture emerges when we include measures of risk to fish.

The variations in our index of environmental performance are largely the result of variation in our index of undesirable outputs. Thus, before discussing the results for individual states, we take a short detour to discuss the trends in the undesirable inputs.

Table 10.1. Average Annual Growth Rates of Undesirable Outputs: 1960-1996

STATE	HR Leach	STATE	HR Runoff	STATE	FR Leach	STATE	FR Runoff
AR	0.003609	AR	0.053439	AR	0.002211	AR	0.065803
AL	0.015061	AZ	0.004246	AL	0.007486	AZ	0.03763
CA	0.019543	AL	0.040374	CA	0.003301	AL	0.059933
CO	0	CA	0.028055	CO	0	CA	0.048709
DE	0	CO	0.030575	FL	0.008947	CO	0.042489
FL	0.015311	CT	0.004619	GA	0.022217	CT	0.008524
GA	0.034995	DE	0.006975	IA	0.006424	DE	0.027501
IA	0.008591	FL	0.026437	IL	0.006701	FL	0.058469
IL	0.017311	GA	0.047323	IN	0.005231	GA	0.067537
IN	0.016384	IA	0.089265	KS	0.010138	IA	0.071506
KS	0.015967	ID	0.012061	KY	0.006399	ID	0.050432
KY	0.007456	IL	0.079649	LA	0.003298	IL	0.063785
LA	0.002535	IN	0.069216	MD	0	IN	0.050814
MD	0.000592	KS	0.053499	MI	0.002249	KS	0.049745
MI	0.016726	KY	0.041559	MN	0	KY	0.046471
MN	0.000554	LA	0.045221	MO	0.003088	LA	0.060388
MO	0.013205	MA	0.003168	MS	0.000772	MA	0.011061
MS	0.004601	MD	0.022161	NC	0.020022	MD	0.023363
MT	0.001617	ME	0.010266	NE	0.018315	ME	0.076338
NC	0.025175	MI	0.057034	NY	0.000665	MI	0.058123
NE	0.025093	MN	0.073593	OH	0.001727	MN	0.073241
NM	0	MO	0.067273	OK	0	MO	0.059444
OH	0.003815	MS	0.053746	PA	0.00052	MS	0.069072
OK	0.000717	MT	0.022729	SC	0.001052	MT	0.03564
OR	0.007617	NC	0.050874	TN	0.003063	NC	0.071705
PA	0	ND	0.03536	TX	0.007698	ND	0.072778
SC	0.008616	NE	0.064959	VA	0.001078	NE	0.058882
TN	0.009349	NH	0	WI	0.005262	NH	0
TX	0.01755	NJ	0.007599	average	0.003737	NJ	0.034895
VA	0.008748	NM	0.00941			NM	0.031302
WA	0.001615	NY	0.043317			NV	0.006609
WI	0.021684	OH	0.065774			NY	0.056977
WY	0	OK	0.02954			OH	0.051001
average	0.001805	OR	0			OK	0.04681
		PA	0.033617			OR	0.021251
		SC	0.040318			PA	0.047449
		SD	0.05841			RI	0.016257
		TN	0.048191			SC	0.058602
		TX	0.05896			SD	0.046308
		UT	0.002924			TN	0.061223
		VA	0.030648			TX	0.081047
		VT	0.011702			UT	0.012921
		WA	0.006081			VA	0.066899
		WI	0.063792			VT	0.006178
		WV	0			WA	0.033732
		WY	0.004089			WI	0.065578
		average	0.032571			WV	0.012153
						average	0.037598

Table 10.2. Average Indices by Year: QB Index Based on Risk to Human Health

Year	QY	QB	E	Year	QY	QB	E
1962	1.03077	1.17374	0.87816	1979	1.01147	0.99655	1.01506
1963	1.02646	1.10788	0.92657	1980	1.02437	0.9941	1.03043
1964	1.0245	1.19253	0.859	1981	1.01907	0.96335	1.05776
1965	1.0131	1.14728	0.88303	1982	1.0016	0.92908	1.07804
1966	1.02717	1.38454	0.73994	1983	1.00201	0.9904	1.01166
1967	1.01723	1.16812	0.87087	1984	1.00607	1.01416	0.99204
1968	1.02372	1.11873	0.91491	1985	1.02787	1.01386	1.01377
1969	1.01107	0.98894	1.02228	1986	1.01005	0.93192	1.0837
1970	1.02939	1.04952	0.98075	1987	0.98993	0.98735	1.00267
1971	1.0216	1.06235	0.96162	1988	1.00709	0.98854	1.01877
1972	1.02225	1.09911	0.92999	1989	1.01535	0.9713	1.04538
1973	0.9971	1.08628	0.91789	1990	1.03577	0.9476	1.09311
1974	1.01144	1.01694	0.99476	1991	1.04226	0.98808	1.05474
1975	1.00852	0.98933	1.01938	1992	1.01109	1.02761	0.98386
1976	1.02791	0.99879	1.02905	1993	1.034	1.03785	0.99618
1977	1.02472	0.97536	1.05065	1994	0.99876	0.99693	1.00189
1978	1.02954	0.96992	1.0615	1995	1.02958	0.9589	1.07387

Indicators of Risk to Human Health

In table 10.2, we present indices of undesirable outputs averaged over states. The indices proxy changes over time in risk from exposure to pesticide in drinking water. Therefore, a decline in the index would indicate a reduction in the risk to human health. The index generally exceeded one in the 1960s and early 1970s, but was less than one more recently. This reduction in risk was coincident with the passage of the Federal Environmental Pest Control Act (FEPCA) of 1972, which significantly increased authority to regulate pesticides. The FEPCA allowed registration of a pesticide only if it did not have an unreasonable adverse effect on human health or the environment. It also required an examination of the safety of all previously registered pesticides using new health and environmental protection criteria. Pesticides with risks that exceed those criteria were subject to cancellation. The result has been a decline in use of pesticides that pose a risk to human health and that are highly mobile in the environment.

To better understand this result, we examine chemical use patterns in corn, soybean, and in cotton production. The use of pesticides in the production of these crops accounted for approximately two-thirds of the total pounds of active ingredients applied. The pesticides that accounted for much of the leaching and runoff in the early 1970s were alachlor, atrazine, aldrin, carbofuran, phroate, trifluralin, and propachlor. The Environmental Protection Agency (EPA) cancelled all uses of aldrin in 1974. In the case of trifluralin, restrictions

Table 10.3. Average Indices by Year: QB Index Based on Risk to Fish

Year	QY	QB	E	Year	QY	QB	E
1962	1.02579	1.01154	1.01396	1979	1.01276	0.88384	1.1459
1963	1.02396	0.96736	1.0584	1980	1.02553	1.00145	1.02398
1964	1.01867	1.02226	0.99648	1981	1.02066	0.94269	1.08274
1965	1.00823	0.99507	1.01326	1982	1.0023	1.01865	0.98403
1966	1.02083	1.05971	0.96348	1983	1.00132	0.97359	1.0285
1967	1.01455	1.00854	1.00594	1984	1.00508	0.98104	1.0245
1968	1.01667	0.95229	1.06777	1985	1.02497	0.97381	1.05242
1969	1.01128	1.06035	0.95215	1986	1.01047	0.94386	1.07072
1970	1.02822	1.01849	1.00946	1987	0.98275	0.85117	1.15433
1971	1.02104	1.19056	0.8576	1988	1.00586	0.85381	1.17862
1972	1.02372	1.12265	0.91181	1989	1.01292	1.04209	0.97191
1973	0.99951	1.09225	0.91493	1990	1.03461	1.08986	0.94937
1974	1.01819	1.21313	0.83931	1991	1.04148	0.89862	1.15897
1975	1.00968	1.16667	0.86545	1992	1.00919	0.94886	1.06363
1976	1.02832	1.17561	0.87472	1993	1.03578	0.90384	1.14598
1977	1.0256	1.06359	0.9643	1994	0.99868	0.93466	1.0686
1978	1.02976	0.9517	1.08216	1995	1.03227	1.02434	1.00788

on product formulation were imposed in 1982. And, most uses of carbofuran were cancelled in 1990 and 1991.

In cotton production, where insecticide use is more important, those pesticides that posed the greatest environmental risk were aldrin and EPN. As noted above, use of aldrin was cancelled in 1974. And the use of EPN was cancelled in 1987, thereby contributing to a decline in risk to human health in recent years.

Indicators of Risk to Fish

Table 10.3 displays our indicator of risk to fish. Note that this series is more volatile than our indicator of risk to human health. The index was greater than one during the 1969 to 1977 period, indicating increased risk to fish from water contamination. There were periods in the 1980s and 1990s when the index dropped below one, suggesting potential environmental improvement. This pattern is consistent with periodic swings in use of such pesticides as endosulfan to control aphids in cotton production.

Table 10.4. Average by State: QB Index Based on Risk to Fish: 1962-95

Year	QY	QB	E	Year	QY	QB	E
AL	1.02463	1.08332	0.94587	MS	1.01825	1.09743	0.92784
AR	1.03438	1.09334	0.94593	MT	1.01454	0.99902	1.0156
AZ	1.02037	1.04721	0.9745	NC	1.02723	1.09267	0.94004
CA	1.02581	1.04794	0.97893	ND	1.02756	0.98308	1.04531
CO	1.02577	1.04181	0.98454	NE	1.0303	0.96529	1.06759
CT	1.00049	0.95523	1.04737	NH	0.98289	1	0.98289
DE	1.03408	0.88034	1.17446	NJ	0.99992	0.94588	1.05716
FL	1.03357	1.01199	1.02125	NM	1.02863	0.95155	1.08092
GA	1.03504	1.10229	0.93891	NV	1.01674	1.12244	0.90588
IA	1.00953	0.97121	1.03961	NY	1.00215	0.94524	1.06033
ID	1.02633	0.98314	1.04389	OH	1.01045	0.9312	1.08512
IL	1.01619	0.98696	1.02962	OK	1.01889	1.03372	0.98567
IN	1.01108	1.03763	0.9745	OR	1.0243	0.97975	1.04535
KS	1.02141	1.12278	0.90811	PA	1.01604	0.96631	1.05151
KY	1.01876	1.06612	0.95538	RI	0.99969	0.90726	1.10184
LA	1.02213	1.14738	0.89076	SC	1.00728	1.02004	0.98752
MA	0.99388	1.02092	0.97344	SD	1.01463	1.03124	0.98387
MD	1.01754	0.87599	1.16237	TN	1.00912	1.00858	1.00053
ME	1.0065	0.97319	1.03426	TX	1.02245	1.0555	0.96871
MI	1.01519	1.00554	1.00958	UT	1.01008	1.04764	0.96416
MN	1.01249	1.05039	0.96383	VA	1.01949	1.02501	0.9945
MO	1.00763	0.97466	1.03376	VT	1.00518	0.94387	1.06494
				WA	1.03211	1.00932	1.02258
				WI	1.01196	0.96874	1.04444
				WV	0.99184	0.89078	1.11351
				WY	1.01392	1.01035	1.00352

Selected State Results

Although the broad trends discussed above and summarized in tables 10.2 and 10.3 are of interest in their own right, they mask considerable variation across states. To illustrate, we present in table 10.6 the indices of environmental performance and the corresponding trends in chemical use for two important corn and soybean producing states — Iowa and Illinois — and two cotton-producing states — North Carolina and Mississippi.

Beginning with the two corn and soybean producing states, we see that Iowa's environmental indices, based on both the risk to fish and to human health, show improvements in recent years. In contrast, our environmental index for Illinois suggests no real improvement in recent years. The difference is most obvious when considering the risk to human health. A glance at table 10.6 suggests why — our index of pesticide runoff declined at an annual average rate of 4.08% in Iowa between 1980/82 and 1994/96, while it increased by

Table 10.5. Average by State: QB Index Based on Risk to Human Health: 1962-95

Year	QY	QB	E	Year	QY	QB	E
AL	1.02463	1.08332	0.94587	MS	1.01825	1.09743	0.92784
AL	1.02463	1.08332	0.94587	MS	1.01825	1.09743	0.92784
AR	1.03438	1.09334	0.94593	MT	1.01454	0.99902	1.0156
AZ	1.02037	1.04721	0.9745	NC	1.02723	1.09267	0.94004
CA	1.02581	1.04794	0.97893	ND	1.02756	0.98308	1.04531
CO	1.02577	1.04181	0.98454	NE	1.0303	0.96529	1.06759
CT	1.00049	0.95523	1.04737	NH	0.98289	1	0.98289
DE	1.03408	0.88034	1.17446	NJ	0.99992	0.94588	1.05716
FL	1.03357	1.01199	1.02125	NM	1.02863	0.95155	1.08092
GA	1.03504	1.10229	0.93891	NV	1.01674	1.12244	0.90588
IA	1.00953	0.97121	1.03961	NY	1.00215	0.94524	1.06033
ID	1.02633	0.98314	1.04389	OH	1.01045	0.9312	1.08512
IL	1.01619	0.98696	1.02962	OK	1.01889	1.03372	0.98567
IN	1.01108	1.03763	0.9745	OR	1.0243	0.97975	1.04535
KS	1.02141	1.12278	0.90811	PA	1.01604	0.96631	1.05151
KY	1.01876	1.06612	0.95538	RI	0.99969	0.90726	1.10184
LA	1.02213	1.14738	0.89076	SC	1.00728	1.02004	0.98752
MA	0.99388	1.02092	0.97344	SD	1.01463	1.03124	0.98387
MD	1.01754	0.87599	1.16237	TN	1.00912	1.00858	1.00053
ME	1.0065	0.97319	1.03426	TX	1.02245	1.0555	0.96871
MI	1.01519	1.00554	1.00958	UT	1.01008	1.04764	0.96416
MN	1.01249	1.05039	0.96383	VA	1.01949	1.02501	0.9945
MO	1.00763	0.97466	1.03376	VT	1.00518	0.94387	1.06494
				WA	1.03211	1.00932	1.02258
				WI	1.01196	0.96874	1.04444
				WV	0.99184	0.89078	1.11351
				WY	1.01392	1.01035	1.00352

1.53% in Illinois. The increase in pesticide runoff in Illinois appears to be driven by the increased use of two herbicides: cyanazine (which is being voluntarily withdrawn by the manufacturer this year) and atrazine.

Our two cotton states also illustrate very divergent patterns, as can be seen in figures 10.1-10.4. Our environmental performance index for Mississippi declined, suggesting more adverse environmental impact. In contrast, North Carolina shows improvement, the result of a successful boll weevil eradication program. Restrictions were placed on the use of some hazardous chemicals. More important, the number of insecticide applications was reduced.

Table 10.6. Risk to Fish and Human Health and Chemical Use: Selected States

	TEUs Fish Mill. Of TEU		Important Pesticides Mill. Of lbs.			
	Leach	Runoff	Triflur.	Carbofur.	Aldicarb	Endosulf.
Iowa						
1980-1982	1.40	469.00	4.80	0.45	-	-
1994-1996	0.51	229.00	2.76	0.21	-	-
% annl chg	-7.21	-5.12	-3.94	-5.29	-	-
Illinois						
1980-1982	4.73	366.93	4.22	0.56	-	-
1994-1996	5.72	454.63	1.11	0.19	-	-
% annl chg	1.36	1.53	-9.96	-7.79	-	-
Mississippi						
1980-1982	0.87	194.00	1.62	0.19	0.04	0.00
1994-1996	0.73	1352.11	1.06	0.03	0.19	0.60
% annl chg	-1.28	13.87	-3.02	-14.29	10.88	34.72
North Carolina						
1980-1982	6.62	1496.86	4.49	0.44	0.07	0.03
1994-1996	4.32	121.74	0.26	0.03	0.37	0.01
% annl chg	-3.04	-17.92	-20.26	-19.46	12.03	-7.70
Iowa						
1980-1982	2.59	1580.82	7.79	13.71	9.12	0.01
1994-1996	2.16	892.76	7.69	2.35	5.47	0.01
% annl chg	-1.30	-4.08	-0.09	-12.61	-3.65	2.37
Illinois						
1980-1982	4.73	366.93	9.92	10.60	2.00	0.04
1994-1996	5.72	454.63	10.74	3.07	6.44	0.11
% annl chg	1.36	1.53	0.57	-8.85	8.34	24.08
Mississippi						
1980-1982	0.87	83.21	0.19	1.07	0.23	0.00
1994-1996	0.73	106.37	0.51	0.15	0.68	0.06
% annl chg	-1.28	1.75	7.27	-13.94	7.91	45.68
North Carolina						
1980-1982	10.11	53.81	2.20	1.71	0.03	0.02
1994-1996	15.64	16.85	1.02	0.83	0.11	0.02
% annl chg	3.12	-8.29	-5.46	-5.19	8.79	-1.61

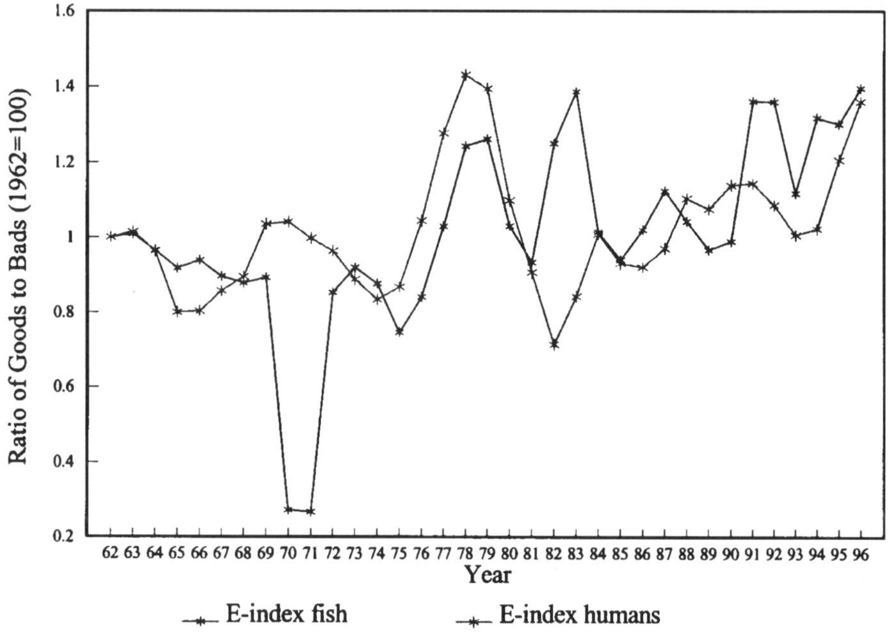

Figure 10.1. Iowa: Goods vs Fish and Human Bads

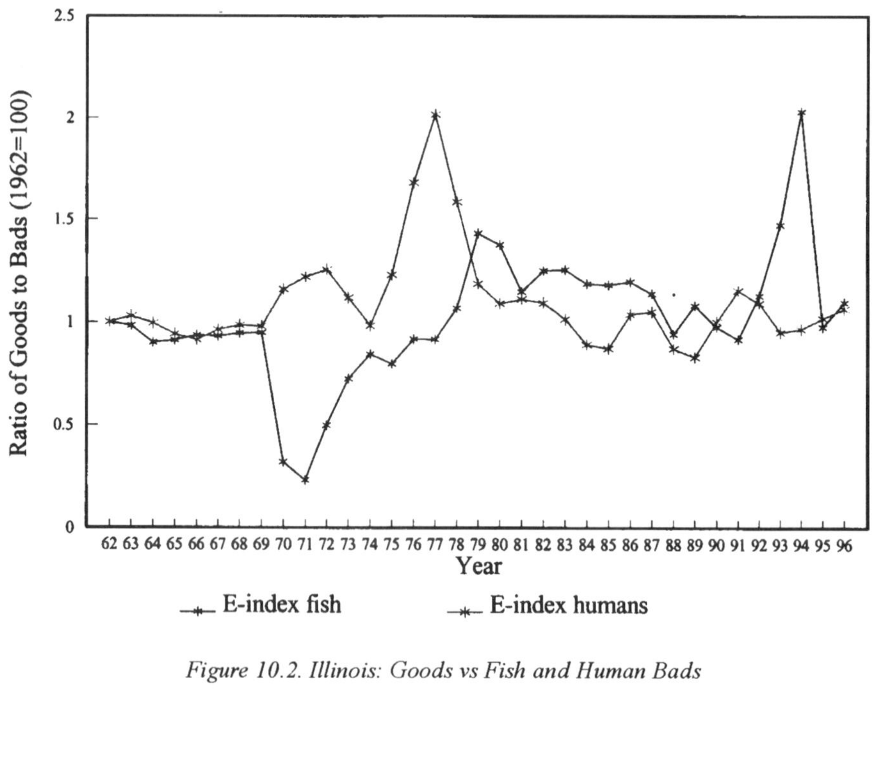

Figure 10.2. Illinois: Goods vs Fish and Human Bads

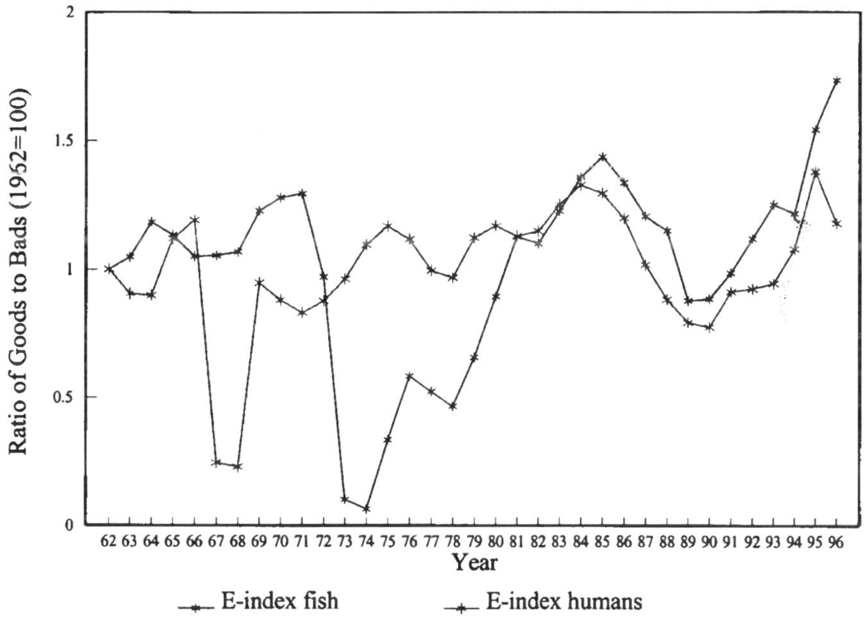

Figure 10.3. North Carolina: Goods vs Fish and Human Bads

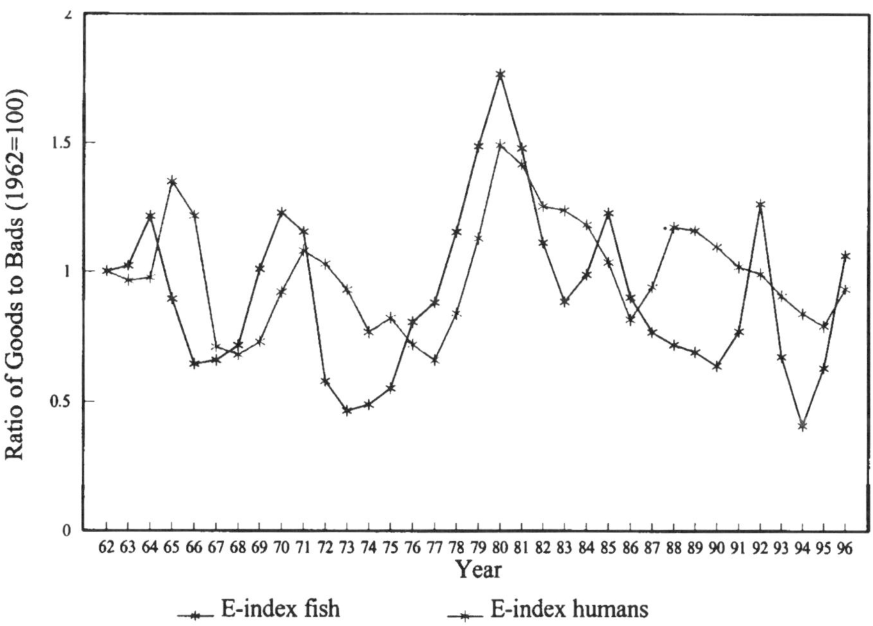

Figure 10.4. Mississippi: Goods vs Fish and Human Bads

Summary and Future Research

Our motivation for developing indices of environmental performance was to assess improvements in the ability of the agricultural sector to produce marketable agricultural outputs while minimizing environmental damage. Since our indices are based on distance functions that are representations of technology, we are in effect measuring "environmental productivity". And since these distance functions explicitly allow us to model joint production of good and bad outputs, and do not require information on shadow prices of bads, our index has fairly minimal data requirements. The distance function allows us to aggregate across multiple undesirable outputs (again without requiring information on shadow prices), allowing us to provide a stand-alone index of bad outputs as well. Our index of environmental performance is then constructed as the ratio of a quantity index of marketed agricultural outputs to the index of undesirable outputs.

Although data requirements are fairly minimal, we nevertheless require information on the production of bad outputs. In this paper, we take advantage of a unique data set recently developed by the USDA's Economic Research Service and Natural Resource Conservation Service, which provides information on the contamination of ground and surface water due to agricultural pesticide use. These data are further refined to reflect the risks posed both to human health and to fish. These risks are the undesirable outputs we use in our analysis. We compute our indices at the state level by year for the 1962 to 1996 period. Our preliminary results suggest that environmental productivity in the agricultural sector may have improved in recent years; however, there is considerable variation by state, which accords with trends in pesticide use.

We provide anecdotal evidence that increased regulator scrutiny of pesticides has played a positive role. Further analysis of the effects of regulation on environmental performance is clearly warranted. A related line of inquiry is the relationship between R&D expenditures and the environmental risk associated with pesticide use. Ultimately, we would also like to parse out the effects of weather, change in regulations, and investments in R & D on environmental performance.

Finally, there are on-going efforts to produce data on the effects of plant nutrients such as nitrogen on water quality, and to extend measures of risk to algae, crustaceans, etc. This extension would provide a richer picture of environmental effects of agricultural chemical use.

Notes

1. In practice, we include a three-year "window" of data in the reference technology; that is, we include data from period *t*, *t*-1, and *t*-2.

2. In practice, we use a moving average of the current period and the two previous periods for each observation under evaluation. As mentioned in the earlier footnote, the reference technology is also expanded to include data from the two previous periods.

3. The observations on *w* consist of average prices. When average data are used rather than actual observations on price, the disturbance terms are likely to be heteroskedastic. Efficient parameter estimates are obtained by applying weighted least squares, where the weights are land areas.

4. For the semi-logarithmic specification used here, a consistent estimate of the parameter δ_i is given by exp (δ_i) - 1 (Halverson and Palmquist, 1980).

5. In addition to being exposed to pesticide runoff, fish are exposed to pesticides that leach into groundwater and then enter surface water through groundwater recharge. Exposure to pesticides that leach is particularly important during low-flow conditions when most surface water originates from groundwater recharge.

6. In some states, we indicate zero environmental risk for some years, because the pesticide concentration in some years did not exceed the water quality threshold. See table 10.1.

7. Although our data begin in 1960, our first reported results are for 1962, due to the fact that we used a 3-year moving average of the inputs and outputs in computing our indices

References

Ball, V. E., R. Färe, S. Grosskopf, and R. Nehring. 2001. "Productivity of the U.S. Agricultural Sector: The Case of Undesirable Outputs." In *New Developments in Productivity Analysis*, ed. C. Hulten, E. Dean, and M. Harper. Chicago: University of Chicago Press, 541-85.

Ball, V. E., F. Gollop, A. Kelly-Hawke, and G. Swinand. 1999. "Patterns of Productivity Growth: Linking State and Aggregate Models." *American Journal of Agricultural Economics*, 81:164-79.

Eltetö, 0., and P. Köves. 1964. "On a Problem of Index Number Computation Relating to International Comparison." *Statisztikai Szemle*, 42:507-18.

Färe, R., S. Grosskopf, and F. Hernandez-Sancho. 2000. "Environmental Performance: An Index Number Approach." Oregon State University, unpublished manuscript.

Färe, R., and D. Primont. 1995. *Multioutput Production and Duality: Theory and Applications.* Boston: Kluwer Academic Publishers.

Fisher, I. 1922. *The Making of Index Numbers.* Boston: Houghton Mifflin.

Halverson, R., and R. Palmquist. 1980. "The Interpretation of Dummy Variables in Semilogarithmic Equations." *American Economic Review*, 70:645-59.

Kellogg, R., R. Nehring, A. Grube, D. W. Goss, and S. Plotkin. 2000. "Environmental Indicators of Nitrogen and Pesticide Leaching and Run-off from Farm Fields." Presented at "Agricultural Productivity: Data, Methods and Measures," ERS, USDA, March 9-10, 2000, Washington, D.C.

Malmquist, S. 1953. "Index Numbers and Indifference Surfaces." *Trabajos de Estadistica*, 4:209-42.

Shephard, R. W. 1970. *Theory of Cost and Production Functions.* Princeton: Princeton University Press.

Szulc, B. 1964. "Indices for Multiregional Comparisons." *Przeglad Statystyczny*, 3:239-54.

11

The Effect of Ground Water Regulation on Productivity Growth in the Farm Sector

Kelly A. Chaston and Frank M. Gollop

Abstract

Proper measures of productivity growth are barometers of how well society is allocating its scarce resources. In this context, there is little difference between labor, capital, and material inputs, on the one hand, and air and water resources, on the other. Consumption of any one entails true opportunity costs. The case for expanding the now familiar Total Factor Productivity (TFP) measures to Total Resource Productivity (TRP) is self-evident and compelling.

This paper adopts a TRP growth framework and demonstrates the relative ease with which changing environmental quality can be incorporated into formal productivity modeling. State-specific TRP growth rates for U.S. agriculture for the 1960-96 period are calculated and compared with corresponding TFP growth rates. Annual TFP growth rates across states decreased by an average 0.50 percentage points between 1960-77 and 1977-96 while annual TRP growth increased by an average 0.45 percentage points. Improved risk management of chemical and water use on U.S. farms wholly explains the differential. It not only offset the 0.50 percentage point reduction in annual TFP growth but also added an additional 0.45 percentage points to TRP growth. Without that improvement in water quality, annual TRP growth in the 1977-96 period would have equaled 1.15 percent instead of its measured 2.10 percent rate. The importance of incorporating environmental quality into measures of productivity growth should not be underestimated.

Introduction

Proper measures of productivity growth are barometers of how well society is allocating its scarce resources. In this context, there is little difference between labor, capital, and material inputs, on the one hand, and air and water resources, on the other. Each is scarce. Consumption of any one entails true opportunity costs. The case for expanding the now familiar Total Factor Productivity (TFP) measures to Total Resource Productivity (TRP) is self-evident and compelling.

This chapter adopts a TRP growth framework introduced in Gollop and Swinand (1998, 2000).The production-based model is applied to state-specific production accounts for the U.S. farm sector (See chapter 2). The 48-state, 1960-96 panel was prepared by the Economic Research Service at USDA in cooperation with the Environmental Protection Agency. The panel includes data for marketable outputs, inputs, prices, and indicators of water quality.

This chapter demonstrates the relative ease with which changing environmental quality can be incorporated into formal productivity modeling, but it has two empirical objectives as well. First, state-specific TRP growth rates are calculated and compared with corresponding TFP growth rates. Because of a rapid rise in the risk-adjusted measure of ground water pollution from U.S. farms through 1977, TRP growth is actually below conventional TFP growth through that year. However, regulatory-induced improvements in risk management after 1977 led to the expected result that annual TRP growth on average exceeds TFP growth in the 1977 to 96 period. Nearly every state shows the same result. This result reflects the farm sector's success in reducing the use of toxic chemicals while continuing to expand production of marketable outputs. Second, the importance of environmental regulation in contributing to TRP growth is assessed. While annual TFP growth rates across states decreased by an average 0.50 percentage points between 1960-77 and 1977-96, annual TRP growth increased by an average 0.50 percentage points. Regulation-induced reductions in chemical use improved water quality and accounted on average for a 0.95 percentage point increase in TRP growth across states between 1960 and 1996. Improvements in risk management not only offset the 0.50 percentage point reduction in annual TFP growth but also added an additional 0.45 percentage points to TRP growth. The importance of incorporating environmental quality into measures of productivity growth should not be underestimated.

MEASURING TOTAL RESOURCE PRODUCTIVITY

The productivity model developed in this chapter is based on a model of producer behavior for a representative farm. The farm produces conventional outputs Y_L (livestock) and Y_C (crops) and an undesirable output S (chemical contamination of neighboring water sources). Assume that the production of Y_C is the only source of S. Government regulations targeting the effluent S constrain and therefore enter the description of the farm's technology. Producers are held accountable; therefore S enters the model of producer behavior.

Developing an index of the farm's aggregate output begins by selecting any arbitrary set of non-negative quantities of outputs Y_L, Y_C, and S. Given this initial product set, the farm's aggregate output can be defined as a proportion of quantities of outputs Y_L,Y_C, and S or, equivalently, as a proportion of conventional outputs Y_L and Y_C holding fixed environmental quality S/Y_C. The maximum value of aggregate output ($\hat{O}$) then can be expressed as a function of Y_L, Y_C, S/Y_C, resources $\underline{X}$, and a time-based technology index T

$$\Phi = H(Y_L, Y_C, S/Y_C, \underline{X}, T) \tag{1}$$

The variables Y_L and Y_C represent separable, aggregate measures of marketable farm outputs. Aggregate Y_L is formed over cattle, sheep, etc.; Y_C represents an aggregate for wheat, corn, fruits, etc. All output and input markets are assumed to be competitive. S/Y_C is an exogenously imposed standard for water quality associated with fertilizers and pesticides used in crop production. Full compliance with environmental standards is assumed. The task of the farm manager is to select the profit-maximizing level of $\hat{O}$ given market prices, market equilibrium conditions for inputs $\underline{X}$ and conventional outputs $\underline{Y}$, and existing societal restrictions (if any) on S/Y_C.

The environmental standard (S/Y_C) enters the farm's production function as a rate rather than as a level. This representation is appropriate for agriculture since environmental restrictions for fertilizers and pesticides are posed in terms of application rates, not in terms of total pounds of applied materials.

The marginal rates of transformation among the arguments in (1) are of note. The function H is increasing in S/Y_C, $\underline{X}$, and T and decreasing in both Y_L and constant quality output Y_C. *Ceteris paribus*, an increase in S/Y_C (holding Y_L and Y_C fixed) frees resources to produce additional aggregate output Φ; an increase in either Y_L or Y_C consumes resources and therefore reduces aggregate output Φ. There is the expected negative rate of transformation between Y_L and Y_C but a positive rate of transformation between either Y_L or Y_C, on the one hand, and S/Y_C, on the other.

The function H exhibits the usual homogeneity properties. H is homogeneous of degree minus one in Y_L, Y_C, and S since, holding S/Y_C, $\underline{X}$, and T constant, any proportional increase in all three outputs definitionally generates an equal proportional decrease in Φ. In addition, H is assumed to be homogeneous of degree one in $\underline{X}$. As a result, H is homogeneous of degree zero in Y_L, Y_C, S, and $\underline{X}$ and exhibits constant returns to scale.

Deriving the model of productivity growth represented by the production function H begins by setting aggregate output Φ equal to unity. Fixing Φ at unity does not imply that output does not or cannot change over time. Given the negative one-to-one relationship between Φ and the scale of conventional outputs Y, output growth can be represented either by an increase in Φ or, equivalently, by an identically proportional increase in Y_L and Y_C. Fixing Φ at unity does nothing more than force growth to be reflected in the Y variables.

Taking the logarithmic differential of (1) with respect to time and solving for $\partial \ln H/\partial T$ yields the following formulation for the representative farm's growth rate of total resource productivity (TRP)

$$\dot{TRP} \equiv \frac{\partial \ln H}{\partial T} = -\sum \frac{\partial \ln H}{\partial \ln Y_j}\frac{d \ln Y_j}{dT} - \frac{\partial \ln H}{\partial \ln\left(S/Y_c\right)}\left[\frac{d \ln S}{dT} - \frac{d \ln Y_c}{dT}\right] - \sum \frac{\partial \ln H}{\partial \ln X_i}\frac{d \ln X_i}{dT} \tag{2}$$

Necessary conditions for producer equilibrium transform the logarithmic partials in (2) into well-defined variables:

$$\frac{\partial \ln H}{\partial \ln Y_j} = \frac{-q_j Y_j}{P_\phi \Phi} \qquad (j = L, C), \tag{3a}$$

$$\frac{\partial \ln H}{\partial \ln S} = \frac{\rho S}{P_\phi \Phi} \tag{3b}$$

$$\frac{\partial \ln H}{\partial \ln X_i} = \frac{w_i X_i}{P_\phi \Phi} \qquad (i=1,\ldots,n), \tag{3c}$$

where P_ϕ is the unit price of aggregate output, q_j is the unit price of marketable output Y_j, the w_i represent input prices, and ρ equals the marginal abatement cost of S. It is important to note that the partial derivative in (3a) is taken with respect to marketable output Y_j, holding constant environmental quality, S/Y_C. This derivative, together with the assumption of competitive markets for Y_L and Y_C, permits necessary condition (3a) to be expressed in terms of q_j, the observed market prices of Y_j — a desirable property of a model destined for empirical application. Note also that the value of aggregate output $P_\phi\Phi$ equals the value of marketable output $\Sigma q_j Y_j$. This equality follows from the assumption of competitive markets and the definitions of H and Φ. The producers' abatement costs are reflected in both q_C and P_ϕ.

The definition of H and the competitive nature of markets for Y_L, Y_C, and $\underline{X}$ also guarantee

$$\sum \frac{w_i X_i}{P_\phi \Phi} = \sum \frac{w_i X_i}{\sum q_j Y_j} = 1. \quad (4)$$

First, abatement requires resources $\underline{X}$ so that $\Sigma w_i X_i$ captures the total cost of producing Y_L and Y_C as well as abating pollution to the given level S/Y_C. Second, the competitive market price q_C reflects the marginal cost of both producing Y_C and engaging in abatement to the level S/Y_C. It follows that $\Sigma q_j Y_j = \Sigma w_i X_i$, as required by (4).

Making the necessary substitutions from (3) into (2) yields the formula for measuring TRP growth conditioned on a model of producer equilibrium

$$\dot{TRP} = \sum \frac{q_j Y_j}{\sum q_j Y_j} \frac{d \ln Y_j}{dT} + \frac{\rho S}{\sum q_j Y_j} \left[\frac{d \ln Y_c}{dT} - \frac{d \ln S}{dT} \right] - \sum \frac{w_i X_i}{\sum q_j Y_j} \frac{d \ln X_i}{dT} \quad (5)$$

TRP growth equals the weighted growth in marketable outputs plus the weighted growth in environmental quality [positive (negative) if $d\ln Y_C/dT > (<) \, d\ln S/dT$] less cost-share weighted input growth. TRP growth also can be expressed in terms of the conventional measure of TFP growth

$$\dot{TRP} = \dot{TFP} + \frac{\rho S}{\sum q_j Y_j} \left[\frac{d \ln Y_c}{dT} - \frac{d \ln S}{dT} \right] \quad (6)$$

since $$\dot{TFP} = \sum \frac{q_j Y_j}{\sum q_j Y_j} \frac{d \ln Y_j}{dT} - \sum \frac{w_i X_i}{\sum q_j Y_j} \frac{d \ln X_i}{dT}.$$

The measure of TRP growth has some very nice properties. If, as outputs increase, either the pollution content per unit of Y_C does not change ($d\ln Y_C/dT = d\ln S/dT$) or marginal abatement cost is zero ($\rho=0$), then TRP growth in (5) collapses to the traditional TFP form. In the former case, the producer receives credit only for the growth in Y_C — exactly as it would in the TFP framework. In the latter instance, though S may be an undesirable byproduct of production, society has chosen to impose no binding restrictions ($\rho=0$) on producer behavior. Producers behave rationally and allocate no resources to abatement. Productivity growth, viewed from the producer's perspective, is properly measured as TFP growth. However, should society impose a binding regulatory constraint on producers, then, since reducing S/Y_C is costly in terms of either $\underline{X}$ or foregone marketable outputs (Y_L or Y_C), $\rho > 0$ and farm managers are induced to consider the S content of their output Y_C. Consequently, when $\rho > 0$, proper productivity measurement cannot ignore changes in S/Y_C. Given $\rho > 0$,

TRP must "grade" producers for changes in environmental quality per unit of output. *Ceteris paribus*, TRP growth > (<) TFP growth when the production sector improves (diminishes) the environmental quality of its product. Ignoring the non-market output term S/Y_C in (5) would lead to a biased measure of total resource productivity growth.

Marginal Abatement Cost

The measurement of TRP growth in (6) relies almost wholly on observable data. The only exception is ρ, the marginal abatement cost corresponding to S. An estimate of ρ can be derived from an econometric specification of a factor-minimal cost function for a representative farm. Farm managers minimize production cost (C) given input prices ($\underline{w}$), marketable farm outputs (Y_L and Y_C), regulatory-imposed environmental standards (S/Y_C), and the state-of-technology variable (T)

$$C = C(\underline{w}, Y_L, Y_C, S/Y_C, T). \tag{7}$$

The assumptions implicit in (7) are consistent with those underlying the model of producer behavior in equation (1). Factor markets are assumed to be competitive. The water quality variable S/Y_C is exogenous, and the farm is expected to comply fully with environmental constraints. Application of (7) is consistent with any market structure for either Y_L or Y_C including the competitive market assumed in section 1. Finally, outputs Y_L and Y_C in (7) are exogenous variables for econometric purposes while endogenous variables in the larger model of producer behavior underlie (7). They are determined outside the cost function.

The translog cost function is selected to give explicit functional form to the implicit cost function (7)

$$\ln C = \alpha_o + \sum \alpha_i \ln Z_i + \frac{1}{2}\sum_i \sum_j \beta_{ij} \ln Z_i Z_j \tag{8}$$

where Z_i represents Y_L, Y_C, S/Y_C, input prices $\underline{w}$, and T. The cost function (8) and its familiar set of input cost-share equations make up the estimating system.

The all-important marginal abatement cost (MAC), stated in logarithmic form, is represented by the logarithmic partial of (8) with respect to S:

$$\frac{\partial \ln C}{\partial \ln S} = \alpha_S + \sum \beta_{Sj} \ln Y_j + \beta_S \ln(S/Y_C) + \sum \beta_{Si} \ln w_i + \beta_{ST} T \tag{9}$$

The second-order nature of the translog form has the attractive property that the resulting marginal abatement cost estimates are not constants but are functions of all the variables entering the cost function. This feature is useful for the particular purposes of this research project. Marginal abatement costs entering the measurement of TRP can vary over states and time.

One modification is made to the cost function (8). There were at least two watershed developments in the evolution of water pollution regulation that directly impacted U.S. agriculture. DDT was banned in 1972 and the Federal Water Pollution Act of 1972 was amended in 1977. The 1977 amendment focused on toxic pollutants. To formally recognize these discrete policy shifts, all first and second-order water quality variables in the cost function (8) are differentiated by three time periods: 1960-72, 1972-77, and 1977-96. All parameters in the marginal abatement cost expression (9) therefore differ by time period, extending further the ability of the estimating model to capture state-specific changes in MAC over time.

DATA

Estimating the cost function (8) and measuring TRP growth described in (6) require not only price and quantity data on conventional farm outputs and labor, capital, land, and material inputs but also quantity data on the industry's pollution of local water sources. The data necessary to support this research project now are available, after many years of effort, from the U.S. Department of Agriculture (USDA). The output, input, and price data were developed by the Environmental Indicators and Resource Accounting Branch of the Economic Research Service (ERS). Measures of various environmental indicator variables were developed by ERS in cooperation with both USDA's Natural Resources Conservation Service and the Environmental Protection Agency. These complementary data sets are described in more detail in chapters 2 and 9. A brief overview should suffice.

The panel data set used in this study includes annual observations for 48 states that span the period 1960-96. The market-based data on conventional outputs and inputs are formed for each state as Tornquist indices over detailed output and input accounts. Hundreds of distinct farm crop and grain product categories, 160 demographically cross-classified labor cohorts, eight distinct capital asset classes, and nine material goods are distinguished. Considerable effort was expended to develop output and input prices that have spatial as well as temporal integrity. A major contribution of the ERS effort is that output, input, and price data reflect a true panel that can be used for both cross section and time series analyses. A full description of the underlying data series, sources, and indexing technique is presented in Ball, Butault, and Nehring (2000).

Each of USDA's environmental indicator variables also exhibits true panel characteristics. They cover all states, span the 1960-96 period, and capture cross-section variation. The panel data sets contain separate toxicity-adjusted measures of chemicals reaching surface and ground water sources. Each indicator variable captures the effects of agricultural use of both pesticides and fertilizers. The panels are constructed from county-specific (3,041) analyses controlling for differing soil conditions, crop and therefore chemical types, application rates per acre, chemical toxicity, and rainfall patterns. The indicator variables also are adjusted for risk. Separate risk-adjusted indicators for ground and surface water are developed for human exposure; a second pair is constructed for chemical exposure by fish life.

The indicator variable for ground water adjusted for risk to fish life is adopted for the purpose of measuring and evaluating TRP growth in this paper. The basis for this choice over the other three is somewhat arbitrary but is premised on the following reasoning. One would expect that toxic chemicals leaching through the soil to ground water sources is likely to be a more widespread problem than chemical runoff into surface water. In addition, inland fish are likely to have a lower tolerance for toxic chemicals than do humans. In terms of both relative importance and regulatory stringency, the risk-adjusted leaching indicator for fish life seems the most appropriate candidate variable for measuring water pollution.

TRP Measurement and the Effect of Regulation

The translog cost function (8) and its derived cost-share equations were estimated as a four-equation system. Coefficient estimates are reported in table 11.1. Since all data were normalized to their respective means before estimation, the reported first-order coefficients provide useful central-tendency estimates of critical economic variables for the "representative farm" (i.e., the observation defined at mean values for all variables).

Estimates of two productivity-related variables are of particular interest — scale economies and technical change. Scale economies in a multiple-output cost function are measured as the sum of the logarithmic partial derivatives of cost with respect to Y_L and Y_C. In the context of the cost function (8), scale economies exhibited by the representative farm equal the sum of the first-order coefficients on Y_L and Y_C less the sum of the first-order coefficients on the three S/Y_C variables, one for each sub-period 1960-72, 1972-77, and 1977-96. All five first-order estimates are statistically significant at no less than the .05 level and their sum (see table 11.1) equals .954, implying economies of scale

Table 11.1. Cost Function Estimation
Nonlinear ITSUR Summary of Residual Errors

Equation	DF Model	DF Error	SSE	MSE	Root MSE	R-Square	Adj R-Sq
LTVCD	36.5	1606	36.54052	0.02276	0.15086	0.9823	0.9819
LBCS	4.667	1637	6.01485	0.0036736	0.06061	0.5676	0.5667
LDCS	5.167	1637	5.60631	0.0034251	0.05852	0.4454	0.4439
KCS	4.667	1637	1.61237	0.0009848	0.03138	0.4819	0.4808

Nonlinear ITSUR Parameter Estimates

Parameter	Estimate	Approx. Std Err	'T' Ratio	Approx. Prob>|T|
ALPHA	0.395958	0.01874	21.13	0.0001
BQCP	0.358264	0.01652	21.69	0.0001
BQAN	0.504167	0.01749	28.82	0.0001
BSFL1	-0.019506	0.0059884	-3.26	0.0011
BSFL2	-0.015318	0.0078709	-1.95	0.0518
BSFL3	-0.056884	0.0098019	-5.80	0.0001
BPK	0.157517	0.0023621	66.68	0.0001
BPLBR	0.287888	0.0042852	67.18	0.0001
BPLND	0.180093	0.0041290	43.62	0.0001
BQCP2	-0.00952957	0.01577	-0.60	0.5458
BQAN2	0.052500	0.02007	2.62	0.0090
BSFL12	-0.00448065	0.0017090	-2.62	0.0088
BSFL22	-0.013432	0.0035415	-3.79	0.0002
BSFL32	-0.010143	0.0020804	-4.88	0.0001
BPK2	0.075942	0.0041652	18.23	0.0001
BPLBR2	0.100474	0.0048082	20.90	0.0001
BPLND2	0.068494	0.0024383	28.09	0.0001
BQCPPK	0.00490478	0.0013522	3.63	0.0003
BQANPK	-0.012488	0.0015040	-8.30	0.0001
BQCPPLBR	0.017723	0.0025813	6.87	0.0001
BQCPLND	0.012805	0.0024504	5.23	0.0001
BQCPSFL1	-0.018205	0.0059298	-3.07	0.0022
BQCPSFL2	-0.012152	0.0076060	-1.60	0.1103
BQCPSFL3	-0.014146	0.0045564	-3.10	0.0019
BQANSFL1	0.010194	0.0065178	1.56	0.1180
BQANSFL2	0.019661	0.0095198	2.07	0.0391
BQANSFL3	0.014679	0.0059121	2.48	0.0131
BQANPLBR	-0.039693	0.0028657	-13.85	0.0001
BQANPLND	-0.010723	0.0027252	-3.93	0.0001
BQCOQAN	-0.039707	0.01690	-2.35	0.0189
BPKPLBR	-0.033326	0.0025781	-12.93	0.0001
BPKPLND	0.021884	0.0015147	14.45	0.0001
BPLPLND	-0.040102	0.0024068	-16.66	0.0001
BPKSFL1	-0.00104853	0.0006166	-1.70	0.0892
BPKSFL2	0.00241023	0.0010384	2.32	0.0204
BPKSFL3	0.00377107	0.0006357	5.93	0.0001

continued

Table 11.1, *continued*

Parameter	Estimate	Approx. Std Err	'T' Ratio	Approx. Prob>|T|
BPLABSFL1	-0.00400409	0.0011846	-3.38	0.0007
BPLABSFL2	-0.00218064	0.0019519	-1.12	0.2641
BPLABSFL3	-0.00182826	0.0012184	-1.50	0.1337
BPLNDSFL1	-0.00017949	0.0045544	-0.04	0.9686
BPLNDSFL2	-0.00987049	0.0018494	-5.34	0.0001
BPLNDSFL3	-0.011586	0.0011470	-10.10	0.0001
BTM	-0.016174	0.0017120	-9.45	0.0001
BTM2	0.000076812	0.00007959	0.97	0.3346
BTMQCP	0.00029689	0.0006377	0.47	0.6416
BTMQAN	0.00102642	0.0007006	1.47	0.1431
BTMS	0.00104424	0.0003111	3.36	0.0008
BTMPK	0.00035108	0.0001003	3.50	0.0005
BTMPLBR	-0.00350659	0.0001796	-19.52	0.0001
BTMPLND	-0.00066546	0.0001705	-3.90	0.0001

where,

QCP = Quantity of Crop Production
QAN = Quantity of Livestock Production
SFL1 = Risk_Adjusted (Fish) Leaching Indicator (1960-1971)
SFL2 = Risk_Adjusted (Fish) Leaching Indicator (1972-1977)
SFL3 = Risk_Adjusted (Fish) Leaching Indicator (1978-1996)
PK = Price of Capital Input
PLBR = Price of Labor Input
PLND = Price of Land Input
TM = Time Index

in U.S. agriculture. The estimate of the average annual rate of technical change is taken directly from the first-order coefficient on the time variable. That coefficient estimate, also significant at the .01 level, from Table 11.1 equals –0.016, suggesting that technical change reduced cost by an average 1.6% per year. The model appears to have generated credible results that are largely consistent with conventional wisdom about traditional sources of productivity growth in U.S. agriculture.

The parameter estimates of interest, of course, are those that enter expression (9) for marginal abatement cost. The relevant first-order coefficients (see table 11.1) are –0.019, –0.015, and –0.056 for the 1960-72, 1972-77, and 1977-96 sub-periods, respectively. The three estimates are statistically significant at the .01, .05, and .01 levels, respectively. Each reveals the representative farm's share of abatement cost in total cost

$$\frac{\partial \ln C}{\partial \ln S} = \frac{\partial C}{\partial S}\frac{S}{C} = \frac{\rho S}{C}. \qquad (10)$$

It is noteworthy that the cost share was approximately constant through 1977 and then rose significantly through 1996. Abatement of toxic chemicals leaching into ground water aquifers appears to have become an increasingly important (costly) task in U.S. agriculture.

Table 11.2 presents the complete decomposition of TRP growth for each state in each sub-period. The source decomposition follows that derived in equation (6)

$$\dot{TRP} = \dot{TFP} + \frac{\rho S}{\Sigma q_j Y_j}\left[\frac{d\ln Y_c}{dT} - \frac{d\ln S}{dT}\right] \tag{6}$$

TFP growth is calculated directly from ERS price and quantity data for inputs and marketable outputs. The differential growth rates of Y_C and S also are calculated from USDA data. Marginal abatement cost shares $\rho S/C$ ($C = \Sigma q_j Y_j$) are estimated from equation (9) for each state and year using USDA data and coefficients from the estimated cost function.

The most striking result from table 11.2 is the sign reversal in the growth of S/Y_C that takes place after 1977. Water pollution per unit of crop output increased in 43 states during 1960-72 and in 44 states in the 1972-77 sub-period. However, it decreased in 42 states after 1977. Not only is the sign reversal ubiquitous but it also is substantial. The average annual growth rates in S/Y_C across the United States, found as simple arithmetic averages over all states, are 30.0, 30.3, and –10.2 percents in the 1960-72, 1972-77, and 1977-96 sub-periods, respectively (see first row of table 11.2). Improvement in the quality of water leaving U.S. farms after 1977 was both widespread and substantial.

This unambiguous pattern has the expected effect on the measured contribution of risk management on TRP growth. Positive (negative) growth in S/Y_C [see equation (6)] necessarily makes a negative (positive) contribution to TRP growth. As expected, S/Y_C growth and risk management's TRP contribution have opposite signs at each entry in table 11.2. Following the growth pattern for S/Y_C, only four states register positive sources of TRP growth through improved water risk management in both 1960-72 and 1972-77 periods. This growth jumps to 42 states after 1977. The average economy-wide annual rates of growth in the source contributions of risk management equaled –0.61, –0.75, and 0.30 percents, respectively, in the 1960-72, 1972-77, and 1977-96 sub-periods.

The implication for the relationship between measured TFP and TRP growth rates is equally clear. In the first two sub-periods, measured TFP growth is an upward-biased estimate of TRP growth in 43 and 44 states, respectively. The simple average annual rates of TFP growth in the 1960-72 and 1972-77 periods are 2.35 and 2.18 percents, respectively. The corresponding TRP growth

Table 11.2. Source Decomposition of TRP Growth in U.S. Agriculture

	Average Annual Rates of Growth (%) 1960 – 1972		Sources of TRP Growth		1972 – 1977	
	S/Yc	TRP	TFP	Risk Mgt.	S/Yc	TRP
United States	29.98	1.74	2.35	-0.61	30.27	1.44
Alabama	8.60	1.88	2.17	-0.28	40.15	-0.54
Arkansas[a]	81.89	-0.69	1.04	-1.73	2.70	2.28
Arizona[a]	34.39	0.94	0.94	0.00	8.68	0.90
California	19.85	1.67	2.32	-0.65	4.22	2.19
Colorado	32.43	1.47	1.54	-0.07	27.92	1.27
Connecticut[b]	26.87	2.25	2.30	-0.05[c]	25.70	0.65
Delaware	14.27	3.79	4.14	-0.35[c]	38.32	-0.28
Florida	16.79	2.31	2.38	-0.07[c]	56.32	0.47
Georgia	5.32	3.14	3.38	-0.24	38.00	0.31
Idaho	11.81	2.83	2.83	0.00[c]	25.86	2.23
Illinois	31.00	1.23	2.24	-1.02	10.69	1.04
Indiana	2.36	2.40	2.40	0.00[c]	120.47	-0.26
Iowa	174.20	-9.59	1.62	-11.21	0.89	1.46
Kansas	81.95	1.32	2.24	-0.92	15.42	2.23
Kentucky	-1.48	2.81	2.78	0.02	55.52	-0.09
Louisiana[a]	77.56	0.79	3.07	-2.28	-6.50	3.41
Maine[b]	14.55	3.39	3.42	-0.03	31.37	0.52
Maryland	-1.75	2.82	2.81	0.01	53.37	0.05
Massachusetts	19.41	3.00	3.02	-0.02[c]	35.56	2.24
Michigan	31.04	3.57	3.67	-0.10	32.29	1.86
Minnesota[a]	75.48	0.17	1.98	-1.81	-4.00	2.05
Mississippi	33.45	3.68	3.86	-0.18[c]	34.67	1.93
Missouri	27.14	1.62	2.03	-0.41	37.02	0.88
Montana	19.98	2.16	2.18	-0.03[c]	22.94	2.52
North Carolina	29.37	3.07	3.07	0.00	67.47	0.71
Nebraska	81.63	1.32	2.21	-0.89	6.54	2.28
Nevada	0.00	1.35	1.35	0.00	74.53	2.00
New Hampshire	24.50	3.46	3.49	-0.03[c]	38.86	2.52
New Jersey	15.36	0.71	0.77	-0.06[c]	27.73	0.16
New Mexico	44.75	2.23	2.30	-0.07[c]	27.09	2.09
New York	24.90	1.38	1.48	-0.10	35.45	0.44
North Dakota	—	0.00	—	—	-12.37	5.27
Ohio	18.57	2.63	2.83	-0.20	45.07	0.48
Oklahoma	54.50	1.45	1.49	-0.04[c]	26.59	2.29
Oregon	26.56	3.59	3.59	-0.01[c]	23.91	3.49
Pennsylvania	25.39	2.50	2.55	-0.04	33.20	1.36
Rhode Island	16.14	2.76	2.80	-0.04[c]	24.34	2.37
South Carolina	27.61	2.32	2.51	-0.19[c]	45.13	-0.58
South Dakota	40.91	1.87	1.98	-0.11	-17.14	2.08
Tennessee	-6.05	2.25	2.12	0.13	52.18	-0.65
Texas	63.46	-0.18	1.64	-1.82	21.08	2.28
Utah	-31.69	1.72	1.72	0.00	23.98	2.38
Vermont	33.64	3.25	3.28	-0.03[c]	45.50	2.15
Virginia	22.26	2.32	2.37	-0.05[c]	61.39	-0.12
Washington	21.32	3.64	3.66	-0.02[c]	21.65	2.92
West Virginia	-12.18	2.17	2.15	0.02[c]	53.01	1.67
Wisconsin	31.47	1.35	1.68	-0.33	-15.61	1.07
Wyoming	45.70	1.37	1.41	-0.04[c]	36.02	1.09

a. Growth rates calculated over subset of the 1960-1972 interval because of missing values in initial years.
b. Growth rates calculated over subset of the 1977-1996 interval because of missing values in later years.

Table 11.2, *continued*

State	Average Annual Rates of Growth (%) 1972 – 1977: Sources of TRP Growth TFP	Risk Mgt.	1977 - 1996: S/Yc	TRP	Sources of TRP Growth TFP	Risk Mgt.
United States	2.18	-0.75	-10/18	2.10	1.80	0.30
Alabama	1.58	-2.12	-9.63	2.81	2.20	0.60
Arkansas[a]	2.39	-0.11	-6.81	1.95	1.61	0.34
Aarizona[a]	0.90	0.00[c]	-12.18	1.58	1.58	0.00
California	2.28	-0.09	0.92	2.11	2.15	-0.04
Colorado	1.53	-0.26	-13.07	1.84	1.41	0.43
Connecticut[b]	1.28	-0.62	-10.79	2.69	2.13	0.56
Delaware	3.62	-3.90	-22.16	4.38	3.02	1.36
Florida	2.15	-1.68	-6.04	2.31	1.91	0.40
Georgia	2.50	-2.19	-4.11	3.76	3.48	0.28
Idaho	2.23	0.00[c]	-2.31	1.79	1.79	0.00[c]
Illinois	1.45	-0.40	-6.43	0.31	-0.08	0.39
Indiana	1.99	-2.24	-32.89	1.73	0.35	1.38
Iowa	1.48	-0.02	-7.65	0.88	0.44	0.43
Kansas	2.41	-0.18	-11.72	2.16	1.68	0.48
Kentucky	1.77	-1.86	-20.72	3.44	2.38	1.06
Louisiana[a]	2.90	0.52	0.71	2.49	2.54	-0.05
Maine[b]	1.95	-1.43	-28.19	1.53	1.53	0.00
Maryland	2.39	-2.34	-8.15	3.11	2.53	0.58
Massachusetts	2.24	0.00[c]	-17.45	2.53	2.46	0.07[c]
Michigan	2.75	-0.89	-16.56	2.72	2.10	0.62
Minnesota[a]	1.96	0.09	-18.60	2.03	1.00	1.03
Mississippi	3.31	-1.38	-6.57	3.35	2.98	0.37
Missouri	1.69	-0.81	-12.04	0.93	0.43	0.51
Montana	2.62	-0.10	-13.86	0.91	0.84	0.07
North Carolina	3.30	-2.59	-5.26	3.21	2.88	0.32
Nebraska	2.32	-0.04	-14.17	1.44	0.88	0.56
Nevada	2.00	0.00[c]	-5.46	0.72	0.72	0.00
New Hampshire	2.52	0.00[c]	2.73	1.26	1.27	-0.02
New Jersey	0.95	-0.80	-19.65	2.26	2.26	0.00
New Mexico	2.30	-0.21	-6.03	1.24	1.07	0.17
New York	1.02	-0.58	-11.69	1.29	0.91	0.38
North Dakota	5.27	0.00[c]	4.97	0.37	0.38	-0.01[c]
Ohio	1.68	-1.20	-35.51	3.06	2.03	1.03
Oklahoma	2.45	-0.16	-0.90	3.23	3.21	0.02
Oregon	3.49	0.00[c]	0.89	3.74	3.74	0.00[c]
Pennsylvania	1.89	-0.54	-9.62	2.22	1.81	0.41
Rhode Island	2.59	-0.23	-16.76	0.96	0.96	0.00
South Carolina	2.48	-3.06	-7.49	4.47	3.86	0.61
South Dakota	1.90	0.18	-15.93	0.94	0.71	0.23
Tennessee	1.49	-2.14	-11.86	2.39	1.67	0.72
Texas	2.42	-0.14	-1.82	2.15	2.09	0.06
Utah	2.38	0.00[c]	-12.49	2.91	2.81	0.10
Vermont	2.56	-0.41	-7.60	2.38	2.11	0.27
Virginia	1.91	-2.03	-5.73	2.41	2.04	0.37
Washington	2.92	0.00[c]	3.82	2.78	2.89	-0.11
West Virginia	1.67	0.00[c]	-14.64	1.95	1.92	0.03[c]
Wisconsin	0.90	0.17	-4.68	0.59	0.42	0.17
Wyoming	1.09	0.00[c]	-7.28	1.31	1.29	0.02

c. State-specific fitted value of MAC is negative. The TRP contribution is calculated using marginal abatement costs derived from the first-order coefficient, a_s (see equation 9), for the relevant sub-period.

rates are only 1.75 and 1.44 percents. In contrast, measured TFP growth is a downward-biased measure of TRP growth in the 1977-96 sub-period. The simple 48-state average TFP growth rate of 1.80 percent is 0.30 percentage points lower than measured TRP growth.

The inference for productivity measurement follows directly. Not only are state and year-specific point estimates of TFP growth biased measures of TRP growth, but trends in measured TFP growth also are biased indicators of trends in TRP growth. TFP growth rates decrease from an all-state average 2.35 percent per year in the 1960-72 period to a 1.80 percent annual rate in the 1977-96 period. Annual rates of TRP growth, in contrast, climbed from 1.75 percent to 2.10 percent. Annual TFP rates declined by 0.55 percentage points while annual TRP growth rates increased by 0.35 percentage points. TFP measures suggest a decline in farm sector productivity growth while TRP measures reveal an increase.

The rise in TRP growth rates is clearly due to improved risk management. That, in turn, must be attributed to the influence of tightened water-related regulations and their enforcement. The widespread and substantial contributions of increasing water pollution, averaging 0.65 percent annually in the 1960 to 77 period, were transformed into a positive 0.30 percent annual contribution in the post-1977 period. Tighter water regulations were responsible for nearly a full percentage point swing in TRP growth between 1960 and 1996.

CONCLUSION

The farm sector has long been heralded as a productivity leader among U.S. industries. Nothing in this paper detracts from that honor. Indeed, when evaluated against a TRP backdrop, the farm sector's productivity performance is even more impressive than that suggested by traditional TFP accounting. Annual TFP growth rates actually declined by 0.50 percentage points over the 1960-96 period while TRP growth rates increased by 0.45 percentage points per year. The difference is wholly explained by improved risk management of chemical and water use on U.S. farms. Environmental regulation of agriculture's impact on local water sources added 0.95 percentage points to TRP growth from the pre- to post-1977 periods. Without that improvement in water quality, annual TRP growth in the 1977-96 period would have equaled 1.15 percent instead of its measured 2.10 percent. It would be interesting to determine whether other water-polluting industries have matched agriculture's record.

NOTES

1. This assumption is not required for the development of the model but follows from the environmental indicator variables developed jointly by USDA and EPA. The S variables applied later in this paper apply to pesticides and fertilizers used in crop production. Consequently, the regulatory constraints are defined on Y_C only.

2. The technology index T enters the cost function as c^T.

3. See Ball, Butault, and Nehring (2000); and Kellogg, Nehring, Grube, Goss, and Plotkin (2000).

REFERENCES

Ball, E., J.-P. Butault, and R. Nehring. 2000. "United States Agriculture, 1960-96: A Multilateral Comparison of Total Factor Productivity." Paper prepared for USDA Economic Research Service conference - Agricultural Productivity: Data, Methods, and Measures, March 9-10.

Gollop, F., and G. Swinand. 1998. "From Total Factor to Total Resource Productivity: An Application to Agriculture." *American Journal of Agricultural Economics*, 80(August): 577-83.

Gollop, F., and G. Swinand. 2000. "Total Resource Productivity: Accounting for Changing Environmental Quality." In *New Developments in Productivity Analysis*, ed. Edwin Dean, Michael Harper, and Charles Hulten. Conference on Research in Income and Wealth.

Kellogg, R., R. Nehring, A. Grube, D. Goss, and S. Plotkin. 2000. "Environmental Indicators of Nitrogen and Pesticide Leaching and Run-off from Farm Fields." Paper prepared for USDA Economic Research Service conference - Agricultural Productivity: Data, Methods, and Measures, March 9-10.

12

Costs of Production and Environmental Risk: Resource-Factor Substitution in U.S. Agriculture

V. Eldon Ball, Ronald G. Felthoven, Richard F. Nehring,
and Catherine J. Morrison Paul

INTRODUCTION

Agricultural production not only generates marketed "good" outputs (primary food commodities), but also non-marketed "bad" outputs (leaching and runoff of agricultural chemicals). In particular, the use of chemicals creates benefits for producers by augmenting the effective output produced from a given amount of other inputs (labor, capital, land, and materials), but these benefits are counteracted by the risk associated with environmental degradation. Decreasing the risk to human health from exposure to pesticides imposes costs on agricultural producers not balanced by direct benefits, since most of the social costs from agricultural chemical use are not translated into private costs borne by farmers.

Reductions in risk from leaching and runoff in the U.S. agricultural sector are thus associated with decreases in marketed outputs at existing input levels, or increases in input costs at prevailing output levels. In a dual production model, such cost effects can be expressed as shadow values associated with risk reduction. These costs arise from input- and output-substitution, potentially including induced innovation – embodied R&D – resulting in increased effectiveness of the chemicals used. The 1st-order shadow-value measures imply 2nd-order input and output substitution, and thus changes in "netput" (output and input) composition, or biases associated with risk reduction.

The shadow value (marginal abatement cost) measures, and their input- and output-specific components, can facilitate evaluating the social benefits required to justify environmental legislation limiting application of agricultural chemicals. Shadow value estimates may also be useful for determining "real" or net productivity growth in the agricultural sector, as risk to human health from leaching and runoff may be thought of as a negative output, or use of the environment for chemical disposal as a costly non-marketed input.

In this study we measure and evaluate such shadow values or costs of reducing the risk to human health from exposure to pesticides, and the changes in input and output characteristics, and intensity and composition, that accompany these reductions. We use a detailed model of U.S. agricultural production processes, founded on a cost-function relationship but incorporating profit-maximizing output choice, to represent both the technological base and the economic behavior underlying input- and output-substitution. We recognize health risks from leaching and runoff as bad outputs inherently associated with producing good outputs, and thus implying input costs for their reduction. And, we model the linkages between risk reductions and *effective* chemical (pesticide) use by incorporating pesticide quality changes from embodied innovation.

We apply this framework to the state-by-year (48 states for 1960-96) panel data set for U.S. agricultural production used by other authors in this volume and described in Ball et al. (1999). Price and quantity data are presented for capital, land, and labor inputs, and for three materials components, separately distinguishing pesticides and fertilizer from other materials inputs. Two "good" outputs (crops and livestock) are identified. Data series are included that represent "bad" outputs, or measures of health and environmental risk, based on the extent to which the concentration of a specific pesticide exceeds a water quality threshold. The two bad outputs are the risk from exposure to pesticide runoff (surface water contamination), and the risk from pesticides leaching into groundwater.[1]

The data were carefully constructed to accommodate quality changes, to the extent possible, to generate effective price and quantity measures. Perhaps the most important data refinement for our purposes was for the pesticide data, which involved a hedonic analysis of effective pesticide abatement prices, based on physical characteristics related to their actual or perceived quality (Nehring and Grube, 1999).[2] The pesticide data are thus quality-adjusted; they reflect embodied R&D or innovation that enhance abatement and reduce the human health risk associated with a given weight of applied pesticides.[3]

We find statistically significant and increasing shadow values of risk from leaching and runoff. The implied private costs from risk reduction are associated with increases in *effective* pesticide use, indicating greater embodied R&D in pesticide inputs. The costs of reducing risk also involve higher levels of most other inputs, except land, for a given level of agricultural output. The associated output composition adaptations are toward animal and away from crop production, but also imply somewhat greater scale and scope economies.

The Methodology: Model and Measures

Representing the structure of U.S. agricultural production, and its linkage to human health risk from leaching and runoff, required characterizing a broad range of output production and input use patterns. This characterization allowed us to capture the various netput-substitution relationships allowed by the existing technological base, and motivated by the economic climate facing agricultural producers.

A cost-based model is desirable for representing the technological and behavioral characteristics underlying cost-minimizing production behavior. From such a model, econometric estimates of the true economic values to agricultural producers of (both bad and good) outputs and inputs may be generated. The benefits and costs of risk "production" associated with agricultural commodity production can then be evaluated in terms of their implications for changing input and output composition.

The cost function we use for our analysis has been augmented for empirical implementation by price determination equations to represent profit-maximizing (good) output decisions, and by fixed effects to accommodate differences across specific states and time periods. This function takes the general form TC = TC(**Y**,**B**,**p**,**D**,t) where **Y** is a vector of outputs (crops, Y_C, and animal products, Y_A); **B** is a vector of "bad outputs" or proxies for human health risk from leaching and runoff (B_{HL}, B_{HR});[4] **p** is a vector of prices for inputs (land, LD, labor, L, capital, K, pesticides, P, fertilizers, F, and other materials, M); **D** is a vector of dummies corresponding to fixed effects for each state, and to specific temporal and spatial divisions; and t is a time trend.[5]

Including **B** in the cost function represents risk essentially as an "environmental factor", as suggested by McFadden (1978) for variables that affect costs but are not necessarily chosen by the economic agent. In our scenario, however, environmental risk is in a sense endogenous to the production process,[6] since bad outputs are produced jointly with **Y**. Production of **B** allows more **Y** production for a given input level, or less input costs for a given **Y**, than would be the case if the bad outputs were restricted. Jointness thus leads to complementarities or scope economies associated with producing both **Y** and **B**.

This treatment facilitates measuring and interpreting the marginal costs of producing good and bad outputs. The marginal cost to farmers of reducing B_k, or marginal abatement cost, can be expressed as a shadow value (SV) of the bad output, $SV_{Bk} = \partial TC/\partial B_k < 0$.[7] For example, the marginal cost of reducing leaching associated with risk to human health (B_{HL}) is $SV_{HL} = \partial TC/\partial B_{HL}$.[8] SV_{HL} thus reflects the farmer's implicit valuation of the ability to "produce" the marginal unit of B_{HL}, in terms of the incremental input costs that would be incurred if a decrease in B_{HL} were pursued.

In our cost function framework, by contrast to a purely technical representation such as a production or distance function, SV_{Bk} incorporates behavioral motivations underlying cost-efficient production choice, as well as technological substitution possibilities. Thus, SV_{Bk} allows us to represent cost-minimizing input and output composition changes associated with reductions in environmental damage, as elaborated below.

The cost derivatives representing these shadow values are similar to other more familiar 1st-order relationships representing input demands and the marginal costs of good outputs. That is, the derivatives of the cost function with respect to components of the **p** and **Y** vectors capture input use decisions through Shephard's Lemma, $v_j = \partial TC/\partial p_j$ (j=LD,L,K,P,F,M), and the marginal cost of Y_m production, $MC_m = \partial TC/\partial Y_m$, (m=A,C). And we can measure the cost impacts of changes in shift factors — the time trend and fixed effects — through the derivatives $\partial TC/\partial t$ and $\partial TC/\partial D_r$.

Such derivatives are often expressed in percentage terms, or as cost elasticities. This convention allows for a more direct comparison of the various cost impacts captured in the model, since the relative changes reflected in elasticity measures are more interpretable than absolute changes when the scale of the outputs and inputs compared to costs differs.[9] Specifically, the proportional cost impacts of changes in the bad outputs, t, and fixed effects (in **D**) can be expressed as $\varepsilon_{TC,Bk} = \partial \ln TC/\partial \ln B_k$, $\varepsilon_{TC,t} = \partial \ln TC/\partial t$, and $\varepsilon_{TC,Ds} = \partial \ln TC/\partial D_r$.[10] Similarly, the elasticities $\varepsilon_{TC,pj} = \partial \ln TC/\partial \ln p_j = v_j \bullet p_j/TC$ and $\varepsilon_{TC,Ym} = \partial \ln TC/\partial \ln Y_m = MC_m \bullet Y_m/TC$ represent the cost-share of each input or output. For the outputs, these elasticities become scale economy measures for each commodity produced, since they can be rewritten as MC_m/AC, as discussed further below.

These first-order cost relationships may in turn be decomposed into their input-specific effects. In particular, the impacts on individual inputs of reducing human health risk, such as the relationships between bad output production and pesticide use, may be established by directly identifying and measuring the associated elasticities. For example, $\varepsilon_{P,Bk} = \partial \ln P/\partial \ln B_k$, where $P = \partial TC/\partial p_P$, represents the pesticide demand response associated with a change in B_k. More generally, input demand; thus composition changes related to adaptations of human health risk can be examined via the elasticities $\varepsilon_{vj,Bk} = \partial \ln v_j/\partial \ln B_k$, where $v_j = \partial TC(\bullet)/\partial p_j$.

Although $SV_{Bk} < 0$ implies that decreases in risk are associated in general (and thus "on average" in some sense) with increases in input use and thus costs, if the underlying substitution patterns are biased, the demand responses $\varepsilon_{vj,Bk}$ will differ, implying that B_k reductions generate adaptations in input composition. Discrepancies may appear in terms of sign (absolute) as well as magnitude (relative), depending on the kinds of interdependencies exhibited among inputs and (good and bad) outputs.

In particular, although the cost elasticity $\varepsilon_{TC,Bk} = \partial \ln TC/\partial \ln B_k = SV_{Bk} \bullet B_k/TC$ should be negative ($SV_{Bk} < 0$), one might think P and B_k would be directly or positively related, so $\varepsilon_{P,Bk} = \partial \ln P/\partial \ln B_k > 0$. This relationship would imply an input bias in absolute terms if other inputs overall, and perhaps each individually, were instead negatively related to B_k. That is, if $SV_{Bk} < 0$ but $\varepsilon_{P,Bk} > 0$, other inputs must on balance increase even more than is implied by the overall cost elasticity to accommodate a B_k decline.

We find empirically, however, that this is not what is exhibited by the data when the pesticide inputs are measured in effective or quality-adjusted units. A primary input adaptation associated with risk (B_k) reduction is instead an *increase* in effective P use that exceeds those for the other inputs. This result, elaborated by Paul et al. (2001) in more detail in the context of discrepancies between the effects on the unadjusted pesticide quantities, and the effective or quality-adjusted pesticide inputs, implies that a desire to reduce B_k induces innovations that become embodied in the effective P measure.

The impacts of reductions in bad outputs on the marginal costs of the good outputs, and the resulting implications for output production and composition and associated cost (scale and scope) economies, may be characterized by $\varepsilon_{m,Bk} = \partial \ln MC_m/\partial \ln B_k$ elasticities, similar to the $\varepsilon_{vj,Bk}$. For example, the marginal cost of crop production might be higher at lower B_k levels, implying increased animal-output intensity. In addition, these elasiticities provide insights about the relationship between B_k and cost economies, which are reflected by the cost-output elasticity $\varepsilon_{TC,Y} = \Sigma_m \partial \ln TC/\partial \ln Y_m = \Sigma_m MC_m/AC_m = MC/AC$ in our multiple output scenario (where AC is average cost), as shown by Baumol, Panzar, and Willig (1982). Since $\varepsilon_{m,Bk}$ represents the response of MC_m to a B_k change, and $\varepsilon_{TC,Bk}$ captures the associated adaptation in TC (and thus AC), the relative changes provide insights on the linkage between human health risk and cost economies.

The $\varepsilon_{TC,Y}$ elasticity represents scope as well as scale economies with multiple outputs, and these two effects may be distinguished. In particular, complementarities among good and bad outputs, or scope economies, involve shifts in the cost curve from output composition changes, whereas scale economies are captured in the slope of the curve. Scope economies may thus be measured as $SC=((\Sigma_m TC(Y_m) - TC(Y))/TC(Y))$, where $TC(Y_m)$ is the minimum cost of producing Y_m, holding other outputs constant. Since the difference between TC for each output separately and combined simply depends on the interaction terms for the outputs, SC is dependent on the second-order derivatives $\partial^2 TC/\partial Y_m \partial Y_n$, or $\partial MC_m/\partial Y_n$. These derivatives imply economies (diseconomies) of scope if $\partial MC_m/\partial Y_n < (>)\ 0$. If the bad outputs are treated in a parallel fashion, relationships between the good and bad outputs are similarly captured by the $\partial MC_m/\partial B_k$ derivatives, or the corresponding elasticities $\varepsilon_{MCm,Bk} = \partial \ln MC_m/\partial \ln B_k$.[11]

The relationships between the B_k and the various cost characteristics captured by the first-order cost elasticities (such as input demands and output marginal costs) may alternatively be motivated in terms of exogenous impacts on the SV_{Bk}. From Young's theorem, the impact of a p_j (or Y_m) change on SV_{Bk} is symmetric to the effect of a B_k change on the demand for $v_j = \partial TC/\partial p_j$ (or $MC_m = \partial TC/\partial Y_m$). For example, $\partial SV_{Bk}/\partial p_P = \partial^2 TC/\partial B_k \partial p_P = \partial^2 TC/\partial p_P \partial B_k = \partial P/\partial B_k$. Thus, the dependence of the $SV_{Bk}(\bullet)$ on any argument of the cost function can be measured via a derivative or elasticity with respect to that argument, such as $\varepsilon_{SVk,pP} = \partial \ln SV_{Bk}/\partial \ln p_P$.

Although this perspective does not give rise to conceptually attractive indicators for the inputs (the $\varepsilon_{SVk,pj}$ elasticities are not as readily interpretable as their $\varepsilon_{j,Bk}$ counterparts), more interpretable relationships emerge for the components of the **Y** and **D** vectors, and the time trend t. The output elasticities, $\varepsilon_{SVk,Ym} = \partial \ln SV_{Bk}/\partial \ln Y_m$, for example, indicate how the cost of risk reduction depends on output levels or composition. That this measure has a somewhat more intuitive implication than its converse, $\varepsilon_{MCm,Bk}$; $\varepsilon_{SVk,Ym}>0$, would suggest that at higher Y_m levels the input cost to farmers of reducing risk is heightened, or that output increases enhance the value of using the environment for disposal of leaching and runoff. Differences in B_k shadow values associated with t (time) or **D** (space) can similarly be represented by the elasticities $\varepsilon_{SVk,t} = \partial \ln SV_{Bk}/\partial t$, and $\varepsilon_{SVk,Ds} = \partial \ln SV_{Bk}/\partial D_s$.

One final point to be addressed before considering the empirical implementation of this model is the potential link between these indicators and appropriate measurement of productivity growth.The estimation of risk shadow values is motivated by the idea that using the environment for disposal of bad outputs has a value to agricultural producers in terms of lower input costs per unit of output. That is, since bad outputs are jointly produced with good outputs, the value of higher B_k levels stems from the resulting greater Y_m output levels produced from a given amount of purchased inputs. Or, from the dual perspective, the environment is used for disposal of leaching and runoff as an input that generates a marginal benefit but has no direct private cost.

Thus, depending on the perspective, measured output may be too high (negative outputs should be subtracted from positive outputs to generate a "net" measure), or measured inputs too low (the contribution of the environment should be added to internalize the externality) in productivity computations that ignore B_k. Since the B_k are not marketed, their weights in such productivity computations would depend on the shadow values SV_{Bk}, which represent private benefits to the farming sector. Note, however, that since productivity measures are typically expressed in terms of growth rates, an understatement of productivity *growth* only occurs if B_k are rising. If B_k are falling, productivity growth measures are instead understated by ignoring risk.

THE RESULTS: ECONOMETRIC IMPLEMENTATION AND ESTIMATES

To econometrically implement our model and generate estimates of the cost-impacts of the risk factors, B_k, we first need to make an assumption about the functional form for $TC(Y_A,Y_C,B_{HL},B_{HR},p_P,p_K,p_L,p_{LD},p_M,p_F,t,D_P,D_F,D_{CT},D_{CN},\mathbf{D_s})$, where, as discussed in the previous section, Y_A and Y_C represent animal and crop outputs; B_{HL} and B_{HR} human risk from leaching and runoff; p_P, p_K, p_L, p_{LD}, p_M, and p_F effective prices of labor, land, materials, and fertilizer; and t time. The fixed effects include dummy variables for structural shifts in pesticides and fertilizer use (D_P, D_F),[12] for the cotton and corn states (D_{CT}, D_{CN}), and for each of the 48 states ($\mathbf{D_s}$, with cross effects for each input price and output quantity, so each estimating equation has state-specific intercepts).

The cost function is assumed to be approximated by a generalized Leontief functional form, with level and shift factors included in quadratic form, as in Paul (1999):

$$
\begin{aligned}
TC(\mathbf{Y},\mathbf{B},\mathbf{p},\mathbf{D},t) = {} & \delta_{P1}\, p_P D_P + \delta_{F1}\, p_F D_F + \Sigma_s\Sigma_i\, \delta_{sj}\, p_j D_s + \Sigma_j p_j \bullet (\Sigma_s\Sigma_m\, \delta_{sm}\, Y_m D_s) \\
& + \Sigma_j\Sigma_{i\,(i\neq j)}\, \alpha_{ji}\, p_j^{.5}\, p_i^{.5} + \Sigma_j\, \alpha_{jDP}\, p_j^{.5}\, p_P^{.5}\, D_P + \Sigma_j\, \alpha_{jDF}\, p_j^{.5}\, p_F^{.5}\, D_F \\
& + \Sigma_j\Sigma_{m\,(j\neq M)}\, \delta_{jm}\, p_j\, Y_m + \Sigma_m\Sigma_r\, \delta_{mPDr}\, Y_m\, p_p D_r + \Sigma_m\Sigma_r\, \delta_{mFDr}\, Y_m\, p_F\, D_r \\
& + \Sigma_j\Sigma_k\, \delta_{jk}\, p_j\, B_k + \Sigma_k\, \delta_{kDP}\, B_k\, p_p D_P + \Sigma_k\, \delta_{kDF}\, B_k\, p_F\, D_F \\
& + \Sigma_j\, \delta_{jt}\, p_j\, t + \Sigma_r\, \delta_{tPDr}\, t\, p_p D_r + \Sigma_r\, \delta_{tFDr}\, t\, p_F\, D_r \\
& + \Sigma_j p_j\, (\Sigma_m\Sigma_n\, \gamma_{mn}\, Y_m\, Y_n + \Sigma_m\Sigma_k\, \gamma_{mk}\, Y_m\, B_k + \Sigma_k\Sigma_l\, \gamma_{lk}\, B_k\, B_l \\
& + \gamma_{tt}\, t^2 + \Sigma_m\, \gamma_{mt}\, Y_m\, t + \Sigma_k\, \gamma_{kt}\, B_k\, t), \qquad (1)
\end{aligned}
$$

where (i,j) denote the inputs, (m,n) the good outputs, (k,l) the bad outputs, s the state, and r the other fixed effects.[13] The estimating model derived from this function comprises a system of factor demand equations, output pricing equations, and the cost function itself.

The factor demand estimating equations are defined via Shephard's lemma; $P=\partial TC/\partial p_P$, $F=\partial TC/\partial p_F$, $K=\partial TC/p_K$, $L=\partial TC/\partial p_L$, $LD=\partial TC/\partial p_{LD}$, and $M=\partial TC/\partial p_M$. Since these demand equations are based on the *effective* price data used as arguments of the cost function, fixity or other effects that might violate Shephard's lemma and invalidate these input demand relationships are accommodated, making these expressions appropriate for estimation purposes.[14]

The output pricing equations are defined according to $p_m=MC_m$ equalities representing optimization over outputs; $p_A = \partial TC/\partial Y_A$ and $p_C = \partial TC/\partial Y_C$, where p_m is the market price of Y_m. These equations were included in the estimating system because scale economy measures were implausible when no structure for output choice was incorporated in the model. A profit function framework was also tried, where the output equations were defined via Hotelling's lemma, but the cost-based framework was empirically more justifiable (in terms of levels and signs of implied input and output optimization relationships). The possibility of imperfectly competitive markets was also recognized in an alternative specification based on $MR_m=MC_m$ equations (where marginal revenue, MR_m, is defined as $p_m+\partial p_m/\partial Y_m \bullet Y_m$), but the estimated deviations between p_m and MR_m were invariably statistically insignificant.

The final equation system, composed of one cost, six input demands, and two output pricing expressions, was estimated by seemingly unrelated (SUR) systems-estimation procedures. We did not invoke instrumental variables for our analysis, which are sometimes used to take into account potential output endogeneity or errors in variables for the input prices. Because of the care taken in the data development, the previously mentioned conclusions regarding endogeneity, and support from results generated by preliminary empirical investigation, the SUR specification seemed empirically justified.

Potential heteroskedasticity was accommodated in the estimation process by using a procedure in TSP that computes White's heteroskedastic-consistent covariance matrix to generate appropriate standard errors. And in response to Durbin-Watson tests that indicated first-order autocorrelated errors were present in the cost and input demand equations, we incorporated an AR(1) error structure directly into the model by appending an AR(1) term into the cost equation, and estimating $TC = TC(\bullet) + \rho_{TC} \cdot \varepsilon_{t-1} + \varepsilon_t$ (where ρ_{TC} is the cost function-specific AR(1) parameter, and ε_t is the period t estimation error for $TC(\bullet)$). The input equations were adapted analogously. Standard statistical tests indicated that the adjustment satisfactorily adapted for autocorrelation in the estimates.

The parameter estimates for this model are presented in table 12.1 (without the state dummies to keep the table manageable). The overall statistical significance of the parameters is very high even with so many estimated coefficients (the state dummies also were almost invariably significant). Also, the R^2's indicate excellent "fits" for the estimated equations, with all exceeding 0.92.

Little intuition can be attributed to most individual parameter estimates (since the derivatives and elasticities of interest are combinations of parameters), so we can move directly to our primary focus — the derivatives and elasticities indicating the production impacts of changes in the B_k. We present the B_k shadow values and their input- and output-specific components in Table 12.2. The reported estimates are (non-weighted) averages of each measure across all states and time periods, and the t-statistics are based on evaluating the measures at the average (mean) values of the data.[16] The SV_{HL} and SV_{HR} measures are both negative (reducing human risk is costly for agricultural producers), and statistically significant at the 5% level for the sample as a whole.

These shadow values are also presented in elasticity form to facilitate comparison with their input- and output-specific counterparts. The $\varepsilon_{TC,HL}$ and $\varepsilon_{TC,HR}$

Table 12.1. Coefficient Estimates

	estimate	*t-stat.*		*estimate*	*t-stat.*		*estimate*	*t-stat.*
δ_{F1}	-0.139	*-5.74*	α_{PFCN}	0.036	*3.55*	δ_{HLDP}	0.0005	*0.77*
δ_{P1}	-0.190	*-6.32*	α_{FDP}	-0.002	*-0.83*	δ_{LDHR}	0.00004	*1.68*
α_{LDL}	-0.013	*-2.09*	δ_{LDA}	-0.971	*-76.43*	δ_{LHR}	0.0001	*3.08*
α_{LDK}	0.082	*8.12*	δ_{LA}	-0.921	*-54.45*	δ_{KHR}	-0.0001	*-5.42*
α_{LDF}	-0.004	*-0.91*	δ_{KA}	-0.900	*-66.55*	δ_{FHR}	-0.0001	*-2.88*
α_{LDDF}	0.012	*3.61*	δ_{FA}	-0.905	*-58.76*	δ_{HRDF}	-0.00001	*-0.89*
α_{LDFCT}	-0.025	*-3.48*	δ_{ADF}	0.015	*4.90*	δ_{MHR}	-0.0006	*-8.24*
α_{LDFCN}	-0.007	*-0.97*	δ_{PA}	-0.931	*-65.03*	δ_{PHR}	-0.0001	*-3.61*
α_{LDM}	-0.034	*-3.74*	δ_{ADP}	-0.001	*-0.56*	δ_{HLDP}	-0.00003	*-2.18*
α_{LDP}	0.006	*1.19*	δ_{APCT}	0.026	*2.43*	γ_{AA}	-0.0003	*-0.43*
α_{LDDP}	-0.012	*-3.21*	δ_{APCN}	-0.014	*-1.29*	γ_{CC}	-0.0011	*-3.13*
α_{LDPCT}	0.001	*0.09*	δ_{AFCT}	-0.013	*-1.07*	γ_{HLHL}	0.00003	*1.86*
α_{LDPCN}	0.013	*1.50*	δ_{AFCN}	-0.029	*-2.44*	γ_{HRHR}	0.00000001	*1.41*
α_{LK}	0.012	*1.65*	δ_{LDC}	-0.623	*-87.22*	γ_{At}	-0.0018	*-26.53*
α_{LF}	-0.022	*-1.93*	δ_{LC}	-0.584	*-55.48*	γ_{Ct}	-0.0018	*-28.78*
α_{LDF}	0.013	*1.30*	δ_{KC}	-0.576	*-88.39*	γ_{HLt}	-0.0001	*-1.54*
α_{LFCT}	0.026	*1.47*	δ_{FC}	-0.571	*-69.82*	γ_{HRt}	0.000001	*1.27*
α_{LFCN}	0.075	*4.46*	δ_{CDF}	0.004	*2.00*	γ_{AHL}	0.0003	*2.20*
α_{LM}	0.234	*8.70*	δ_{PC}	-0.613	*-84.63*	γ_{CHL}	0.0005	*3.62*
α_{LP}	0.016	*2.30*	δ_{CDP}	-0.001	*-0.49*	γ_{HLHR}	0.000005	*2.20*
α_{LDP}	-0.001	*-0.15*	δ_{CPCT}	0.021	*4.48*	γ_{AHR}	0.00001	*2.43*
α_{LPCT}	0.014	*1.21*	δ_{CPCN}	0.009	*2.43*	γ_{CHR}	-0.00001	*-2.85*
α_{LPCN}	-0.0004	*-0.04*	δ_{CFCT}	-0.011	*-1.81*	γ_{AC}	-0.0016	*-2.60*
α_{KF}	0.019	*3.40*	δ_{CFCN}	-0.006	*-1.23*	ρ	0.835	*85.77*
α_{KDF}	0.034	*10.25*	δ_{LDt}	0.004	*5.67*	ρ_{L}	0.607	*54.85*
α_{KFCT}	0.024	*2.08*	δ_{Lt}	-0.009	*-10.71*	ρ_{F}	0.786	*59.89*
α_{KFCN}	0.016	*1.30*	δ_{Kt}	-0.013	*-9.63*	ρ_{M}	0.883	*114.40*
α_{KM}	-0.066	*-5.01*	δ_{Ft}	-0.003	*-4.17*	ρ_{P}	0.967	*294.84*
α_{KP}	0.031	*3.90*	δ_{tDF}	0.005	*4.62*	ρ_{LD}	0.896	*108.39*
α_{KDP}	0.018	*4.09*	δ_{Mt}	-0.023	*-8.54*	ρ_{K}	0.954	*284.05*
α_{KPCT}	0.028	*1.68*	δ_{Pt}	0.003	*2.29*			
α_{KPCN}	-0.004	*-0.20*	δ_{tDP}	0.005	*4.22*			
α_{FM}	0.047	*4.07*	δ_{tPCT}	0.010	*4.84*			
α_{MDF}	0.0004	*0.03*	δ_{tPCN}	0.022	*10.52*			
α_{MFCT}	0.161	*6.27*	δ_{tFCT}	0.004	*4.75*			
α_{MFCN}	0.169	*6.16*	δ_{tFCN}	0.007	*8.89*			
α_{MP}	-0.019	*-2.16*	δ_{LDHL}	0.002	*1.18*			
α_{MDP}	0.043	*4.15*	δ_{LHL}	-0.009	*-3.52*			
α_{MPCT}	0.088	*4.46*	δ_{KHL}	-0.002	*-1.14*			
α_{MPCN}	0.042	*1.87*	δ_{FHL}	0.0001	*0.06*			
α_{FP}	0.008	*2.25*	δ_{HLDF}	-0.001	*-1.60*			
α_{PDF}	0.013	*3.38*	δ_{MHL}	-0.012	*-2.56*			
α_{PFCT}	0.007	*0.79*	δ_{PHL}	-0.002	*-1.85*			

Equation	R-squared
TC	**0.989**
L	**0.974**
F	**0.932**
M	**0.970**
P	**0.966**
LD	**0.999**
K	**0.996**
MC_A	**0.942**
MC_C	**0.920**

Table 12.2. Shadow Value and Elasticity Measures (overall averages)

measure	*estimate*	*t-stat.*	*measure*	*estimate*	*t-stat.*	*measure*	*estimate*	*t-stat.*
SV_{HL}	-0.0164	*-2.118*	$\varepsilon_{P,HL}$	-0.0243	*-1.887*	$\varepsilon_{MCA,HL}$	0.0039	*1.731*
SV_{HR}	-0.0004	*-1.951*	$\varepsilon_{P,HR}$	-0.0644	*-2.472*	$\varepsilon_{MCA,HR}$	0.0053	*1.884*
$\varepsilon_{TC,HL}$	-0.0090	*-2.118*	$\varepsilon_{F,HL}$	0.0035	*0.175*	$\varepsilon_{MCC,HL}$	0.0055	*1.295*
$\varepsilon_{TC,HR}$	-0.0077	*-1.951*	$\varepsilon_{F,HR}$	-0.0151	*-1.273*	$\varepsilon_{MCC,HR}$	-0.0001	*-1.943*
			$\varepsilon_{LD,HL}$	0.0086	*2.157*			
			$\varepsilon_{LD,HR}$	0.0107	*2.348*	$\varepsilon_{SVHL,A}$	-0.1905	*-1.732*
			$\varepsilon_{L,HL}$	-0.0205	*-1.867*	$\varepsilon_{SVHL,C}$	-0.1208	*-1.295*
			$\varepsilon_{L,HR}$	0.0144	*1.961*	$\varepsilon_{SVHR,A}$	-0.1557	*-1.884*
			$\varepsilon_{K,HL}$	-0.0017	*-1.063*	$\varepsilon_{SVHR,C}$	0.1807	*1.943*
			$\varepsilon_{K,HR}$	-0.0074	*-2.563*	$\varepsilon_{SVHR,t}$	0.0175	*1.058*
			$\varepsilon_{M,HL}$	-0.0141	*-1.850*	$\varepsilon_{SVHR,t}$	-0.0104	*-0.556*
			$\varepsilon_{M,HR}$	-0.0255	*-2.875*			

measures indicate that a 1 percent reduction in B_{HL} requires an approximately 0.009 percent increase in production costs. Similarly, about an 0.008 percent cost increase is associated with a fall in B_{HR} of 1 percent. These measures are much closer in magnitude than are the SV_{Bk} measures, due to differences in leaching and runoff risk levels unaccounted for in the (nominal) derivatives, but accommodated in the (proportional) elasticities.

The input composition implications underlying these marginal costs of reducing risk — or marginal abatement costs — in the agricultural sector are biased in absolute as well as relative terms. In particular, risk reduction is clearly pesticide-using in the sense that lowering risk requires augmenting *effective* pesticide demand, implying the development of more and better pesticides.

That is, changes in human health risk are negatively related to effective pesticide use ($\varepsilon_{P,HL}<0$ and $\varepsilon_{P,HR}<0$), and are significant both statistically and in terms of magnitude (especially for runoff). Thus, even if B_k reductions are associated with declines in the *pounds* of applied pesticides, increased embodied innovation in the quality-adjusted or effective pesticide input must increase substantively to allow productivity to be maintained concurrently with lower risk levels.

By contrast, the only input insignificantly related to both risk factors is the other agricultural chemical input, fertilizer. The negative but insignificant $\varepsilon_{K,HL}$ suggests a similar relationship for capital, although the significantly negative $\varepsilon_{K,HR}$ measure indicates that capital has a tendency to substitute for the environment, in the sense that additional capital is required to reduce human risk from runoff. The M elasticities with respect to B_{HL} and B_{HR} are also negative, large, and significant; lower risk levels for given amounts of good outputs are associated with higher levels of materials inputs.

Comparing the magnitudes of these elasticities with the overall cost measures — the bad-output cost elasticities $\varepsilon_{TC,HL}$ and $\varepsilon_{TC,HR}$ — provides indications of relative input changes or biases, and thus implies input composition changes associated with adaptations in human health risk levels. The increase in effective P demand corresponding to a one percent B_{HL} reduction is the greatest for all inputs, exceeding 0.024 percent, whereas labor requirements rise by just over 0.02 percent, M demand by 0.015 percent, and K use by 0.003 percent. L and M demands climb relative to other inputs, and capital changes on average less than does overall input use.

These results contrast with the relationship of both types of risk to land (LD), and of B_{HR} to labor (L). Risk reductions seem biased toward land saving in absolute terms. The relationship between B_{HR} and L demand is also positive, suggesting that reductions in human risk from runoff are associated with stagnating agricultural labor demand. These values are all statistically significant at the 5% level, although the implied LD changes are quite small in proportionate terms. A one percent decline in B_{HL}, for example, suggests a 0.006 percent drop in land use.

For the outputs, the $\varepsilon_{MCm,Bk}$ elasticities are small and largely positive, a result that suggests that risk reduction is consistent with augmented scale economies, since average cost is instead higher at lower B_{HL}, B_{HR} levels (so MC/AC falls if a decrease in risk is associated with a decline in MC_m). This result also implies that scope economies between the risk factors and outputs are limited (marginal costs do not change substantively with risk reduction, even though average costs are significantly affected). This conclusion makes intuitive sense especially for the animal inputs, which might not be thought to have much connection with leaching and runoff from agricultural chemical use.[17]

Some interesting implications may also be generated from the elasticities of SV_{Bk} with respect to the outputs and the time trend – the $\varepsilon_{SVBk,m}$ and $\varepsilon_{SVBk,t}$ elasticities in table 12.2. These measures indicate that SV_{HL} and SV_{HR} are lower (in absolute value, so reductions in risk are less costly) when output becomes more Y_A-intensive; leaching and runoff abatement costs are inversely related to animal production. SV_{HR} is instead larger in absolute value when crop production is higher; abatement costs are positively related to crop production. And over time (with changes in t), SV_{HL} appears to be rising (in absolute value, so the costs of reducing B_{HL} are greater over time), whereas SV_{HR} is falling.[18] But most of these measures, except $\varepsilon_{SVHR,C}$, are statistically insignificant.[19]

Further insights emerge from comparison of the shadow value and input- and output-specific risk cost estimates across time periods and regions. To more explicitly evaluate the time dimension, we have separated the measures into the four decades covered in the analysis — the 1960s, 1970s, 1980s, and 1990s. The spatial or regional division is into 10 regions (for the 48 contiguous states) according to the breakdown used by the USDA to report regional productivity data: the Northeast, the Corn Belt, the Lake states, the Northern Plains, the Appalachian region, the Southeast, the Delta states, the Southern Plains, the Mountain states, and the Pacific states. [20]

First, the temporal division for the Table 12.3. Bad Output Shadow Value and Elasticity Measures (temporal) $\varepsilon_{TC,HL}$ and $\varepsilon_{TC,HR}$ values reported in table 12.3 indicates an upward trend in the agricultural-sector cost-saving benefits (in proportional terms) of allowing B_{HL}, but the reverse for B_{HR} , as indicated by the elasticities. This result is also consistent with the implications for the associated pesticide costs; the contributions of more and better pesticides (increasing effective P) to reductions in B_{HL} ($\varepsilon_{P,HL}$) are increasing over time, although those associated with B_{HR} ($\varepsilon_{P,HR}$) are declining. And virtually all estimates of input-demand responses are greater in absolute value in the later periods (although $\varepsilon_{L,HR}$ changes little), suggesting greater substitutability and thus flexibility of production processes over time. For the outputs, the same relationships are apparent again as for the overall averages, but all impacts except that of B_{HR} on MC_A (represented by $\varepsilon_{MCA,HR}$) are stronger in the later time periods.

For the regional division, in table 12.4 we find very different trends for the Corn Belt states compared to others. Leaching and effective pesticide use are clearly positively correlated for these states ($\varepsilon_{P,HL}$>0, as they are for the Northern Plains and Pacific states as groups), with $\varepsilon_{P,HL}$ larger than for other regions, while the associated (proportional) marginal cost of reducing B_{HL} is somewhat lower.

The states that exhibit the highest proportional cost of reducing risk from leaching are those in the Southeast, followed by those in the Appalachian region (although the nominal values are greater in the Northeast and Mountain areas, with the lowest values, except for the Corn Belt, in the Lake states). The implied proportional changes in pesticide use follow closely, with the Appalachian and Southeast states requiring more augmentation of effective pesticide use than the Northeast to reduce leaching risk, and the Mountain states requiring significantly less.

Table 12.3. Bad Output Shadow Value and Elasticity Measures (temporal)

	overall	*sixties*	*seventies*	*eighties*	*nineties*
$\varepsilon_{TC,HL}$	-0.0090	-0.0033	-0.0099	-0.0092	-0.0148
$\varepsilon_{TC,HR}$	-0.0077	-0.0131	0.0084	-0.0047	-0.0039
$\varepsilon_{P,HL}$	-0.0243	-0.0073	-0.0294	-0.0299	-0.0312
$\varepsilon_{P,HR}$	-0.0644	-0.1708	-0.0434	-0.0222	-0.0180
$\varepsilon_{F,HL}$	0.0035	0.0090	0.0185	-0.0064	-0.0111
$\varepsilon_{F,HR}$	-0.0151	-0.0245	-0.0130	-0.0130	-0.0089
$\varepsilon_{LD,HL}$	0.0086	0.0037	0.0110	0.0088	0.0110
$\varepsilon_{LD,HR}$	0.0107	0.0103	0.0120	0.0105	0.0096
$\varepsilon_{L,HL}$	-0.0205	-0.0051	-0.0200	-0.0238	-0.0361
$\varepsilon_{L,HR}$	0.0144	0.0134	0.0160	0.0131	0.0152
$\varepsilon_{K,HL}$	-0.0017	-0.0001	-0.0011	-0.0025	-0.0038
$\varepsilon_{K,HR}$	-0.0074	-0.0121	-0.0074	-0.0047	-0.0050
$\varepsilon_{M,HL}$	-0.0141	-0.0063	-0.0179	-0.0151	-0.0173
$\varepsilon_{M,HR}$	-0.0255	-0.0363	-0.0286	-0.0189	-0.0164
$\varepsilon_{MCA,HL}$	0.0039	0.0012	0.0039	0.0048	0.0061
$\varepsilon_{MCA,HR}$	0.0053	0.0053	0.0056	0.0055	0.0047
$\varepsilon_{MCC,HL}$	0.0055	0.0018	0.0056	0.0066	0.0087
$\varepsilon_{MCC,HR}$	-0.0001	0.0000	-0.0001	-0.0001	-0.0002

By contrast, the cost of reducing risk from runoff overall, as well as from pesticide runoff, is by far the highest in the Corn Belt. The states with the next highest input costs of reducing runoff are in the Lake, Southeast, and Delta regions; although for pesticide use the Northern Plains are next in line, with less than one third the required adjustment of the Corn Belt states. In the Pacific region virtually no change in effective pesticide use seems to be associated with reductions in runoff.

Table 12.4. Bad Output Shadow Value and Elasticity Measures (spatial)

	Northeast	Corn Belt	Lake states	North Plains	Appalachian	Southeast	Delta	South Plains	Mountain	Pacific
$\varepsilon_{TC,HL}$	-0.0088	-0.0007	-0.0077	-0.0031	-0.0177	-0.0399	-0.0079	-0.0020	-0.0022	-0.0015
$\varepsilon_{TC,HR}$	-0.0031	-0.0264	-0.0134	-0.0068	-0.0080	-0.0116	-0.0111	-0.0031	-0.0009	-0.0004
$\varepsilon_{P,HL}$	-0.0534	0.0406	-0.0035	0.0130	-0.0691	-0.0827	-0.0153	-0.0001	-0.0173	0.0119
$\varepsilon_{P,HR}$	-0.0200	-0.3059	-0.0817	-0.1004	-0.0595	-0.0461	-0.0334	-0.0204	-0.0085	-0.0020
$\varepsilon_{F,HL}$	-0.0131	0.0252	0.0316	0.0211	-0.0090	0.0104	-0.0056	0.0142	-0.0076	0.0193
$\varepsilon_{F,HR}$	-0.0075	-0.0419	-0.0226	-0.0312	-0.0117	-0.0086	-0.0287	-0.0062	-0.0055	-0.0013
$\varepsilon_{LD,HL}$	0.0013	0.0150	0.0261	0.0083	0.0092	0.0346	0.0034	0.0027	0.0003	0.0026
$\varepsilon_{LD,HR}$	0.0033	0.0514	0.0356	0.0077	0.0039	0.0073	0.0089	0.0015	0.0004	0.0001
$\varepsilon_{L,HL}$	-0.0147	-0.0116	-0.0176	-0.0151	-0.0294	-0.0941	-0.0187	-0.0075	-0.0050	-0.0053
$\varepsilon_{L,HR}$	0.0030	0.0654	0.0270	0.0188	0.0067	0.0128	0.0210	0.0061	0.0017	0.0003
$\varepsilon_{K,HL}$	-0.0045	0.0027	0.0004	0.0023	-0.0043	-0.0072	-0.0034	0.0011	-0.0012	0.0031
$\varepsilon_{K,HR}$	-0.0023	-0.0264	-0.0120	-0.0077	-0.0065	-0.0107	-0.0124	-0.0032	-0.0011	-0.0005
$\varepsilon_{M,HL}$	-0.0102	-0.0091	-0.0176	-0.0057	-0.0267	-0.0595	-0.0098	-0.0046	-0.0023	-0.0047
$\varepsilon_{M,HR}$	-0.0081	-0.1030	-0.0503	-0.0194	-0.0252	-0.0338	-0.0300	-0.0088	-0.0022	-0.0009
$\varepsilon_{MCA,HL}$	0.0005	0.0064	0.0080	0.0055	0.0054	0.0123	0.0020	0.0040	0.0004	0.0040
$\varepsilon_{MCA,HR}$	0.0004	0.0322	0.0135	0.0045	0.0017	0.0017	0.0029	0.0025	0.0002	0.0002
$\varepsilon_{MCC,HL}$	0.0007	0.0082	0.0106	0.0076	0.0074	0.0195	0.0027	0.0059	0.0005	0.0054
$\varepsilon_{MCC,HR}$	0.0000	-0.0002	-0.0002	-0.0002	-0.0002	-0.0004	-0.0001	-0.0001	0.0000	-0.0001

Concluding Remarks

This study uses a detailed structural model to investigate the costs to U.S. agricultural producers of reducing human health risks associated with leaching and runoff, and the netput substitution underlying these costs. Our cost-based measures represent both technological possibilities and behavioral responses. They indicate significant costs of risk reduction, which involve higher levels of embodied innovation in the pesticide inputs, but also imply increased use of other inputs, except land and possibly labor, and greater animal-output intensity.

The findings presented here are only indicative of the many results that might be generated from this type of framework to facilitate interpretation and analysis of U.S. agricultural production structure. In particular, many issues have been raised about compositional, temporal, and spatial patterns of agricultural production and productivity, and their association with non-marketed production aspects, including innovation/quality and environmental issues. Further exploration of these patterns will provide useful insights about the causes and effects of agricultural sector performance, and its contribution to overall economic performance in the U.S. food sector.

Notes

1. These indices are based on ratios of simulated concentration in runoff or leaching water divided by a threshold concentration, generated for each year and state from site-level data, based on the amount of applied pesticides, and conditioned by soil, weather, crop and other variables.

2. These characteristics included pesticide application rates as a proxy for pesticide potency, LD50 scores as a proxy for toxicity, and persistence dummies are based on solubility, vapor, sorption, and half lives.

3. This transformation of the data into true effective or quality-adjusted pesticide prices and implicit quantities accommodates the problem, emphasized by studies such as Lichtenberg and Zilberman (1986) and Chambers and Lichtenberg (1994), that standard pesticide quantity measures based on pounds of applied chemicals do not reflect the input actually demanded — abatement.

4. Fish stock risk factors from leaching and runoff were also included in these data, and used in preliminary investigation, but were dropped from the estimation because when included along with human risk measures they were invariably insignificant. On their own, however, they behave similarly to the human risk measures, suggesting that the independent impacts on the human and fish risk factors are not well identified. As noted by a referee, such a result might be expected since the effects are derived from the same basic data, with the only difference being the threshold concentration.

5. These dummy variables were determined *a priori* to be likely representative of structural changes and differences, and were retained since they were almost invariably significant in preliminary investigation, although they did not have a substantive impact on the elasticity measures.

6. It may not be completely endogenous, however, due to spillovers across boundaries.

7. Note that such a shadow value should be interpreted in the context of a *private* value to the agricultural sector, since it represents the amount private expenditure on other inputs would have to increase (given output levels) if the environment could not be used freely.

8. This cost can also be thought of as the marginal cost savings from using the environment to dispose of such wastes.

9. That is, cost and revenue shares of the inputs and outputs vary substantially.

10. Note that, as is conventional for the cost-based productivity measure $\varepsilon_{TC,t}$, these "elasticities" are not expressed in terms of the logarithms of t or D_s since these are just trend (time counter) or qualitative (0-1) variables.

11. This use of the term *scope economies* is only strictly valid if the B_k are incorporated as outputs in the definition of ε_{TCY}. Note also that the interpretation of the B_k as bad outputs means $SV_{Bk} = \partial TC/\partial B_k<0$, by contrast to $MC_m = \partial TC/\partial Y_m>0$ for the Y_m. But the presence of scope economies implies that the cross terms $\partial MC_m/\partial B_k = \partial 2TC/\partial Y_m\partial B_k = \partial 2TC/\partial B_k\partial Y_m = \partial SV_{Bk}/\partial Y_m$ are negative, similarly to $\partial MC_m/\partial Y_n$. Thus, increases in the B_k should reduce the MC_m, and higher Y_m levels raise the SV_{Bk} in absolute value.

12. D_P represents a 1984 break in the pesticide data found with the hedonic research to indicate when most cropping sectors switched from or reduced use of many of the old-line chemicals to the new. D_F represents the post-1979 time period, motivated from Chow test results that indicate this to be a point of structural change in the fertilizer input. These and the corn and cotton state dummies were initially included to establish the sensitivity of the specification to potential shifts or differences in the technological base that otherwise might not be captured in our measures. Although they had negligible impact on the substantive results, and tended to be small in magnitude, they were retained due to their overall statistical significance.

13. Linear dependency arises without the $j\neq m$ requirement for the cross p_j-Y_m terms, and $i\neq j$ for the input price cross-terms, due to the $\Sigma_j p_j$ term incorporated to impose linear homogeneity in prices.

14. Alternative models allowing for land, capital, or labor to be quasi-fixed (separately or together), incorporating imperfect competition in the output markets, or explicitly addressing the deviations between measured and effective pesticide price and quantity, were also evaluated in preliminary empirical investigation. However, the results from the model invoking Shephard's lemma were plausible and comparable, so this specification was retained.

15. This justification could potentially be due to the time-frame of agricultural production, where output choices are made initially and prices are determined after goods are brought to market.

16. The measures were constructed for these data using a generalized Wald test, through the ANALYZ command in TSP. This procedure computes the constraints underlying the hypothesis that the measure is equal to zero, and the associated covariance matrix, evaluated at the estimated parameter vector for a given data point.

17. Of course risk issues from animal waste are also a major issue, at least in some states. Although we do not currently have measures of such risk factors, work is proceeding to generate measures that will be used in later research to establish these relationships.

18. Note that a negative value for the $\varepsilon_{SV,r}$ elasticity implies a positive measure of $\partial SV_{Bk}/\partial r$, since the derivative is multiplied by the (negative) SV_{Bk} value to construct the elasticity.

19. The elasticities with respect to the fixed effects were similarly insignificant.

20. The states included in these divisions are: [1] Northeast (CT, ME, MA, NH, RI, VT, DE, MD, NJ, NY, PA); [2] Corn Belt (IL, IN, IA, MO, OH); [3] Lake States (MI, MN, WI); [4] Northern Plains (KS, NE, ND, SD); [5] Appalachian (KY, NC, TN, VA, WV); [6] Southeast (AL, Fl, GA, SC); [7] Delta (AR, LA, MS); [8] Southern Plains (OK, TX); [9] Mountain (AZ, CO, ID, MT, NV, NM, VT, WY); and [10] Pacific (CA, OR, WA).

References

Ball, V. E., F. Gollop, A. Kelly-Hawke, and G. Swinand. 1999. "Patterns of State Productivity Growth in the U.S. Farm Sector: Linking State and Aggregate Models." *American Journal of Agricultural Economics*, 81(1)(February): 164-79.

Ball, V. E., R. Färe, S. Grosskopf, and R. Nehring. 2000. "Productivity of the U.S. Agricultural Sector: The Case of Undesirable Outputs." Unpublished manuscript.

Baumol, W. J., J. C. Panzar, and R. D. Willig. 1982. *Contestable Markets and the Theory of Industry Structure*. New York: Harcourt Brace Jovanovich.

Chambers, R. G., and E. Lichtenberg. 1994. "Simple Econometrics of Pesticide Productivity." *American Journal of Agricultural Economics*, 76(3)(August): 407-17.

Lichtenberg, E., and D. Zilberman. 1986. "The Econometrics of Damage Control: Why Specification Matters." *American Journal of Agricultural Economics*, 68(2)(May):261-73.

McFadden, D. 1978. "Cost, Revenue and Profit Functions." In *Production Economics: A Dual Approach to Theory and Applications*, vol. 1, ed. M. Fuss and D. McFadden. Amsterdam: North Holland.

Nehring, R., and A. Grube. 1999. "Quality-Adjusted Quantity Indices for Pesticides: A Comparison of Corn/Soybean and Cotton States." USDA/ERS, unpublished manuscript.

Paul, C. J. M. 1999. "Cost Economies and Market Power in the U.S. Beef Packing Industry: A Plant-Level Analysis." Unpublished manuscript, November.

Paul, C. J. M., V. E. Ball, R. G. Felthoven, A. Grube, and R. Nehring. 2001. "Effective Costs and Chemical Use in U.S. Agricultural Production: Benefits and Costs of Using the Environment as a 'Free' Input." Manuscript, resubmitted to the *American Journal of Agricultural Economics*, January.

13

Conclusion: The Usefulness of Productivity Measurement

Robert E. Evenson, Bruce L. Gardner, Dale W. Jorgenson,
and C. Richard Shumway

Productivity measurement for the U.S. agricultural sector has long been the subject of research. The early studies of Barton and Cooper (1948) and Loomis and Barton (1961) provided important guidelines for the practical aspects of productivity measurement. Subsequent studies have focused on aggregation procedures and on methods for adjusting input measures for quality changes (see, for example, Jorgenson and Griliches, 1967; Ball, 1985; Jorgenson, Gollop, and Fraumeni, 1987; Craig and Pardey, 1996; and Ball et al., 1997). More recently, Jorgenson and Stiroh (2000) have attempted to quantify agriculture's contribution to aggregate productivity growth. The present volume extends productivity measurement in agriculture to the individual states, focusing on relative levels and growth rates of productivity across states. It also considers the economic costs of addressing undesirable environmental impacts of agricultural production.

In this concluding chapter, we offer comments on the "user community" for measures of total factor productivity (TFP), on current measurement issues, on intersectional spillover issues, on state TFP indexes, and on the usefulness of TFP measures with "incomplete" quality adjustment. Finally, we offer suggestions for further research.

THE USER COMMUNITY

The "user community" for measures of TFP includes the following groups:

a) Policy-makers concerned with "competitiveness"
b) Policy-makers concerned with structural adjustment costs
c) Policy-makers concerned with sectoral and regional income inequalities
d) Policy-makers concerned with public sector policies and investments that affect rates of TFP change and the directions of change
e) Policy-makers concerned with environmental and resource quality issues associated with agriculture

The chapters in this book attest to the usefulness of productivity measures to each of these user communities. The chapters in Part I reflect the continued pursuit of disaggregate data suited to better measurement of inputs and outputs in more quality-homogeneous categories. These data provide measures of competitiveness. They also enable insights into sources of TFP changes through the association of policies and inventions with the production of factor quality.

Perhaps more important, TFP measures can also be interpreted as measures of Real Cost Reduction (RCR), and RCRs are related to changes in supply of agricultural goods. The agricultural sector has achieved TFP (or RCR) gains that have exceeded those of other sectors of the economy over most of the 20th century. This achievement has important implications for structural adjustment costs. The agricultural sector and rural communities have undergone major structural changes over the last few decades. An understanding of TFP (or RCR) trends can lead to informed policy regarding future change.

The chapters in Part II of this volume provide measures of TFP that are suited to statistical analyses of sources of TFP change. Part III provides measures of TFP when environmental externalities are considered.

Current Measurement Issues

It is difficult to be critical of an enterprise that has a record of responsiveness to general criticisms regarding methods and to user-community demands. The responsiveness in productivity measurement in agriculture has probably not been as rapid or as timely as desired by some critics, but there is little doubt that USDA/ERS has been responsive, most notably to the AAEA task force on agricultural productivity measurement chaired by Gardner (1981). Advances in methodology developed by Dale Jorgenson and others (with Eldon Ball providing leadership in ERS) have been incorporated into productivity measurement, as well as the increasingly important concerns over resource use and environmental effects. Nearly all of the recommendations of the AAEA task force have now been implemented in construction of the official data series. Regular criticism and advice have been sought to determine additional ways to both improve the quality of the data and to capitalize on their analytical potential. The productivity of the productivity group at ERS has been extraordinary, and could be one of the most highly leveraged programs in USDA. The work of many people has benefited from this group's example, support, and careful attention to developing high quality data. It is an example of a successful government-university partnership. Agricultural economists everywhere will be reaping the benefits of this work, and using it as a springboard for further investigation, for years to come.

The general principle of attempting to adjust for input quality change, and the approaches that ERS follows in attempting to quality-adjust inputs, make sense for the most part. But there may be a possible problem with the quality adjustment used for labor. The approach uses market wage rate differences for people with different characteristics, notably age and education, as quality indicators. But we don't have wage rates for farm labor provided by farm operators or unpaid family members. Recent work in ERS and elsewhere, using both individual farm data or state aggregates, has had difficulty finding significant positive effects of farm operators' education on net income or efficiency measures. This fact casts doubt on the quality adjustment coefficients used.

What is the cost of using an inappropriate adjustment? If in fact the increase over time in farm operators' schooling is causing the quality-adjusted index of farm labor to increase at a greater rate than the real value of schooling warrants, then we overstate the growth of inputs and hence understate productivity growth. The evidence is not sufficient to conclude that such understatement in occurring, but the possibility that it is occurring merits further exploration. In the meantime, it is important that users of USDA's productivity statistics be aware of the issue. Indeed it is important to be cognizant of the labor quality adjustment even if the adjustment is appropriate. Griliches (1964) estimated that education increased output, with inputs held constant, implying that education contributed to productivity growth. If one were to try to confirm Griliches' findings by explaining recent productivity growth as measured by USDA as a function of farmers' schooling, the coefficient would not estimate the effect of schooling, but rather the extent to which schooling's effect is greater or less than the effect already built into the productivity statistics through USDA's labor quality adjustment.

A more serious and substantive problem, however, is with USDA's capital input series. Note first that there are differences in measured capital input series available that are sufficiently large to make a difference in productivity measurement. A comparison of the capital input series from Acquaye, Alston, and Pardey (AAP) (Chapter 3) to the USDA index for durable equipment (Chapter 2) shows that capital in agriculture has been falling since about 1980 (peak at 1979 in AAP and 1981 in USDA). But the rate of decline is 3.9% annually in USDA and 0.9% annually in AAP. This 3% difference in the rate of decline, with capital accounting for about a sixth of agricultural inputs, means use of the AAP index would reduce the measured rate of multifactor productivity growth by about 0.5%, for example from 2.0% to 1.5% annually over the 1980-1992 period.

The USDA capital-input index may have been growing at too slow a rate (or declining too fast) for the period since 1960. One hypothesis is that the chief culprit is the treatment of depreciation. As mentioned by Ball et al. in chapter 2, USDA uses a concave decay function for capital as contrasted with the more often used geometric decay. The concave function means a given capital item decays slowly early in its life but then falls rapidly to quite low levels. In contrast, with geometric decay an item loses most of its effectiveness early, but retains more as it gets older (as compared to the USDA approach). While Ball et al. cite good reasons for believing that capital holds on to its effectiveness better in its early life than geometric decay allows, one can question the rapid fall-off to low levels for "middle-aged" capital. Anecdotal evidence for this — take it for what it is worth — is found on many small dairy farms that still have stanchion barns and silos built decades ago, milking parlor equipment and pipelines from the 1960s, barn cleaners from the 1970s, and feed handling equipment, tractors, combines, and other implements almost all 20 years old or more. In short, according to USDA measurement, the capital input on that farm is about zero. In economic management terms, that may be about right. The equipment is depreciated and/or obsolete, and little of it would be acquired by someone investing in dairy farming today. The net income of the farm may be properly treated for both taxes and managerial accounting purposes as largely a return to the family's labor and management. But — and here is where the production-theory foundations of productivity measurement becomes important — those facts are not the relevant ones. What is relevant is how that depreciated capital contributes to output. And in this respect that capital is far from depreciated. If you doubt that, just consider how much output the cows would produce, keeping the labor input intact, in the absence of that capital.

The treatment of older capital is important because so much of the capital equipment in U.S. agriculture is quite old. ERS data on sales of tractors indicate that about 1 million have been sold over the last 15 years (adding up reports of annual sales). The implication is that of the 4.5 million tractors reported on farms in the 1997 Census of Agriculture, at least 3.5 million must be more than 15 years old. Similarly startling percentages of old equipment can be inferred from ERS data on sales compared to Census inventories for combines and other machinery. Probably many of these old tractors do not carry out much useful work, and thus are properly depreciated down. But other old equipment is providing valuable service, and the existence of so many older tractors, with the adoption of a concave rather than geometrically declining depreciation approach by ERS, could be substantially understating the capital stock. And, since recent-year sales relative to the Census inventory figures are lower in 1997 than in earlier Censuses, such bias as it exists has been increasing over time. Another source of possible understatement of growth (or overstatement of decline) in the farm capital stock is that machinery that is less well covered statistically (and not counted in the Agriculture Census) includes machinery, such as corn planters and pesticide applicators, that has become more sophisticated and likely a larger fraction of the farm capital stock in recent years.

These comments represent an important reminder to users of productivity measures that the user needs to be aware of differences between data series where few quality adjustments have been made and data series where such adjustments have been made. When a labor quantity series is adjusted for changes in schooling, experience, and gender composition, it cannot realistically be compared with a labor quantity series where no adjustments have been made. In fact, the relevant question to be asked of a labor series for agriculture is whether the implied quality differentials for education differences based on economy-wide data are actually appropriate for agriculture.

Intersectoral Spillover Issues

The growing importance of the industries (machinery, chemical, veterinary medicine, seeds) supplying inputs to agriculture has implications for agricultural TFP measurement. These industries conduct R&D and generally bring new products onto the market on a regular basis. These new product introductions have a special property that produces a "pecuniary R&D spillover" to agriculture that takes the form of high agricultural TFP growth. This special property is that new products are not priced on a par with old products. That is, the first price observed for a new product typically understates the real productive value of the new product relative to the old product that it typically replaces.

Consider a company producing a new model of a planter. Suppose the new model provided ten percent more productive services in a real sense than an existing model. This improved performance is the product of a private R&D program by the manufacturer. Will the new planter be first priced at a ten percent premium over the existing planter? If so, it would be incorporated into the agricultural capita stock at the correct price. But this is not the typical pricing strategy of suppliers. In order to establish market share and in anticipation of the introduction of new models by competing firms, it will first price at something less than the ten percent real services difference. Suppose it is priced at a 6 percent difference. The agricultural accounts will then regard it as having a 6 percent higher service flow when in reality it has a 10 percent higher service flow. This practice understates the real growth in the capital services stock and results in TFP growth in agriculture.

The company developing the new model captured a private return to its R&D from the 6 percent premium, but the R&D also produced a social rate of return, which showed up in the agricultural sector (embodied in the 4 percent uncaptured premium). Recent reviews of rate-of- return studies of public sector research (Evenson, 2001; Alston et al., 2000) show relatively high social rates of return (around 40 to 50 percent internal rates of return). These rates are often compared to private rates of return to company R&D of roughly 25 percent. But the social productivity realized in the agricultural sector when added to the private rate of return produces total social rates of return to this industrial R&D of roughly the same order of magnitude as those recorded for public sector R&D. This difference in social and private return is a dimension of company R&D that is quite important. With the growing importance of private company R&D to agriculture, we need to more carefully measure this spill-in.

State TFP Consistency

Part I of this volume is concerned with properties of TFP indices by State. TFP indices can be seen in two dimensions, levels and rates of change. For level comparisons, the issues of transitory and base invariance are important. Are they important for rate of change comparisons?

One of the difficulties in level comparisons is that technology has strong environmental interactions. In the 1950s, soybean varieties performed best under conditions in Iowa. By the 1970s, soybean breeders had developed varieties that performed well under the conditions encountered in Mississippi. Because of this plant breeding work, the implicit value of sunlight and humidity in Iowa and Mississippi was changed. What does transitivity mean in this context?

A rate of change comparison would measure productivity change in Iowa independently of measures for Mississippi. An arbitrary base period would be chosen and, for comparison purposes and for statistical analysis purposes, this base period would be the same in different states but it would be arbitrary.

Can one meaningfully compare output/input indices between geographic regions? Presumably, a well-functioning land market would result in land rents (or values) — which are implicitly sunshine, humidity, and related climate rents — that equilibrate the cost of production in two regions producing the same product. Thus, when the soybean breeders in Mississippi modified varieties for the sunshine and humidity of Mississippi, they changed the value of these resources.

This value should be reflected in land prices; but how does one incorporate it into an output/input ratio? The art of making state-by-state TFP or output/input comparisons remains complex at this date, with many unresolved issues. The art of measuring rates of TFP change in different states is more straightforward and is suited to source and consequence analysis.

Factor Quality Adjustments

Are "raw," unadjusted TFP measures useful? As noted above, adjusted TFP measures bear a different interpretation than do unadjusted measures. For many purposes, partially adjusted measures are less useful for "source identification" than are crude unmodified measures or measures with reasonably complete adjustment.

In a sense, the adjustment process itself constitutes identification of and provides insight into sources of TFP production. The alternative to quality adjustment directly is to use statistical techniques to relate TFP to variables defined to identify contribution or sources.

The statistical decomposition of TFP measures has been used most extensively to identify the contribution of agricultural research progress (both public and private) to TFP (and hence to supply) changes. These techniques can be used with unadjusted TFP measures if a reasonably complete set of source variables (research, extension, schooling, institution) can be assembled and defined in meaningful ways.

Judicious use of TFP measures adjusted for schooling and other quality changes can enhance statistical analyses. However, it is important that such adjustments be carefully done and accurately reported.

Conclusion

Having said the above, three principal questions remain: 1) Are the data presented in chapters 2 and 3 ready for prime time? 2) Can the data be improved upon and if so how? 3) How can the data be used? The answer to the first question is an unequivocal yes. It is important not to undervalue the benefit of incremental improvements in the data, but one way to get those improvements is to use them as they exist so that others value the effort that goes into revising them. The other chapters in this volume attest to the variety of useful analyses that can be performed.

Clearly the data can be improved upon for reasons noted above, particularly the capital and labor variables and to some extent thc land variable. The USDA is the logical player to lead the revision effort, drawing on the work of others in the economics profession.

The data need to be used and put through the paces that econometricians can do. Having said this, it is important not to overreach and try to extract more from the data than what is really there. One should remember that there is a limit to what you can extract from residuals.

Much progress has been made, but the following are five suggestions for future work that would help to move productivity-related research forward another step:

1. Decide how to do multilateral spatial comparisons in order to most appropriately index across states, geographic and/or production units.
2. Carefully compare the data construction methods underlying the state-level data presented in Chapter 2 with those in Chapter 3, and with those of Huffman and Evenson. Incorporate the best methods among all three in the development of the next generation of state data. Consider whether they could also improve the U.S. data series. Of particular importance are the quality adjustments for labor and capital.
3. Critique alternative assumptions in analysis that are equally plausible *a priori*. Consider functional form, nature of technical change, time-series properties of the data, behavioral objective, observed vs. shadow prices, nonparametric versus parametric techniques, maximum likelihood vs. Bayesian techniques. Where an unambiguous choice is not forthcoming, conduct sensitivity analyses to determine the range of viable implications.
4. Develop alternative productivity series for different uses. For example, one should account for water and air quality effects of production techniques.
5. Focus on the intended use of the data. Efficiency measurement may not be a particularly important use of state-level data, but measurement of differences in resource endowments and production technologies probably is.

REFERENCES

Alston, J. M., C. Chan-Kang, M. C. Marra, P. G. Pardey, and T. J. Wyatt. 2000. *A Meta Analysis of Rates of Return to Agricultural R&D: Ex Pede Herculuem*. IFPRI Research Report 113. Washington, D.C.

Ball, V. E. 1985. "Output, Input, and Productivity Measurement in U.S. Agriculture, 1948-79." *American Journal of Agricultural Economics,* 67:475-86.

Ball, V. E., J-C Bureau, R. Nehring, and A. Somwaru. 1977. "Agricultural Productivity Revisited." *American Journal of Agricultural Economics,* 79:1045-63.

Barton, G., and R. Cooper. 1948. "Relation of Agricultural Production Inputs." *Review of Economics and Statistics,* 30:117-26.

Craig, B. and P. Pardey. 1996. "Productivity Measurement in the Presence of Quality Change." *American Journal of Agricultural Economics,* 78:1349-54.

Evenson, R. E. 2001. "Economic Impact Studies of Agricultural Research and Extension." in *Handbook of Agricultural Economics,* volume 1, ed. B. L. Gardner and G. Rausser. Amsterdam: Elsevier Science/North Holland.

Gardner, B. L. 1981. "Changes in the Quality of Agricultural Statistics — Inputs, Farm Income, Output, and Prices." In *Economic Statistics for Agriculture: Current Directions, Changes, and Concerns,* ed. R.K. Perrin and E. Reinsel. Proceedings of a Symposium at the American Agricultural Economics Association meetings, Clemson, S.C. July.

Griliches, Z. 1964. "Research Expenditures, Education, and the Aggregate Agricultural Production Function." *American Economic Review*, 54:961-74.
Huffman, W. E, and R. E. Evenson. 1993. *Science for Agriculture*. Ames: Iowa State University Press.
Jorgenson, D., and Z. Griliches. 1967. "The Explanation of Productivity Change." *Review of Economic Studies*, 99:249-82.
Jorgenson, D., and K. Stiroh. 2000. *Raising the Speed Limit: U.S. Economic Growth in the Information Age*. Brookings Papers on Economic Activity, Vol. 1, pp. 125-211.
Jorgenson, D., F. Gollop, and B. Fraumeni. 1987. *Productivity and U.S. Economic Growth*. Cambridge, Mass.: Harvard University Press.
Loomis, R.A. and G.T. Barton. 1961. *Productivity of Agriculture*. Economic Research Service, U.S. Department of Agriculture, Washington, D.C.

Index

F

G

H

I

K

L

T

U

V

W

Y